As per UGC Model Syllabus and Guidelines

ENVIRONMENTAL BIOLOGY

[For B.Sc. and B.Sc. (Hons.) Classes of All Indian Universities]

Dr. H.R. SINGH
M.Sc., Ph.D.
Former Professor and Head
Department of Zoology
University of Allahabad

S. CHAND
PUBLISHING

S Chand And Company Limited

(ISO 9001 Certified Company)

S Chand And Company Limited

(ISO 9001 Certified Company)

Head Office: Block B-1, House No. D-1, Ground Floor, Mohan Co-operative Industrial Estate, New Delhi – 110 044 | Phone: 011-66672000

Registered Office: A-27, 2nd Floor, Mohan Co-operative Industrial Estate, New Delhi – 110 044

Phone: 011-49731800

www.**schandpublishing.com**; e-mail: **info@schandpublishing.com**

Branches

Ahmedabad : Ph: 27542369, 27541965; ahmedabad@schandpublishing.com
Bengaluru : Ph: 22354008, 22268048; bangalore@schandpublishing.com
Bhopal : Ph: 4274723, 4209587; bhopal@schandpublishing.com
Bhubaneshwar : Ph: 2951580; bhubaneshwar@schandpublishing.com
Chennai : Ph: 23632120; chennai@schandpublishing.com
Guwahati : Ph: 2738811, 2735640; guwahati@schandpublishing.com
Hyderabad : Ph: 40186018; hyderabad@schandpublishing.com
Jaipur : Ph: 2291317, 2291318; jaipur@schandpublishing.com
Jalandhar : Ph: 4645630; jalandhar@schandpublishing.com
Kochi : Ph: 2576207, 2576208; cochin@schandpublishing.com
Kolkata : Ph: 23357458, 23353914; kolkata@schandpublishing.com
Lucknow : Ph: 4003633; lucknow@schandpublishing.com
Mumbai : Ph: 25000297; mumbai@schandpublishing.com
Nagpur : Ph: 2250230; nagpur@schandpublishing.com
Patna : Ph: 2260011; patna@schandpublishing.com
Ranchi : Ph: 2361178; ranchi@schandpublishing.com
Sahibabad : Ph: 2771238; info@schandpublishing.com

First Edition 2005
Second Revised Edition 2014; Reprint 2017
Reprint 2020

ISBN: 978-81-219-2476-4 **Product Code:** H6ENB68ENVS10ENAB14O

PRINTED IN INDIA

By Vikas Publishing House Private Limited, Plot 20/4, Site-IV, Industrial Area Sahibabad, Ghaziabad – 201 010 and Published by S Chand And Company Limited, A-27, 2nd Floor, Mohan Co-operative Industrial Estate, New Delhi – 110 044

This book is respectfully dedicated to
Professor Dr. S. S. KHANNA
my teacher in B.Sc., M.Sc. and Ph.D. guide
and to all my students who worked for their
doctorate degrees under my supervision and
guidance and immensely contributed to making
me what I am today

PREFACE TO THE SECOND EDITION

The first edition of this book was published in 2005. During the last ten years, the book has been used both by teachers and students and I am grateful to all of them. Since the publication of the book many changes have taken place. New information and new data is now available. Hence it was necessary to revise the book.

The original plan of the book is maintained in this edition. All the chapters of the book have been revised and new information added. Maximum changes have been done in respect of chapters on biodiversity and wildlife. In view of the new forest policy of the country, now we have changed data in respect of national parks, wildlife sanctuaries, tiger reserves and biosphere reserves, etc. The same has been incorporated to make the subject matter up-to-date. The number of MCQs has been increased.

I am sure, this new edition will be more useful for its users. As usual, the help of Dr. Neeraj Kumar, Head of the Zoology Department, Meerut College Meerut, Mr. Bhuvnesh Kumar, General Manager, NTPC, Noida and Dr. Nishant Kumar, Associate Professor in MBA, Lucknow University and Management of S. Chand & Company Pvt. Ltd., New Delhi is thankfully acknowledged.

Dr. H.R. Singh
3/33, Shraddhapuri, Phase 1
Meerut-250001
email:profsinghhr@yahoo.co.in
Mobile : 09410035821

PREFACE TO THE FIRST EDITION

Under the scheme "Preparation of University Level Books by Invitee Authors on Grey Areas in July 1995, the University Grants Commission, New Delhi had requested me to prepare a book on "Environmental Biology". I had accepted the offer; chapter plan was submitted to UGC and I had consulted some senior colleagues of Jawaharlal Nehru University, New Delhi; Banaras Hindu University, Varanasi ; and Delhi University, Delhi regarding the subject matter of the book. Among others, valuable suggestions were received from Prof. J.S.Singh, Prof. T.R.Rao and Prof. Brij Gopal. But due to my teaching, administrative and other engagements I could not complete the project within the stipulated time given by UGC.

After a lapse of about five years, I have worked on this project during 2002 and 2003 and now the book is ready for publication. In the treatement of the subject, I have adopted a holistic approach "everything is connected to everything else" (Barry Commoner). Keeping this in view, the chapters of the book are designed. All the chapters of the book are planned in such a way that they introduce the reader from simple to difficult to more difficult topics. As a result, these chapters automatically flow into each other and, thus, coherence and connectivity is maintained. The first few chapters of the book are meant to prepare a foundation required to build up the remaining more difficult chapters. But for the benefit of students in all the chapters an attempt is made to discuss all the interrelated things at one place. Thus the book presents a balanced overview of environmental biology with a new and holistic approach and with examples drawn from aquatic and terrestrial environments. To make learning interesting and more relevant, the book covers all the global environmental problems in detail and also the practical applications of environmental biology, wherever possible. Some important features of the book are: crisp language, lucid style and clear illustrations. The specific areas covered in the book are: Biodiversity, Environmental Toxicology and Environmental Biotechnology.

These days environmental biology and environmental science are a necessary component of the syllabi of several disciplines including Botany, Chemistry, Geology, Geography, Zoology, and Social Sciences and Medical, Engineering and Tourism and Hoteliering. As such, the book will be found useful by the students and teachers of all these disciplines. Since it is based on UGC syllabus, it will also be useful for all levels of competitive examinations including IAS, IFS, PCS, NET and All India Engineering and Medical Entrance Examinations.

In the preparation of this book, I have consulted a large number of original research papers, many standard books on ecology and environmental science, and I am indebted to their distinguished authors. All such papers and books are quoted in References. If some students and teachers may like to consult these papers and books, they will be further benefited. I have tried my best to give up-to-date information and my own interpretations, wherever possible. But I have my limitations and I regret the errors that might have crept anywhere in the text.

Besides the names mentioned above, I have been immensely benefited by the brilliant ideas of Dr. Neeraj Kumar, Department of Zoology, Meerut College, Meerut; Mr. Bhuvnesh Kumar, Geologist, J.P. Industries Private Limited; Mr. Nishant Kumar, Lecturer in MBA, Chaudhary Charan Singh University, Meerut; Prof. Trivikram Pati, former Head of the Mathematics Department and former Vice Chancellor, University of Allahabad; Prof. Sarvesh Kumar, Head of the Zoology Department, Kumaun University, Nainital; Dr. P.K.Gupta, Reader in Zoology, Kumaun University, Nainital; and Prof. W.S. Lakra, Principal Scientist and Head of the Fish Genetics and Biotechnology Division, Central Institute of Fisheries Education(Deemed University), Mumbai. I am thankful to UGC, New Delhi for their offer to prepare the book. I am thankful to Mr. Ved Prakash Singh for typing the manuscript. I am also thankful to Mr. Prashant Mishra for his help in properly arranging the computer typed matter. Last but not least, I am thankful to the proprietor of S.Chand & Company Ltd., New Delhi and their staff for publishing the book in a record time.

Author

CONTENTS

CHAPTER 1

Introduction

1.1 ECOLOGY

An organism has its identity as an individual and also as a component of a population, community, ecosystem, and biosphere. But at all the levels of organization (Fig. 1.1), there is sufficient interaction between the organism and its environment. The study of this interaction (relationship) between the organism and its environment is called *ecology* (originally in German: *Oekologie*). According to Ricklefs and Miller (2000), ecology is the study of interactions of organisms with their physical and biological environment. This interaction is multi-dimensional and many features of the environment affect this interaction.

The ecological study of one species (organism) is called *autecology*. This single species may be a living plant or an animal of any habitat. The ecological study of a number of species (population, community, ecosystem) is called *synecology* (Fig. 1.1). Synecology is that branch of ecology which deals with the study of a group (or groups) of organisms which are associated together as a unit. An example of synecological study may be the study of tiger population, the study of aquatic communities of a river or lake and the study of a river, lake or forest ecosystem.

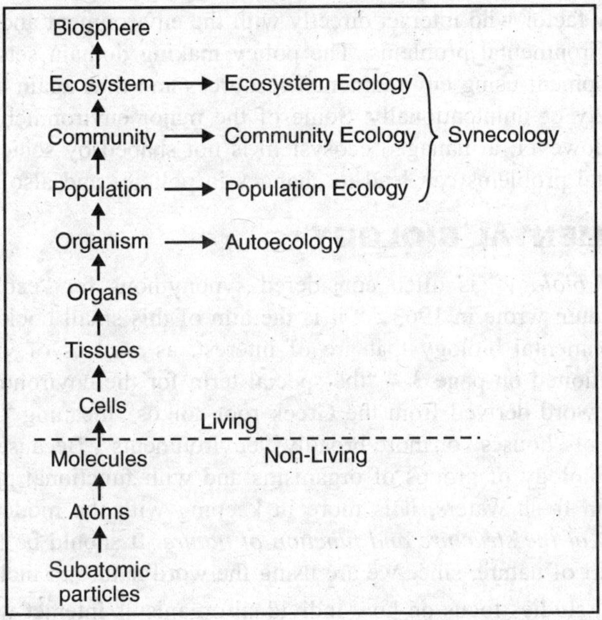

Fig. 1.1 Levels of organisation and two main subdivisions of Ecology

Ecology is both a pure and applied science and ecologists work on a variety of spatial and time scales. Three of the great founders of ecology are A.G. Tansley, F.E. Clements and Charles Elton. In India, Vedic literature contains frequent references to ecological thoughts, particularly of conservation and ecological balance; but as a science subject it developed after 1910, made rapid strides in the late 1940s and assumed an exponential growth in the 1970s (Singh and Singh, 1995). However, almost everywhere in the world the words "ecology", "ecologist" and "environment" have spread from the science of ecology to the press, to the general public and to pressure groups, and some people say that these words have been hijacked and misused. The fact is that the twin threads of pure and applied ecology have developed side by side, each depending on the other. The fundamental ecology deals with four higher levels of organization– *individual organisms, populations* (individuals of the same species), *communities* (a large or small number of populations) and *ecosystems* (the community together with its physical and biotic environment). Applied ecology attempts to make practical application of ecological understanding to predict, manage and control. As such, an ecologist has to depend on Mathematics, Chemistry, Physics and other disciplines of science and social science as tools essential to the understanding of ecology.

In ecology, the unit of study is a *natural ecosystem*. All ecological systems are dynamic and governed by well-defined laws. From a descriptive viewpoint, a stream, a forest, and an ocean may have almost nothing or very little in common, yet as ecological systems they function in a similar way. But we should also know what are human ecosystems where, in ecological terms, humans are the dominant species. In human ecosystems, there is a mutual interaction between society and the environment and this interaction is the very basis for the existence of society. The environment provides with the resources that are used by people according to their needs. Thus, human ecosystems can be easily visualized in terms of domains or interacting parts. Three of these domains are – the *environmental,* the *individual management* and the *policy making* (Clapham Jr.,1981). The environmental domain comprises those components that react according to the laws that govern natural ecosystems. The individual- management domain has the human behaviour as the key factor who interact directly with the environment and manipulate or change it, thus creating environmental problems. The policy making domain sets goals for the society for economic development using environmental resources and thus again creating environmental problems intentionally or unintentionally. Some of the major environmental problems are dealt with in this book. However, a managed ecosystem is not shaped by society management alone. In fact, environmental problems can lead to changes in policies and also in human behaviour.

1.2 ENVIRONMENTAL BIOLOGY

"Environmental biology" is often considered synonymous for "ecology". Describing the scope of ecology, Odum wrote in 1963 : " it is the aim of this small book to outline those basic principles of environmental biology that are of interest, as well as of vital concern, to all of us". He further mentioned on page 3 – "the special term for the environmental biology field of interest is *ecology,*a word derived from the Greek root "oikos" meaning "house". Thus literally, ecology is the study of "houses" or more broadly, "environments". Because ecology is concerned especially with the biology of groups of organisms and with functional processes on the lands, in the oceans, and in fresh waters, it is more in keeping with the modern emphasis to define ecology as *the study of the structure and function of nature.* It should be thoroughly understood that mankind is a part of nature, since we are using the word nature to include the living world."

Thus, ecological studies focus on how individual organisms interact with their environment, or how ecological populations or communities interact with their environment. Study of how organisms are adapted to respond to temperature or other factors of their physical environment is an example of such a focus. This approach is called *ecophysiology* or *physiological ecology.* Organisms of the same species living in the same place and time constitute an *ecological population.* All the populations of a natural habitat are linked to one another by several types of interactions including feeding relationships. These interacting populations form *ecological communities* and their study is called *community ecology.* A community plus its environment forms an interacting entity called an *ecosystem,* which is considered as a unit in the study of ecology. Although ecology has shown tremendous progress during the past 50-60 years, its main focus has been on various types of ecosystems – terrestrial, freshwater, marine and on how human activity has influenced these ecosystems.

The *ecosystems ecology,* for obvious reasons, is even now the most important subfield of ecology. Some other subfields are *landscape ecology* and *restoration ecology* (Fig. 1.2). The former focuses on how ecological processes operate on large spatial scales. A landscape may be defined as a large area that includes one or more ecosystems. Restoration ecology focuses on restoring or reestablishing species and ecosystems after human perturbation. However, some questions in ecology can be better answered in evolutionary terms and this subfield of ecology, referred to as *evolutionary ecology*, focuses on interpretation of the structure and function of organisms, populations, communities and ecosystems in the context of evolutionary theory.

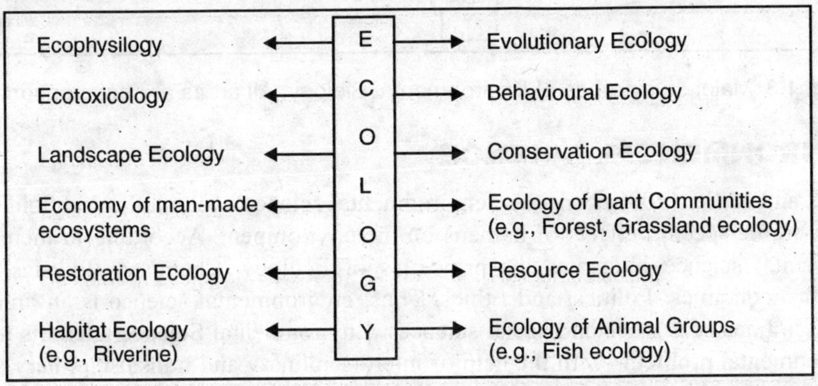

Fig. 1.2 Some important subfields of Ecology (All subfields are not mentioned)

As stated earlier, "environmental biology" is a term often used for "ecology". Thus, these two words depict the same thing and they appear like the two sides of the same coin. When the coin is tossed many times, there are chances that each side (head/tail) comes almost 50:50. In other words, "environmental biology" represents another shade (may be slightly lighter or darker) of the colour of "ecology". However, ecology is deeply rooted in biology and in classical ecology often the focus is mainly on the "ecosystem studies" (energy flow and productivity) but in environmental biology the focus is on the "biological organisms of environment" and their ecophysiological adaptations and interactions. For example, if we use the phrase "environmental biology of fish", it will mean how the fish species of a particular environment (pond, lake, river, estuary, ocean) are interacting with the environmental factors, both biotic and abiotic; which of these factors govern their distribution and abundance, and how these factors have influenced their adaptations and life history strategies. We also know that it is not only the environment and its physical and biotic factors which effect the organism but the organism is also able to modify the environment to a certain extent (Gaia Hypothesis). The hypothesis states that the earth's atmosphere would not support life without regulation by the totality of life in the biosphere (Lovelock and Margulis, 1974; Lovelock, 1988; Mann, 1991). Some major study areas os environmental biology are indicated. (Fig. 1.3)

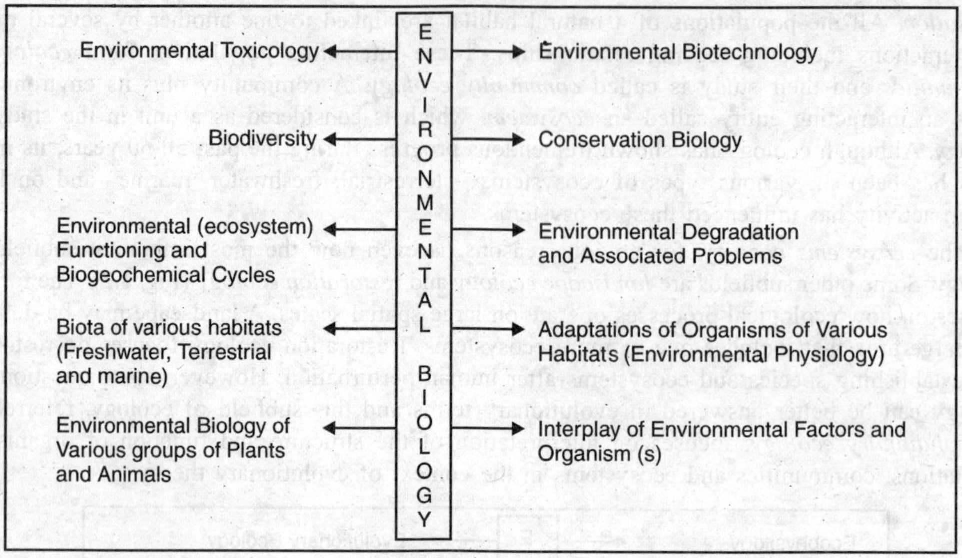

Fig. 1.3 Major study areas of Environmental Biology. All areas are not mentioned

1.3 ENVIRONMENTAL SCIENCE

Ricklefs and Miller (2000) consider "environmental science" yet another subfield of ecology that focuses on the specific effects of humans on the environment. According to them, the study of environmental science integrates the principles of Ecology, Chemistry, Physics, and other sciences with Economics, Politics, and Ethics. Thus, environmental science is an amalgamation of all the disciplines of science and social sciences. Environmental Science attempts to solve the major environmental problems with the help of interdisciplinary and transdisciplinary approaches (Jantsch, 1970), *i.e.*, the entire knowledge of all the disciplines of science such as Mathematics, Physics, Chemistry, Biology, Geology, Geography, Computer Science, Medical Science and Biotechnology, as well as of social sciences such as Economics, Sociology, and Psychology. Environmental science is a subject which draws heavily from environmental biology, but depends more on transdisciplinary approach.

Environmental scientists often study the action and fate of man-made substances, such as detergents and chemical pesticides, in the biosphere. This study is called *ecotoxicology*. Ecotoxicologists often use fish and other animals as models to study the action of toxic compounds in bioassays. (Misra, 1982; Mukherjee and Pankajakshi 1995; Feisthaur *et al* 2002). Since all the three terms– *ecology, environmental biology*, and *environmental science* cannot be defined without reference to environment, they show too much overlapping which is inevitable and unavoidable. Some major problems and disciplines which provide input to environmental science are mentioned in Fig. 1.4.

1.4 THE BOOK

The book is designed to serve as a comprehensive textbook of environmental biology at University level and covers all the topics that are expected in any good book of Ecology. However, the approach throughout is interdisciplinary and transdisciplinary. Chapters covered in the first section of the book are: Organism and environment, Population, Biotic community, and Ecosystem. The second section has chapters on Biosphere degradation and major environmental problems, and Biodiversity. Since the three components of biosphere – hydrosphere, lithosphere and atmosphere are interlinked, any problem or perturbation on anyone part automatically affects the remaining two components (parts) and their living organisms (biodiversity). Therefore, factors responsible for biodiversity crisis and measures for biodiversity conservation are described. The third section of the book deals with Environmental toxicology and Environmental Biotechnology. Main emphasis is paid to the movement of toxic substances in biosphere, and toxicity of pesticides and heavy metals to living organisms. A chapter on environmental biotechnology, often not given in most of the books on ecology, throws light on how the emerging biotechnological tools may be helpful for environmental protection and restoration. Among other things, this chapter deals with microbiological degradation of xenobiotics, bioremediation, and pollution control by genetic engineering.

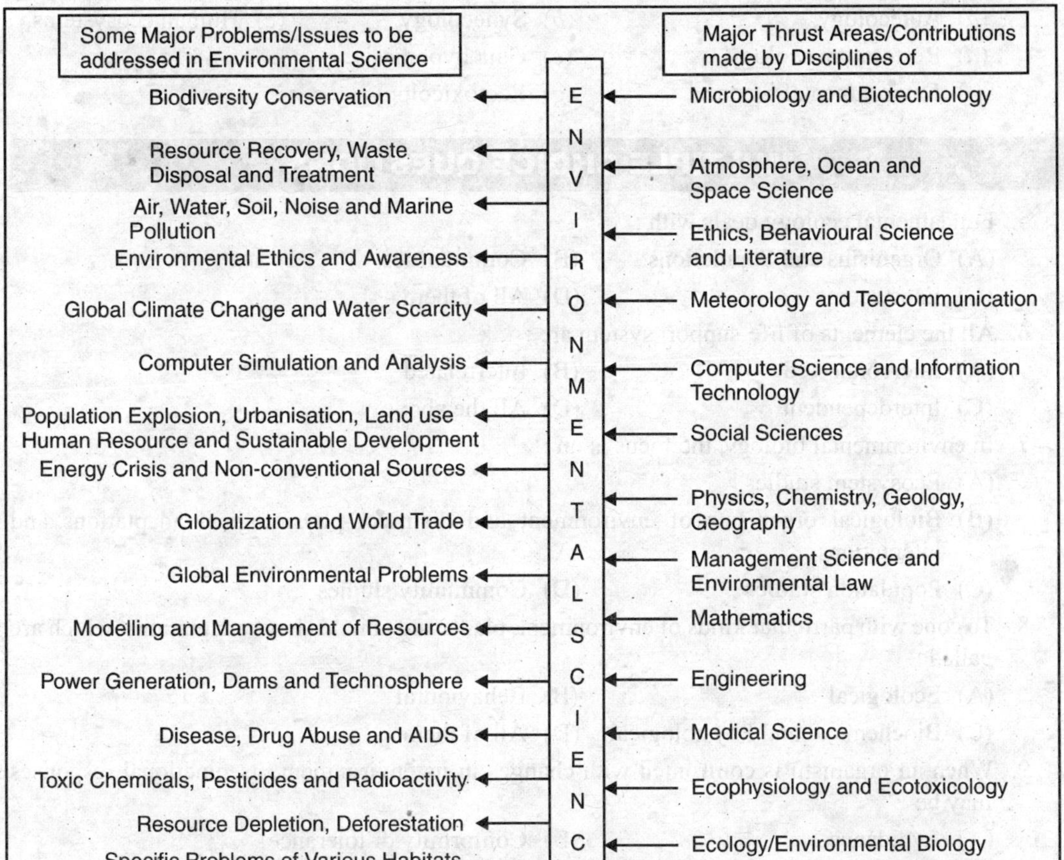

Fig. 1.4 Some major problems and issues and disciplines which provide input to solving these problems in environmental science

Thus, dealing with the related headings under one chapter, the book exposes the readership gradually from simple to complex aspects of environmental biology. The main aim of the book is to throw light on rather grey areas of environmental biology in a lucid style and crisp language so that the book may provide a broad knowledge basis for deeper understanding of the subject and also for achieving sustainable development. The aim also teaches us that we should be sensitive towards conservation of nature and natural resources including wildlife and biological diversity. As such the book is useful for students of all disciplines, teachers, planners and administrators and last but not least to those who prepare for competitive examinations.

MODEL QUESTIONS

1. Define 'ecology', 'environmental biology', and 'environmental science' and write a note on their scope and applications.
2. Justify the statement that in environmental biology the focus is on "the biological organisms of the environment".
3. Write the difference between adaptation, acclimation and acclimatization.
4. Write notes on the following:
   (*a*) Autecology, (*b*) Synecology, (*c*) Human ecosystems,
   (*d*) Restoration ecology, (*e*) Gaia hypothesis,
   (*f*) Environmental Science, (*g*) Ecotoxicology

MULTIPLE CHOICE QUESTIONS

5. Fundamental ecology deals with :
   (A) Organisms and Populations (B) Communities
   (C) Ecosystems (D) All of them
6. All the elements of life-support system are :
   (A) Interconnected (B) Interrelated
   (C) Interdependent (D) All the above
7. In environmental biology, the focus is on :
   (A) Ecosystem studies
   (B) Biological organisms of environment and their eco-physiological adaptations and interactions
   (C) Population studies (D) Community studies
8. To cope with particular kinds of environment, plants and animals have adaptations, which are called :
   (A) Ecological (B) Behavioural
   (C) Biochemical and Physiological (D) All of them
9. When an organism is confronted with changes in its environment, its functional responses may be :
   (A) Avoidance (B) Conformity or tolerance
   (C) Regulation (D) All the above
10. Environmental stress includes:
   (A) Abiotic factors (B) Biotic factors
   (C) Habitat modification (D) All the above

2 CHAPTER Organism and Environment

2.1 INTRODUCTION

In philosophical and poetic terms, life is manifestation of the creative power of nature, but there is a molecular basis of life. Like the non-living, living forms are also made up of molecules and compounds, which are mostly formed by carbon, hydrogen, oxygen and nitrogen. However, the presence of certain molecules at one place does not make life. Life results due to a relationship of molecules and is not a property of anyone molecule (Linus Pauling). A living being, called organism (any living plant or animal) represents a characteristic level of organization; it shows responsiveness and adaptability, it acquires materials and energy; it reproduces, and shows growth, development, ageing and death. These attributes distinguish living from the non-living. In the living, the organizational levels consist of molecules and compounds, living cells, tissues, organ systems, individual (organism), population, community, ecosystem, and biosphere. Thus, living forms, although made up of simple molecules, represent a rather complex organization.

2.2 BROAD CLASSIFICATION OF ORGANISMS

All the organisms require energy to live and obtain energy in different ways. There are two major divisions of organisms — *autotrophs and heterotrophs*. An autotroph (Gk. *aut,* self, and *troph,* feeder) is an organism that can make organic molecules from simple inorganic nutrients in its environment. Green plants use light (as the source of energy), water, carbon dioxide, and inorganic ions to manufacture their food in the process of photosynthesis. Hence, they are called *photoautotrophs.* Some organisms make their food by using energy derived from the oxidation of inorganic substrates in the environment. They are called *chemoautotrophs.* Many prokaryotic organisms are chemoautotrophs (Table 2.1).

Table 2.1 The five kingdom classification of organisms

S.No	*Kingdom*	*Cell complexity*	*Mode of nutrition & role*
1	ANIMALIA	Multicellular, Eukaryotes	Heterotrophic (Consumers)
2	PLANTAE	Multicellular, Eukaryotes	Autotrophic (Producers)
3	FUNGI	Multicellular, Eukaryotes	Heterotrophic (Decomposers)
4	PROTISTA	Unicellular, Eukaryotes	Several (Photosynthetic, Predatory, Parasitic, Saprobic)
5	MONERA	Prokaryotes (Bacteria & Cyanobacteria)	Several (Photoautotrophic, Chemoautotrophic, Heterotrophic)

A heterotroph (Gk. *hetero,* different, and *troph,* feeder) is an organism that cannot synthesize organic compounds from inorganic nutrients in its environment and therefore must take in preformed food. All animals and fungi depend on heterotrophy. A *saprotroph* (Gk. *sapro*, putrid, and *troph*, feeder) is an organism, usually a bacterium or fungus, that digests dead organic matter; secretes digestive enzymes and absorbs the resulting nutrients back across the plasma membrane. Heterotrophic nutrition of animals involves ingestion, digestion, absorption and assimilation of food. The undigested food is excreted.

2.3 ENVIRONMENT AND ITS FACTORS

The term *environment* means surroundings. All the levels of organization have their immediate environment, the so-called microenvironment, and a larger environment, the so-called macroenvironment. The planet earth is covered by various types of environments, which may be broadly classified into aquatic and terrestrial environment. Air, water and soil are the three environments, which support life on this earth. A number of environmental factors such as light, temperature, water, humidity, currents and pressures, general weather conditions, soil and fire (all physical factors), and atmospheric and dissolved gases, pH, nutrients and food (all chemical factors), and biotic factors (plants and animals of the same and other species) effect the distribution and abundance of organisms in different habitats. The physical environment of an organism depends on location, topography (the height and shape of the land), geology of the area (*i.e.*, rock and soil types), and climate. The latitude, altitude and proximity to a coast contribute to the climatic conditions of an area. Climate has several important variables and weather may be defined as the climate at a particular time or place.

The idea that environmental factors could be limiting at their maximum and minimum levels was incorporated in the *law of tolerance* presented by Shelford. The law states that each environmental factor to which an organism responds has maximum and minimum limiting effects between which lies a range that is now described as the *limits of tolerance.* Organisms may have a wide range of tolerance for one factor and a narrow range for another. To describe a species with a narrow range of tolerance for a particular factor, we use the prefix 'steno'; for those with a wide range, we use the prefix 'eury'. The terms stenothermal- eurythermal and stenohaline-euryhaline refer to temperature and salinity tolerance respectively.

The *Liebig's law of the minimum* states that the growth of a plant is dependent upon the amount of foodstuff which is presented to it in minimum quantity. This statement is now generally combined with the *law of limiting factors* first proposed by Blackman who investigated the factors affecting the rate of photosynthesis and concluded that the rate is governed by the level of one factor operating at a limiting intensity. The two concepts combined form the so-called *Liebig-Blackman law.* Any condition or factor that affects an organism by exceeding the limits of tolerance is called a *limiting factor.* Among others, the temperature is quite often a limiting factor. The oxygen may also be a limiting factor and limits the distribution of organisms in land and aquatic environments. A combination of Liebig-Blackman law of minimum and Shelford's law of limits of tolerance gives us a combined concept of limiting factors which is a better expression of the environmental conditions affecting the organisms in natural ecosystems.

2.3.1 Light as an Environmental Factor

Light is defined as the radiation or brightness that enables things to be seen. The sun is the major source of light on earth. Solar radiation can be described in terms of its energy content and its wavelength. The energy comes in discrete packets called *photons* (Gk. *photo,* light). So we can think of radiation as photons that travel in waves. The solar radiation (electromagnetic spectrum) can be divided on the basis of its wavelength. The energy content of photons is inversely proportional to the wavelength of the particular type of radiation, *i.e.,* short wavelength radiation has photons of a higher energy content than long wavelength radiation. High-energy photons, such as those of the short wavelength *ultraviolet radiation* are dangerous to organisms because they can break down their cells and organic molecules. However, low energy photons, such as those of infrared radiation, do not damage cells. The *visible light* is considered most important because it is utilized in photosynthesis. Thus, visible light is only a small portion of the electromagnetic spectrum (380 n m to 750 n m). It is made up of a number of different wavelengths of radiation. When it is passed through a prism, we see these wavelengths as different colours of light. The quality, intensity and the duration of light are important.

2.3.1.1 Light and plants

Light affects plants of all habitats in many ways. The relationship of light intensity to photosynthesis in both terrestrial and aquatic plants follow the same general pattern of a linear increase up to an optimum or light saturation level. The intensity of light at which the rate of photosynthesis is just sufficient to meet the requirement of respiration is called *compensation point.* The orientation of a plant towards light is called *phototropism.* Plants may be classified ecologically according to their requirements of light and shade; for example, long- day plants, short- day plants, and day- neutral plants. *Sciophytes,* plants growing in shady places, usually have large leaves, which are thin in texture and are sparsely distributed on the stem. In *heliophytes,* plants that can grow well in the light; leaves are small, thick and crowded together on the stem. Variations in light and rate of transpiration produce marked differences in leaf structure. In sun exposed plants, leaves have thicker cuticle and cell walls. In sea, plants are restricted to euphotic zone due to their dependence upon light for energy. Phytoplankton exhibit diurnal movement in response to light, moving down during bright light of the day. Thus, light as an environmental factor affects the pattern of distribution of plants through its effect on photosynthesis, chlorophyll production, leaf structure, stomatal movement, transpiration, and production and growth of flowers, seeds and fruits.

2.3.1.2 Light and animals

Light affects metabolism of animals. The absence of light results in the loss of colour among the cave animals. Formation of pigments depends on light. The role of pigmentation and protective colouration is well known in many terrestrial animals such as chameleon, leaf insect *(Phyllium),* stick insect *(Caracisus),* and deadleaf butterfly *(Kallima paralecta).* Many inhabitants of caves and deep sea, where light has no ecological significance, have vestigial eyes or are blind. In certain lower animals, the speed of locomotion is regulated by light. This phenomenon is called *photokinesis.* The response of an organism to the duration of light and day-length is known as *photoperiodism.* It plays an important role in the life cycle of many animals.

Metamorphosis in insects and wing production in aphids are affected by day-length. *Salmon* larvae undergo normal development only when they get sufficient light. Some fish can be induced to breed earlier by artificially changing day-length. Light also affects the movement and migration of honeybees, locusts and birds. The phenomenon of *diapause* in insect eggs and pupae, which means spontaneous arrest of development at any stage, is generally triggered by changing photoperiod. The fact that numerous species of insects interrupt the development of the young stages in summer, despite the apparently favourable high temperature and abundance of food, and enter a resting phase (diapause) can definitely be attributed to the influence of light. In diapausing state insects have highly reduced respiratory rates, viscous body fluids, and antifreeze substances for winter diapause. Some herbivorous insects may be in diapause for about 10 months of the year, only becoming active when their host plant is available. Thus, diapause serves to synchronise the whole life cycle of the insect with the environmental conditions.

The endogenous rhythms of animals are also affected by light. In many countries now light therapy is being used to cure endogenous depression and the use of light therapy may even prove useful in controlling reproduction.

2.3.2 Temperature as an Environmental Factor

Temperature is defined as the degree or measure of heat or cold. Temperature is the most important physical factor of the environment and exerts its influence on plants and animals in various ways. *In poikilothermic* (Gk. *poikilos,* changeable) animals, the body temperature (Tb) depends on the temperature of environment. They have a variable Tb, whereas in *homeothermic* animals, such as birds and mammals, a constant Tb is maintained (37 to 40ºC). E*ctotherms* have a Tb, which is mainly dependent on external heat sources, whereas *endotherms* have a Tb, which depends on their internally generated metabolic heat. However, ectotherms can be *thermoregulators* or *thermoconformers,* but endotherms are always thermoregulators and use both behavioural and physiological mechanisms to regulate their Tb. The terms *stenothermal* and *eurythermal* are used to describe narrow and wide range of temperature tolerance respectively.

2.3.2.1 Temperature and metabolism

All metabolic processes are controlled by temperature. In many organisms, especially poikilothermic animals, the rate of metabolism increases with increasing temperature up to a certain limit. Each animal has a lower and upper limit of temperature tolerance. An 'optimum' temperature is required for the activity of most enzymes. Sudden changes in temperature affect the rate of respiration and other activities in crustacea and other animals.

2.3.2.2 Temperature and development

An optimum temperature is necessary for the growth and development to proceed at a normal rate. The life history of the dragonfly, *Tetragoneuria,* is so regulated by temperature that at one place two years are required to complete the development, whereas at another place only one year. The ambient temperature also affects the absolute size as well as the relative proportions of certain body parts in many animals. For example, variation in the size of ears of arctic fox *(Alopex lagopus)*, red fox *(Vulpes vulpes)* and desert fox *(Megalotis zerda)* are indicative of the role of temperature in different climatic conditions. Since the heat is lost through the surface, the small ears of arctic fox help to conserve heat, whereas the large ears of desert fox help in heat loss and evaporation.

2.3.2.3 Temperature and reproduction

In many animals, the sex ratio is affected by the temperature of environment. In *Daphina,* under normal condition parthenogenetic eggs are produced, which develop into females. But when the temperature of environment is raised, they give rise to sexual eggs, which after fertilization may develop either into males or females. In certain cases, the temperature exerts a more severe limiting effect on organisms when the moisture is either very high or very low, than when it is moderate. Similarly, moisture plays more critical role in the extremes of temperature. For example, the cotton boll weevil cannot develop if the relative humidity is less than 40% or more than 88%, no matter how favourable the temperature may be. Similarly, the animal remains dormant, regardless of the humidity, if the temperature is lower than 10º C or higher than 39º C. Thus, interaction of temperature and humidity decide the number of days required for development in cotton ball weevil..

2.3.2.4 Other Effects of temperature

Temperature affects animal behaviour and structure. Experiments show that *Drosophila* may undergo structural modifications at high temperatures. The vestigial wings of one of the mutants will develop into normal wings at high temperatures. In *Drosophila,* the temperature may affect the mechanism of heredity by affecting the position of genes and the behaviour of chromosmes. In higher plants, the development of cold tolerance is triggered by environmental signals such as temperatures of 0-5º C usually accompanied by decreasing daylength. Acclimatization may also increase the freezing tolerance in plants. During frost hardening many species of plants and animals accumulate compounds such as free amino acids, glycerol, sorbitol, that protect their membranes from the effects of dehydration. Insects also have strategies that allow them to survive during the low temperatures of winter. However, the ability of plants to acclimatize to higher temperatures appears to be less than to low temperatures.

2.3.3 Soil (Edaphic) Factor

Soil as an ecological factor is of great significance, as it affords a medium for anchorage of plants and a depot of water and minerals. Soil also serves as habitat for numerous animals. It can be defined as the weathered layer of the earth's crust with living organisms and their products of decay intermingled. Thus, soil is not only a factor of the environment of organisms but they produce it as well. Soil of any area is affected by climate as also by the vegetation growing on it. As a result, soils are distributed on a latitudinal pattern corresponding approximately to vegetation zones (Chapman and Reiss, 1995).

2.3.3.1 Soil formation

Soil is formed by disintegration and decomposition of rocks due to weathering (action of rain water, running streams, glaciers, wind, and temperature) and the action of soil organisms such as bacteria, fungi, protozoa, and earthworms, and also interactions of various chemical substances present in the soil.

2.3.3.2 Soil profile

Soil profile consists of several distinct horizontal layers often called soil horizons. The upper or A- horizon is the *topsoil.* It contains most of the organic matter consisting of litter and humus.

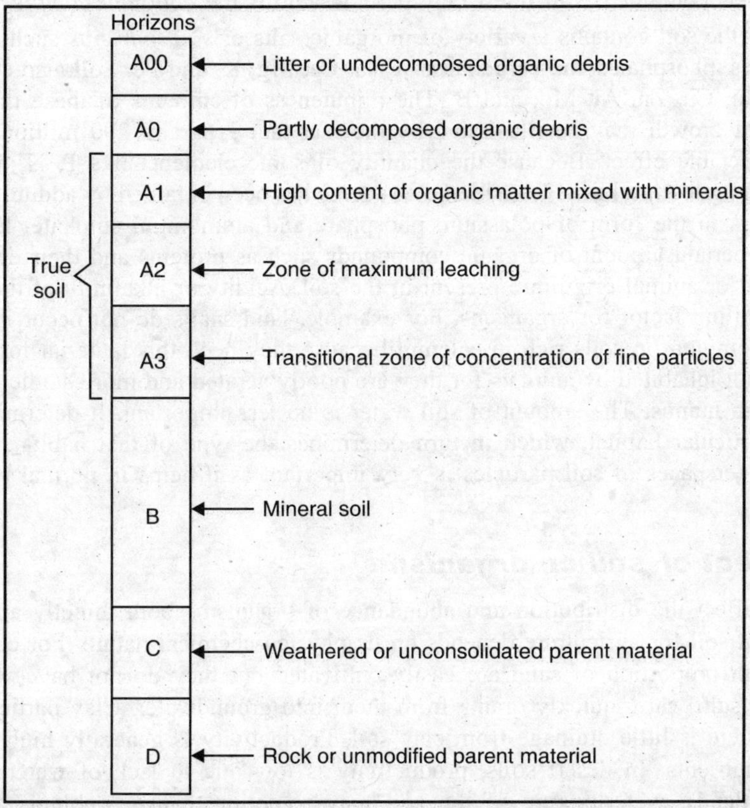

Fig. 2.1 Soil profile. The thickness of soil horizons varies according to the nature of soil

It also has a zone of leaching through which dissolved materials move downward. The roots of small plants are embedded in topsoil. The next or B- horizon is composed of mineral soil (subsoil). The third or C- horizon represents the unconsolidated parent material. The last or D-horizon represents the rock or unmodified parent material. (Fig. 2.1).

The exact nature of the soil profiles depends upon the interaction of climatic and biotic factors, contour of the land, and the type of parent rock in the given area. Hence, the relative thickness of horizons varies from soil to soil. For example, soils formed in grasslands have a deep A- horizon made up from decaying grasses over many years. In forest soils, both the A and B horizons have enough nutrients to allow root growth. In tropical rain forests, the A- horizon is more shallow and the B- horizon is deeper, signifying more extensive leaching.

2.3.3.3 Physico-chemical nature of soil

The physico-chemical composition of soil is an important factor, which determines the presence or absence of particular plants and animals. Physically, soils may be classified into (*a*) *clay soil,* containing more than 30 per cent of clay (size of particles less than 0.002 mm in diameter) ; (*b*) *sandy soil,* containing 10 per cent each of clay and silt (particle size 0.05-0.002 mm) with a large amount of sand (particle size 0.05-1.00 mm); and (*c*) *loam,* containing about 30 per cent each of clay and silt, the rest being sand. Loam is considered the best soil for plant growth and is exploited for agriculture.

Humus of the soil is of considerable importance to both plants and animals. It consists of organic matter in various stages of decomposition, usually present in the topsoil. Soil microorganisms act upon the nitrogenous organic compounds of humus, converting them into nitrates, which are absorbed by plants. Thus, humus is the source of plant food and the seat of most of the bacteria in the soil. Humus combined with finest clay particles forms the *colloidal complex* of soil.

Chemically, the soil contains a variety of inorganic salts or *soil nutrients* such as chlorides, nitrates, sulphates, phosphates and carbonates of Na, Ca, Mg, K, and Fe. Soil also contains trace elements like Mn, Cu, Zn, Al, Mo, and B. The minuteness of amounts of these trace elements involved in plant growth may be illustrated by the fact that 1 part in 200 million parts of Zn will have a detectable effect. Because the quantity of some elements like P, N, and K is the limiting factor in agricultural soils, considerable success has been obtained by adding the mixtures of these elements in the form of potassium phosphate and ammonium sulphate. Besides these nutrient salts, a certain amount of organic compounds such as proteins and their decomposition products of plant or animal origin are present in the soil. Acidity or alkalinity of the soil is also an important limiting factor for organisms. For example, land snails do not occur in acidic soil; they are more common on soils rich in calcium because they need this material for their shells. Peaty soils are not inhabited by animals for they are poorly aerated and more acidic. Earthworms prefer soil rich in humus. The amount of soil water is no less important. It determines the type of flora in a particular habitat, which in turn determines the type of that habitat. The soil air present in the interspaces of soil particles is very important as it helps in normal respiration of soil organisms.

2.3.3.4 Effect of soil on organisms

Soil types affect the distribution and abundance of organisms both directly and indirectly. The suitability of soil for agriculture depends on its physico-chemical nature. For example, soils containing a high proportion of sand are easily cultivated but they do not have water-holding capacity. As a result, water quickly drains from them into groundwater. Clay particles can hold water well, so there is little drainage from clay soil. Productivity is generally high in soils rich in humus and nutrients. In desert soils, productivity is low due to lack of water and organic material. In wetlands, waterlogging of the soil may sometime cause problems as anaerobic conditions are produced. Mangroves get rid off this problem with the help of pneumatophores and aerial roots, which can take in oxygen from the air. The nature of soil in any habitat also affects the distribution of animals. Animals such as worms and termites mix up the different horizons. Many soil organisms like bacteria, fungi, protozoa, earthworms and rodents, are useful agents in altering the soil. Some burrowing animals make the soil loose for better aeration and movement of water.

Soils are also affected by the vegetation growing on them as also by soil pollution. Soil is polluted by a number of inorganic and organic wastes including sewage, fertilizers, pesticides, detergents and dead animals. Pesticide pollution of agricultural soils kills many nontarget organisms and disturbs the ecological balance. Persistent chemicals may remain in the soil for a long time and in turn several food chains may be adversely affected. Soil erosion also affects productivity adversely.

2.4 BIOTIC ENVIRONMENT

Biotic or living environment is constituted by relationships both within and between species. A relationship within the organisms of the same species is called *intraspecific relationship,* whereas a relationship between two different species is known as *interspecific relationship.* These relationships can be found in any community.

2.4.1 Interspecific Relationships

Interspecific relationships between two (or more) species can be discovered in any community and belong to two main categories —symbiosis and antagonism.

2.4.1.1 Symbiotic relationships

Symbiosis means 'living together'. It is a beneficial relationship between two different species in which one or both the species are benefited and neither species is harmed. Symbiotic relationships include *commensalism* (one species benefited, another called host not affected), *protocooperation* (both the species are benefited, the relationship is favourable to both but not obligatory) and *mutualism* (both the species or symbionts are benefited, the relationship is favourable to both and obligatory).

Commensalism occurs when one species is benefited from a symbiotic relationship. All commensalistic relationships are facultative, as the commensals neither harm nor help their hosts and the hosts also appear neither to resist nor to foster the relationship in any way. Examples of commensalism showing more or less continuous contact with the host are offered by a great variety of epiphytes and epizoans. All *epiphytes* use trees only for attachment and manufacture their own food by photosynthesis. In *Vanda,* an epiphytic orchid, special kind of aerial roots hang freely in the air and absorb moisture with the help of their special absorptive tissue called velamen. Sessile invertebrates that grow on plants or other animals represent many permanently fixed commensals. For example, hydroids like *Hydractinia* live as commensals on the gastropod shells occupied by crabs.

Barnacles attached to the skin of whales provide another example, an association, which provides wider distribution and feeding opportunities for the sessile crustaceans. Examples of commensalism without continuous contact also occur. An interesting example of a commensal living within its host is that of a small tropical fish, *Fierasfer.* This species finds shelter in the cloaca of a sea cucumber, moving out for food and returning to the cloacal cavity at its own will. The suckerfish, *Echeneis,* attaches to the underside of a shark, thereby securing protection, wide geographical dispersal and scrapes of food. The shark neither benefits nor suffers in any respect.

Protocooperation is a relationship between two species, which is favourable to both but not obligatory. The association of a crab and a coelenterate shows an interesting example of protocooperation. The sea anemone, *Adamsia palliata,* grows on the back of the hermit crab, *Eupagurus prideauxi,* or is sometime 'planted' there by the crab. It protects the crab with the help of its nematocysts that prevent the approach of predatory fish, which feed on the crab. The sea anemone is transported by the crab from place to place and obtains particles of food when the crab captures and eats another animal. In this case the crab is not absolutely dependent on the coelenterate, nor *vice versa*. Hence the association, through favourable to both, is not obligatory for their existence.

Mutualism occurs when both the species benefit from a symbiotic relationship. Mutualism may be facultative, where the species involved in the association can exist independently. It may be obligatory, where the relationship is imperative to the existence of one or both the species. Mutualism may occur between two animal species, between two plant species, or between an animal and a plant species. An example of obligatory mutualism between two animal species, without continuous contact, is the association between aphids and dairy ants (Fig. 2.2). Dairy ants keep the tiny green aphids (plant lice) as food suppliers. Aphids secrete honeydew, a sugar and protein mixture, on which the ants depend. A common species of garden ant, for example, places aphids on the roots of corn. Aphids feed there and the ants thereafter "milk" these "antcows" by gently stroking them. At the approach of winter the aphids are carried into the ant nest and are put back on corn roots the following spring. Thus, ants obtain food from the aphids, and the aphids in turn secure protection, food and care from the ants.

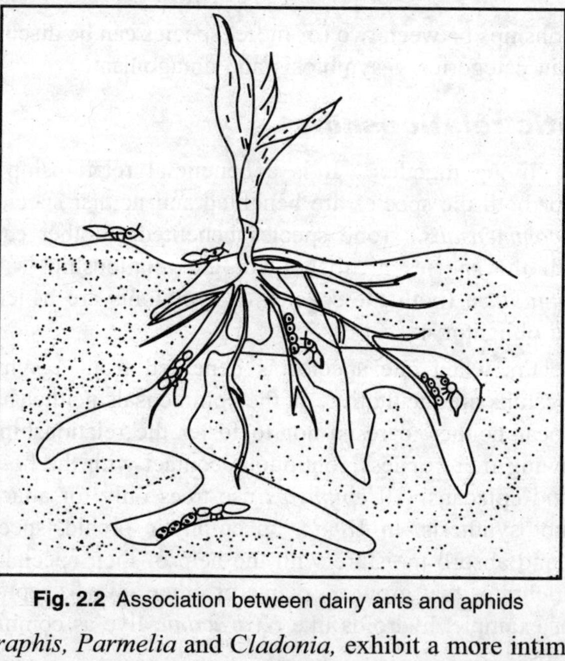

Fig. 2.2 Association between dairy ants and aphids

Lichens such as *Graphis, Parmelia* and *Cladonia,* exhibit a more intimate form of mutualism between two plant species. Each *lichen* is a symbiotic association between a fungus and an alga. In many species of lichens, the algal symbiont is *Trebouxia.* The algae manufacture food for themselves as well as for the fungus. The fungus in turn contributes water and carbon dioxide that enable the alga to synthesize food. If they are separated from their association, they lead a precarious life, more particularly the fungus. Nitrogen fixing mutualistic bacteria of the genus *Rhizobium* infects the roots of many leguminous plants. They enter the roots through the root hair and the infected root cells respond by developing *root nodules* (Fig. 2.3). In these root nodules, the host plant provides nutrients for the bacteria and the bacteria in turn contribute fixed nitrogen to the host plant, which becomes independent of supplies of nitrogen from the soil.

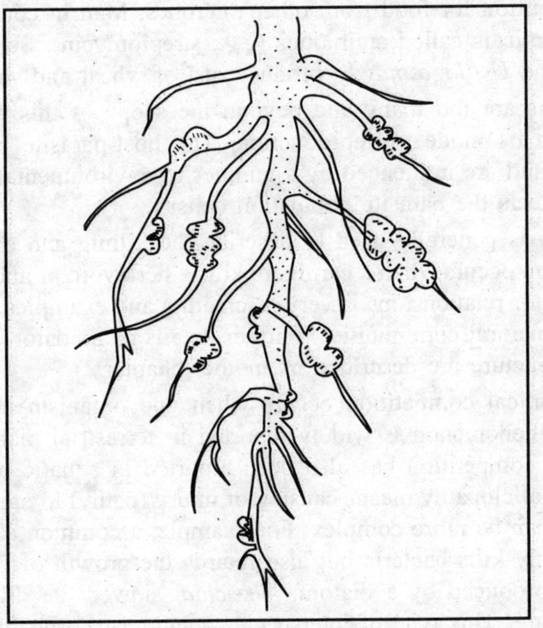

Fig. 2.3 Root nodules of a leguminous plant containing nitrogen-fixing bacteria

There are many examples of symbiotic mutualism between an animal and a plant species. In green hydra, *Chlorohydra viridissima,* the cells of gastrodermis contain many unicellular algae, *Zoochlorellae*, which are transferred by the host eggs from one generation to another. The algae supply food and oxygen, and in turn receive protection, water and other materials essential for continued photosynthesis. Pollination of flowers by insects is yet another manifestation of plant-animal mutualism. In many cases, the relationship between the insect mouth parts and the shape of the flower is complimentary with an adaptation to each other. For example, bee- pollinated plants, such as Orchidaceae, usually have zygomorphic flowers. These are tubular, so the bee can adjust itself to the flower shape and then push its head into the flower. Similar adaptations occur in butterfly and moth- pollinated flowers and pollinator's mouth parts. Interdependence of the yucca moth, *Pronuba yuccasella,* and the yucca flower is very fascinating. These are co-evolutionary relationships. In some cases flower species have coevolved to avoid competition among themselves by flowering at different times of the year, or by attracting different species of insect as their pollinator.

2.4.1.2 Antagonistic relationships

The antagonistic relationships are manifested through parasitism, predation, competition and antibiosis between two different species.

Parasitism is a relationship in which one species (parasite) is always benefited at the cost of the other species (host). A parasite derives nourishment from the host and in many cases finds protection and living space on or inside the host. The host is usually larger than the parasite and an efficient parasite does not kill its host. Many parasites require a secondary host for dispersal or completion of their life cycle. For example, mosquito species serve as vectors for the protozoan malarial parasite, *Plasmodium.* Animals that become infected and serve as a source from which other animals can be infected are known as reservoir hosts. Animals may also be parasitic on plants. Nematodes infest the roots of plants. Wasps or gnats, especially on oak, roses form galls. A variety of insect larvae are leaf miners, wood borers, cambium feeders, and fruit eaters. Plants themselves may be parasites either on other plants or on animals. Bacteria and fungi are among the most important disease-producing organisms in animals. Filamentous bacteria, *Actinomycetes* are chiefly decomposers in the soil. Some of them produce chemicals to reduce severe competition for food from other microbes. Man in combating several bacterial diseases uses these chemicals called antibiotics (*e.g.,* streptomycin). Some basidiomycetes such as *Puccinia graminis* and *Ustilago maydis* produce rust of wheat and smut in corn respectively.

Parasitic adaptations are too many and beyond the scope of this treatise. The size of the parasite also influences its mode of reproduction. The host-parasite relationships have great ecological significance and are influenced by a number of environmental factors. The density of host population also affects the parasite population density.

The term *predation* is generally used to describe the killing and eating of one species by another. Typical predation occurs when a carnivore kills a herbivore or another carnivore for food. The study of prey-predator relationships is very fascinating and examples of this relationship may be discovered in many natural communities. Further details of predator- prey interaction and its effect on community structure are described in another chapter.

Allelopathy or chemical competition occurs when one organism uses chemicals to cause harm to another. This phenomenon is widely reported in terrestrial plants (Seigler, 1996). The occurrence of chemical competition has also been reported in aquatic plants (Gopal and Goel, 1993). In general sense, allelopathy means causing of injury (pathy) to other organisms (allelo) by chemical means, but it can be more complex. For example, a common alga, *Chlorella,* produces a bactericide that not only kills bacteria but also retards the growth of *Daphnia,* which feed on *Chlorella.* A chemical produced by a diatom, *Nitzschia,* slowed the division rate of *Chlorella* grown in the same culture. This type of interspecies antagonism probably exerts control on the abundance of different phytoplankton species in water bodies, and in some cases may influence the seasonal succession of species, so common in nature.

There are many examples of chemical interaction among crop plants. Barley, *Hordeum vulgare* may inhibit germination and growth rate of several weeds, even in the absence of competition for nutrients or water. Allelopathic effects of crop plants against weeds, called heterotoxicity, may be of great benefit in agricultural systems. Some researcher suggested that the abundant oils in the leaves of eucalyptus trees in Australia promote frequent fires in the leaf litter, killing the seedlings of competitors. The study of allelopathic agents is currently a very active field in plant ecology, and it is too early to say how much the distributional patterns of plants are determined by interactions involving toxins or antibiotic substances.

Competition is another manifestation of antagonistic relationship. Competition may occur among species and between species. However, in nature competition may not be always apparent, although it is occurring. Increased growth rate of some tree species after the removal of other species provides direct evidence of competition for water, light and nutrients. Individuals of the same species may compete for food, living space and mate. Competition for food also applies to animals of different species that depend on the same type of food. Competition may result in death of some competitors, but this is usually from fighting or being deprived of food rather than being killed for food as in predation, or by disease as in extreme parasitism. The severity of interspecific competition depends on the extent of similarity or overlap of resource requirements of different organisms and the shortage of supply in the habitat. Such a competition may have several effects on the populations of competing individuals.

Schoener (1983) divided competition into six categories. These are *consumptive competition* (based on the utilisation of some renewable resource), *pre-emptive competition* (based on the occupation of open space), *overgrowth competition* (occurring when one organism grows over another thereby depriving it of light, water, or some other resource), *Chemical competition* (by production of a toxin acting at a distance), *territorial competition* (defence of territory), and *encounter competition* (involving transient interactions over a resource resulting in loss of time or energy, physical harm, or theft of food). These competition mechanisms are defined in terms of the capabilities of organisms and the habitats in which they occur (Ricklefs and Miller, 2000). In terrestrial environments, consumptive competition is most common. However, pre-emptive and overgrowth competition are more common in marine habitats. Territorial and encounter competitions occur only among animals of all habitats, predominating in terrestrial ones.

Darwin emphasized that competition is usually most intense between closely related species or organisms. As they have similarity in structure and habits, the competition is more severe among species of the same genus than between species of different genera. Substitution experiments, developed by Wit (1960), are helpful for studying plȧnt competition. In such experiments, ratios of two plant species are varied, but their total density is kept constant. The results are portrayed on replacement series diagrams showing the relative strengths of interspecific and intraspecific competition. Experimental work with agriculture crops suggested these competitive interactions are very acute in field populations. Experiments with oats *(Avena)* and other plants have demonstrated strong asymmetry in interspecific competition. It is the most critical factor confining a species to a particular niche. According to the competitive exclusion principle of Gause (1934), stabilised populations of more than one species cannot simultaneously and completely occupy an ecological niche. Thus, interspecific competition results in the segregation of species into different niches.

When two populations compete, it is likely that one of them is more strongly affected by competition than the other. This is called *asymmetrical competition.* In plants, root competition for nutrients and water is generally symmetrical, whereas shoot competition for light is asymmetrical. Examples of asymmetrical competition are available in animal kingdom also (Resetarits, 1997). All interspecific interactions are shown in Table 2.2.

Table 2.2 Interspecific interaction of organisms

S.No.	*Types of interaction*	*Species*		*Explanation of relationship*	*Result of interaction*
		A	*B*		
1.	Neutralism	0	0	A and B independent	Neither population affects the other
2.	Competition	–	–	A and B competitors	Eliminated from niche
3.	Amensalism	–	0	A, amensal; B, inhibitor in allelopathy, or antibiotic in antibiosis	A inhibited, B not affected
4.	Parasitism	+	–	A, parasite; B, host	A benefited, B inhibited
5.	Predation	+	–	A predator; B, prey B, inhibited	A benefited (obligatory for A),
6.	Commensalism	+	0	A, commensal; B, host	A benefited (obligatory for A), B not affected
7.	Protocooperation	+	+	A and B cooperators	Favourable to both, but not obligatory
8.	Mutualism	+	+	A and B partners, or symbionts	Favourable to both and obligatory

Types 2 and 3 may be called as "negative interactions", Types 6, 7 and 8 as "positive interactions" and 4 and 5 as both.

0 no net effect; – inhibition (or loss); + benefit (or gain)

2.5 ADAPTATIONS TO ENVIRONMENTAL CONDITIONS

Organisms try to adapt to the prevailing environmental conditions. Thus adaptation may be defined as a process by which the organism becomes fitted to its environment. In other words, adaptation enhances the ability of an organism to cope with its environmental conditions. In an attempt to cope with their environment organisms may have to undergo changes in their morphology, behaviour and physiology, thus showing structural (or morphological), behavioural, and physiological and biochemical adaptations. Sometime an organism may be required to adjust to laboratory condition. Such slight adjustments are through *acclimation.* It is defined as a physiological compensatory mechanism induced under experimental conditions. For example, a fish in the laboratory may be acclimated to different temperatures, say from 10 or 15°C to more than 35°C. Sometime a physiological compensatory change is induced over a longer period of time under natural environmental conditions. This is called *acclimatisation.* If a person is suddenly taken to high attitude and required to live there for a longer time, say a few months, he will acclimatise himself to low concentration of oxygen in the air and accordingly there will be marked physiological changes in his blood parameters.

Adaptations are sometimes divided into *individual* and *population* adaptations. *Individual adaptations* are the morphological or physiological adjustments on the part of an individual. For example, in summer we perspire more than in winter. *Population adaptations* indicate adaptive changes in the whole population. These adaptations help in the survival of the population just as adaptive adjustments of an individual help for his survival. Morphological adaptations in different groups of animals living in the same environment or habitat are called *convergent adaptations.* They are the result of convergent evolution or adaptive convergence; for example, shape of the body of a fish, *Ichthyosaur* (extinct reptile) and *Porpoise* (a mammal). On the other hand, there are *divergent adaptations,* which are the result of adaptive radiation or divergent evolution. Certain features of adaptive radiation found in the evolution of reptiles are principal characteristics of macroevolution in all groups, from tree ferns and trilobites to seed plants and insects. These common elements provide the key to understanding the process of evolution above the species level (Savage, 1966).

2.5.1 Adaptations of Desert Animals

Two characteristics of the desert *i.e., high temperature* and *scarcity of rainfall* (about 25 cm of rainfall per year) determine the occurrence, distribution and adaptations of desert animals. In addition to these the highly characteristic "spaced distribution" of the desert vegetation also affects the desert fauna. The characteristic animals of the desert are insects, small rodents, and reptiles. Birds and mammals are comparatively rare or absent. Poikilothermic animals, which match their internal temperature to the external, are adapted to live easily in the desert.

Insects and reptiles of the desert have impervious integument and excrete nitrogen in the form of uric acid (uricotelic animals). In desert insects, the evaporation from respiratory surface is reduced to the minimum by the internally invaginated spiracle system. Rodents of the desert can live on dry seeds, succulent cacti and other plants that store water and do not require drinking water. They stay in burrows during the day to avoid evaporation and water loss and conserve water by excreting highly concentrated urine and by not using water for temperature regulation. Hibernation is necessary for many ectothermic animals of the desert. Desert snakes and lizards hibernate 0.5 m or more in sand, under rocks or in burrows of other animals. Some ants and crickets burrow deeply into the ground. Burrows of kangaroo rat penetrate 50-65 cm below the surface. Protection of eye, ear and nostril against the sand is an important adaptation. In the burrowing snake, *Typhlops,* the eyes are covered by minute shields. In camel, the eyes are well protected by long eye lashes and are kept high above the ground by long neck. The ear opening of desert animals are also well protected by hair or scales. The presence of poison glands is also an adaptive feature of desert animals. Protection against natural enemies is achieved through protective colouration or spiny covering. Their colour generally matches with their surroundings. The so-called horned lizard, *Phrynosoma,* of the Western American deserts and the spiny devil, *Moloch horridus*, of Australia are classical examples of the desert animals having a spiny covering on their body.

Physiological adaptations of desert animals are no less interesting. Some of them, for example the desert lizard, *Sauromalus obesus,* have the mechanism for selective cooling of blood to the brain. In some African gazelles and ungulates the brain is supplied with cool blood. Some desert reptiles, when exposed to heat, increase their respiration frequency and breathe with an open mouth, a situation reminiscent of panting. Other lizards use saliva regurgitated from the mouth over the throat region to achieve cooling. A camel *(Camelus dromedarius)* in the desert uses several methods to reduce heat gain. Firstly, the heat is stored by an increase in body temperature. In a camel deprived of water the body temperature in the morning may be about 34°C, which rises to 41°C in the late afternoon. This 7°C increase in body temperature corresponds to about 29000 kcal of heat, which equals a saving of 5 litres of water. The increase in body temperature decreases the heat flow from the environment, the fur reducing the heat gain from the environment. A camel can tolerate a greater degree of water depletion of the body, and when water is available, it may drink more than one third of its body weight (Schmidt- Nielsen, 1964). Further adaptations to desert life are splayed hooves, which are ideal for walking on sand and the hump which stores fat.

Thus, adaptations of desert animals are actually the adjustments to protect themselves against high temperatures, to live without water, and to conserve water as far as possible.

2.5.2 Adaptations of Desert Plants

Extreme desert is without any vegetation and rainfall. However, some deserts receive less than 5 cm of rain per year. Deserts include arid regions, which contain considerable vegetation, commonly called *xerophytes,* in the form of desert bushes and shrubs, succulents such as cacti which store water, and other small plants which avoid drought by growing only when there is adequate moisture. In some deserts, plants survive as seeds for several years until a little rainfall provides conditions suitable for growth and flowering.

The *ephemeral plants* of the desert can complete their entire life cycle in a few weeks. *Geophytes* of the desert avoid periods of drought by surviving as underground bulbs or corns. *Succulents* such as cacti *(Opuntia)* and *Euphorbia* have adaptations enabling them to survive above ground for the whole year. They possess thick cuticles, a very low surface area to volume ratio and sunken stomata, which open during night so as to minimise water loss in transpiration. In succulents, the stem is flattened and the function of leaf is performed by stem. In *Opuntia,* leaves are reduced to spines. The stem epidermal surface is covered with waxy material, which prevents water loss, by cuticular transpiration. In some desert plants like *Aloe* and *Agave* the leaves are thick leathery or succulent. Absence of broad leaves and abundance of spines further protect desert plants from being eaten by animal consumers. The roots in perennial xerophytes are very deep. The stem is usually covered with corky layers, which reduce water loss. Many of these plants produce gums and resins as in *Acacia;* their presence also helps in formation of layers, which prevent water loss.

2.5.3 Deep sea Adaptations

Deep sea is characterised by a set of environmental conditions, which in turn determine the adaptations of deep-sea forms. Of all the oceanic zones, light penetrates only into the euphotic zone; the remaining zones are aphotic or devoid of light (bathyal, abyssal and hadal zones). The term hadal zone is used to designate the perpetually cold and dark supreme depths of the ocean. It is characterised by high pressure, low temperature, absence of light, calmness of water, absence of phytoplankton and other producers, scarcity of food and resulting competition and soft bottom. Apart from containing bacteria and perhaps fungi, the community of aphotic zones is characteristically animal. Because of lack of cues like light and temperature change, it was once thought that there was no seasonality in the deep sea, but this is not so. The amount of detritus coming into deeper waters depends on seasonal production cycles at the surface. Due to extreme pressure the bodies of deep-sea fish and other animals are very much compressed. Some *bathypelagic fishes* are 'economy' designs showing reduction of bony skeletons except for their jaws and contain watery muscles. In some of them *e.g.,* angler fish *(Melanocetus)* and gulper eel *(Eurypharynx)* the water content is very high, about 95%. Even without gas-filled swim bladders, they are near neutral buoyancy. Their hearts are very small, they have very little red muscle and low haematocrit values. Some deep-sea fishes exhibit greatly enlarged eyes or the so-called teleoscopic eyes, which are highly effective in visioning lights of very low intensity. In some deep-sea fishes the retina is composed of a number of tiers of rods, presumably arranged to absorb all the limited light that enters the eye (Nicol, 1989). In some other deep-sea fishes, eyes are very small as they are of little apparent use, and still others are without eyes.

Many deep-sea animals produce their own light by means of luminous organs (bioluminescence), *e.g.,* lantern fish. In anglerfish, *Linophryne* (Fig 2.4), the light is used as a bait to attract prey. Besides attracting and seeing the prey, luminescent display may also serve for species and sex recognition. Another interesting adaptation of deep-sea fish is the enormous mouth enabling them to swallow prey larger than themselves (*e.g.,* the gulper, *Eurypharynx,* whale fish, *Cetomimus*). All benthic fishes lack swim bladders and rest on the bottom, sometimes like tripod fishes *(Bathypterois* spp.) on stiff elongate fin rays (Fig 2.4). Correlated with soft substratum, many of the deep-sea animals have long appendages, abundant spines, stalks or other means of support, as illustrated by tripod fish, lampshells and crinoids. Perhaps these appendages are very useful in the darkness and serve for contact reception, or compensate for the difficulties of vision.

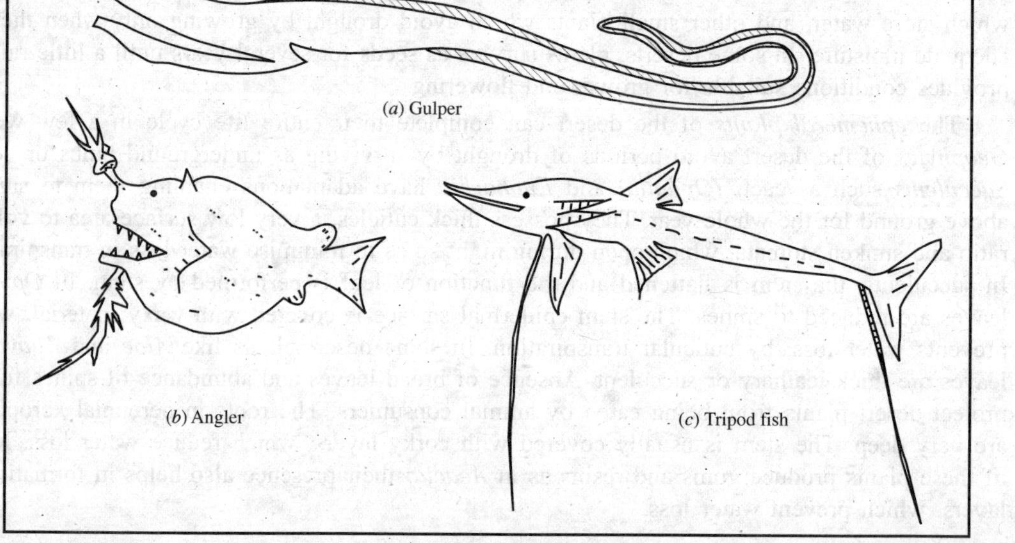

Fig. 2.4 Some deepsea fish. (not to scale)

2.5.4 Adaptations of Hydrophytes

Hydrophytes are plants normally growing in water and swampy or marshy habitats. They are broadly classified as *free-floating* hydrophytes and hydrophytes *attached to the substratum.* They are further divided into several ecological groups. These are : (*i*) free-floating hydrophytes such as *Salvinia, Pistia, Eichhornia* (water hyacinth); (*ii*) submerged floating such as *Utricularia, Najas, Ceratophyllum*; (*iii*) submerged rooted such as *Vallisneria,Sagittaria, Isoetes, Hydrilla*; (*iv*) rooted with floating leaves such as *Nymphaca, Marsilia, Potamogeton natars*; (*v*) rooted emergent such as *Typha, Limnophila.* Land plants, which live permanently in moist environment are called *hygrophilous* and may have considerable resemblance to hydrophytes. Important morphological, anatomical and physiological adaptations of hydrophytes are as follows:

2.5.4.1 Morphological adapatations

Roots are often poorly developed (*e.g. Wolffia, Salvinia*) or completely absent. Root hairs have completely disappeared in some species of *Ceratophyllum.* However, many hydrophytes have well developed root systems. For example, *Eichhornia* and *Pistia* have well developed adventitious roots. In these free-floating rosette plants, the roots are at least partly responsible for preserving the stability of their rosette leaves. The stem may be well developed (*e.g., Ceratophyllum, Hydrilla*), reduced (*e.g. Wolffia ,Spirodela*), or modified into rhizome (*e.g., Vallisneria*). Stem is spongy due to well developed *aerenchyma.* The spongy and elongated petioles of water hyacinth exhibit the development of so-called aerenchyma. Many hydrophytes show *heterophylly, i.e.,* production of different forms of leaves in the same plant. In submerged aquatics with free floating aerial leaves, the submerged leaves are generally linear, ribbon-shaped or finely dissected while the aerial leaves are complete and rounded or lobed. In *Sagittaria,* the heterophylly seems to be due to the difference in the intensity of light in the submerged and aerial parts. More light intensity in the aerial parts favours formation of entire leaves (Fig. 2.5). The floating leaves have waxy surface so that water may not wet the surface and block stomata. The presence of mucilage on the aerial organs seems also an adaptation for protecting them from getting wet. The propagation of most hydrophtes is vegetative.

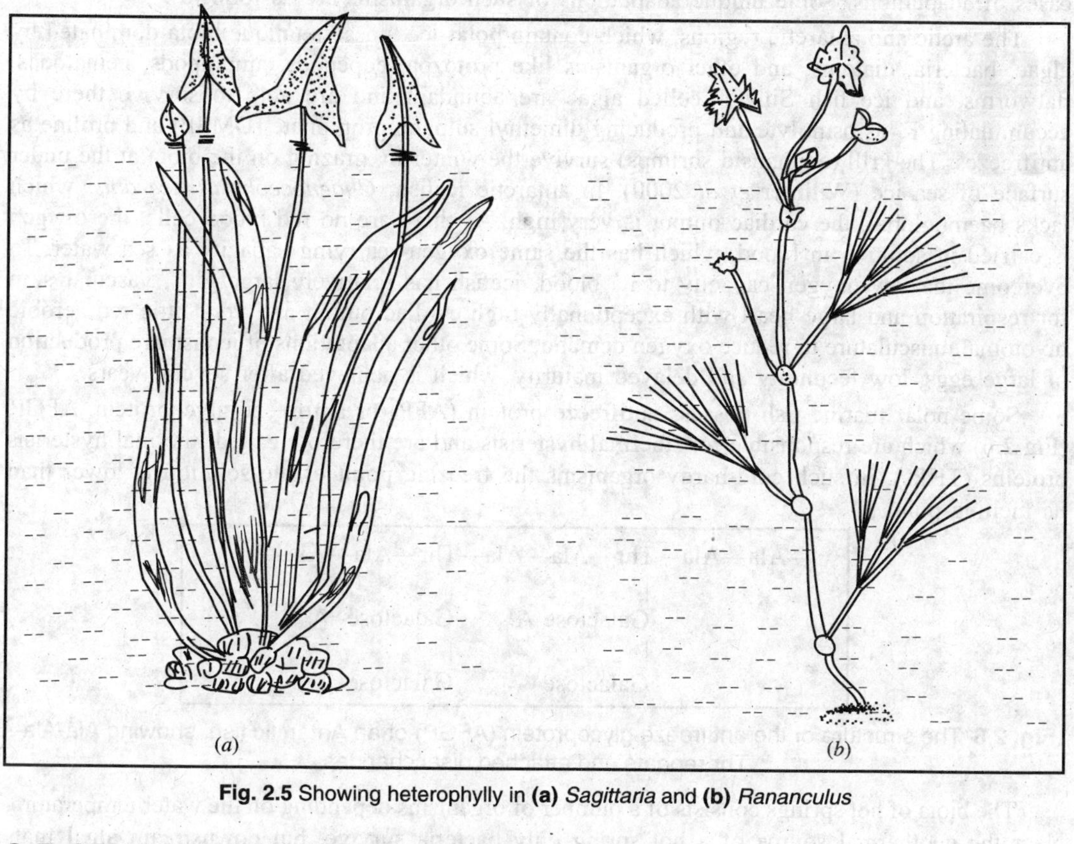

Fig. 2.5 Showing heterophylly in **(a)** *Sagittaria* and **(b)** *Rananculus*

2.5.4.2 Anatomical adaptations

An important anatomical feature of all hydrophytes is the *sponginess* of their tissues. They have extensive air-spaces in their leaves, stems and roots. This helps in keeping the buoyancy of plants and facilitates exchange of gases. The epidermis usually lacks cuticle or periderm and stomata are not present in the submerged leaves. Hydrophytes have reduced vascular elements. The absence of secondary growth in thickness of stems and roots is also an important characteristic of hydrophytes.

2.5.4.3 Physiological adaptations

Petioles of floating- leaved hydrophytes have a great capacity for renewed growth, which is perhaps regulated by *auxins* (phytohormones). In lotus, the long petioles seem to adapt themselves to the depth of water, thus keeping the leaf lamina on the surface of water. Many hydrophytes maintain active photosynthesis. Some carbon dioxide evolved during respiration is stored in the air spaces and utilized during photosynthesis. In hydrophytes, the osmotic concentration of the cell sap is equal to or slightly higher than the surrounding water. In some hydrophytes, the initiation of the flowering phase may depend upon nutrition. The *Utricularia,* a carnivorous plant, flowers when grown aseptically in inorganic nutrient medium but only when supplemented with organic nitrogenous compounds, *e.g.*, mixture of peptone and meat extract.However, in *Eichhornia,* flowering seeems to be influenced by temperature rather than photoperiod. In some hydrophytes(*e.g.*, water-lily) food is stored in the rhizome, in others such as *Sagittaria,* tubers are formed.

2.5.5 Examples of Extreme Cases of Adaptations

Organisms living in polar areas, hot springs, hot deserts, and deepest seas exhibit extreme cases of adaptations. Some unique adaptations of such organisms are as follows :

The arctic and antarctic regions, which contain polar ice, possess unique biota dominated by algae, bacteria, diatoms, and other organisms like protozoa, copepods, amphipods, nematodes, flatworms, and ice fish Single- celled algae are abundant and are able to survive there by accumulating K^+ as osmolyte and producing dimethyl sulphonipropionate (DMSP) and proline as antifreezes. The krill (euphausid shrimps) survive the winter by grazing on the biota at the under surface of sea ice (Willmer *et al* 2000). In antarctic icefish, *Chaenocephalus aceratus,* which lacks haemoglobin, the cardiac output is very high. As there are no red blood cells, the oxygen is carried in solution in blood, which has the same oxygen- carrying capacity as sea water. To overcome the low oxygen capacity to its blood, icefish has relatively large gills, vascular skin for respiration and large heart with exceptionally high cardiac output and very little red aerobic myotomal musculature to reduce oxygen demand. Some other adaptations of icefish are production of large eggs, low fecundity and delayed maturity, which is achieved after several years.

Some polar marine fish possess antifreeze protein (AFP) or antifreeze glycoprotein, AFGP, (Fig 2.6) which are responsible for 'thermal hysterisis and are therefore called 'thermal hysterisis proteins (THPs). In such cold-hardy organisms the freezing point of the solution is lower than its melting point.

```
—Ala—Ala—Thr—Ala—Ala—Thr—Ala—Ala—
           |             |
      Galactose A   Galactose- A
           |             |
      Galactose     Galactose
```

Fig. 2.6 The structure of the antifreeze glycoprotein (AFGP) of an Antarctic fish, showing Ala-Ala-Thr repeats and attached disaccharides

The biota of hot springs consists of a number of organisms depending on the water temperature. Near the geothermal source of a hot spring only bacteria survive, but downstream algal mats appear and are sometimes colonised by protozoans at about 65°C. Some ostracods, larval flies, rotifers and nematodes may appear as the water further cools down to about 50°C. In some hot springs, midge larvae and chironomids may be found. In India, hot springs are scattered in many parts of the country. Recently Saha and Datta Munshi (1982,1983) and Saha (1993) have described limnology of some thermal springs of Bihar. Higher thermal tolerance of bacteria, algae and some other organisms of hot springs may be attributed to their physiological adaptations. For example, *Thermoplasma acidophilus* occur at 59°C and pH 1-2. It has the smallest genome so far known in non-parasitic bacteria. Certain anaerobic thermoproteales have been isolated from hot springs and volcanoes. They are extremely thermophlic, growing optimally at 85-105°C and they depend on 'sulphur respiration'(Schlegel, 1995). They oxidise hydrogen and reduce elemental sulphur to hydrogen sulphide. *Pyrodictium occulatum* is a bacterium that grows at temperatures up to 110°C. It is anaerobic and grows autotrophically with H_2 as H-donor and sulphur as hydrogen acceptor.

Desert adaptations have already been discussed. Desert species have higher body temperatures (desert beetle, 33°C; desert iguana, 41°C; camel, 34°C and 41°C in the morning and afternoon respectively). Extreme desert conditions are very harsh and certain organisms survive there due to their behavioural, physiological and biochemical adaptations. In hyper-arid deserts, there is no rain, wind speed is very high, and night is extremely cold. The soil is sandy, stony and salty. All these conditions make life almost impossible for most kinds of organism. However, some organisms survive as they live in burrows. For example, the peak day temperature on the surface may be 45°C to 65°C, but in the burrow of the scorpion, *Hadrurus,* it is only 32°C to 40°C (Hadley, 1970). Besides low temperature, burrows may also give a constant high humidity, which helps in osmotic regulation. In some naked mole rats the burrows are about 75 cm below surface, providing not only a low temperature (25-28°C) but also high humidity which helps minimize evaporative water loss. Certain locomotory tricks as in sidewinder snakes and dune spider are adopted to minimize body contact with the surface (Pough *et al*, 1999; Henschel, 1990). Endogenous rhythms of desert invertebrates further help in their survival. Some ants and arthropods of hyper-arid deserts possess raised thermal tolerance and extremely low metabolic rate. Mechanisms for water vapour uptake in desert mites and desert cockroach, *Arenivaga,* are quite interesting and many desert insects use discontinuous respiration. In the kangaroo rat, *Dipodomys,* the expired air is cooler than the inspired air, which is a device for nasal heat exchange.

In the deepest seas organisms live in pitch dark, under high pressure and low temperature. In deep sea hydrothermal vents, the water is rich in silicates, hydrogen sulphide and sulphide minerals containing magnesium and iron, which crystallise as the hot water meets the cold sea water, forming 'black smokers', 'chimneys' and other mineral deposits. The vent water contains no oxygen, and high sulphide and ammonium levels. The organisms, especially prokaryotes, which occur in deep-sea thermal vents, are extremophiles having highly specialized enzymes. The vent fauna consists of annelid and pogonophoran tubeworms, bivalve molluscs and some decapod crustaceans. The 'Pompeii worm', *Alvinella pompejana,* occurs on vent chimneys. It is so far the most eurythermal organisms on record; at the tube mouth the water temperature is around 22°C but in other parts of the tube it may be up to 82°C, showing a gradient of 60°C along the length of the worm's body. The deep-sea crustacean, *Rimicaris exoculata,* which also occurs around vents, has no eyes, but has a 'thoracic eye' used to detect the radiation emitted by the very hot vent water (up to 350°C). The vent animals also possess other physiological adaptations in nutrition and respiration to cope with toxic hydrogen sulphide. Their sulphide-binding proteins play an important role by drawing free sulphide into general circulation. In pogonophoran worm, *Riftia pachyptila,* the symbiotic bacteria in trophosome possess adaptations to protect the worm from sulphide toxicity. These bacteria can probably use nitrate in addition to oxygen, producing ammonia and nitrate as end products. The nitrate respiration enables the bacteria to function in an environment which is very low in oxygen (Willmer *et al*, 2000).

In deepest parts of the ocean the effects of depths and pressure are very obvious. In all deep-sea forms, the rate of oxygen consumption decreases rapidly with increasing depth and they use various methods to increase buoyancy. Some of the benthic forms and bacteria are anaerobic or can live in very little oxygen condition. Some other adaptations of deep-sea animals are chemoreception, bioluminescence, electroreception, and echolocation.

MODEL QUESTIONS

1. How organisms are affected by their environment ? Give an account of the limiting factors of the environment.
2. Give an account of light as an environmental factor.
3. Give an account of temperature as an ecological factor.
4. How physico-chemical and biological characteristics of soil affect soil organisms? Why 'humus' of the soil is important?
5. Define 'symbiosis'. Give an account of symbiotic relationships.
6. Give a brief account of antagonistic relationships of organisms.
7. Give an account of ecophysiological adaptations of desert animals and plants.
8. Give an account of adaptations of hydrophytes.
9. Describe the adaptations of organisms living in polar areas and deepest oceans.
10. Write notes on the following :
    (*a*) Liebig-Blackman law, (*b*) Diapause, (*c*) Acclimatization,
    (*d*) Soil profile, (*e*) Soil nutrients, (*f*) Parasitism,
    (*g*) Competition, (*h*) Adaptive convergence, (*i*) Desert adaptations.
    (j) Limiting factor

MULTIPLE CHOICE QUESTIONS

11. Who gave binomial system of naming organisms ?
    (A) Charles Darwin (B) Pat Willmer
    (C) W.B. Clapham Jr. (D) Carolus Linnaeus
12. In the following names, which kingdom is missing ?
    (A) Monera and Protista (B) Fungi and Plantae
    (C) Animalia (D) None
13. How fungi are different from other plant organisms ?
    (A) They are decomposer
    (B) Digestion extracellular and absorption of organic matter
    (C) Many are parasitic on other multicellular organisms
    (D) All the above
14. To which group cyanobacteria belong ?
    (A) Protista (B) Prokaryotes
    (C) Bacteria (D) Filamentous bacteria
15. Cyanobacteria are also called :
    (A) Eubacteria (B) Archaebacteria
    (C) Blue-green algae (D) True bacteria
16. How man is different from other organisms in the biosphere ?
    (A) Tool-making capacity (B) Articultated speech and language
    (C) Ingenious brain power
    (D) As a dominant species with all the above characteristics.
17. Extremities such as tail and ears of mammals are shorter in colder regions. What is the name of this rule ?
    (A) Liebig's rule (B) Bergmann's principle
    (C) Allen's rule (D) Shelford's law

18. In all the organisms, the death at subzero temperture is caused by :
    (A) Freezing of the body tissues (B) Slow metabolic rate
    (C) Low oxygen consumption (D) Hormonal imbalance
19. Which is the most fundamental diurnal rhythm that fluctuates due to daily change of light?
    (A) Food intake (B) Respiration
    (C) Photosynthesis (D) Reproduction
20. Humus of the soil is defined as :
    (A) Organic matter
    (B) Undecomposed organic debris
    (C) Colloidal complex
    (D) Organic matter in various stages of decomposition
21. In which of the following organisms, body temperature(Tb) is independent of ambient temperature(Ta) ?
    (A) Fish (B) Reptile
    (C) Bird and mammal (D) All the above
22. Which of the following is considered the most eurythermal organism ?
    (A) Round worm (B) Flat worm
    (C) Earthworm (D) Pompeii worm
23. Antifreeze glycoprotein is present in :
    (A) Marine fish (B) Marine algae
    (C) Antarctic fish (D) Antarctic krill
24. Sponginess of tissues is a characteristic of :
    (A) Hydrophytes (B) Some hydrophytes
    (C) Xerophytes (D) Land plants living in moist environment
25. Plants growing in shady places are called :
    (A) Hydrophytes (B) Sciophytes
    (C) Heliophytes (D) Day-neutral plants
26. Which of the following plants shows heterophylly?
    (A) *Eichhornia* (B) *Hydrilla*
    (C) *Sagittaria* (D) *Vallisneria*
27. Molluscs are abundant in the soil which is rich in:
    (A) Iron (B) Nitrate
    (C) Aluminium (D) Calcium
28. Which of the following interaction is beneficial to both the partners?
    (A) Parasitism (B) Commensalism
    (C) Competition (D) Mutualism
29. Mutualism between two plant species is shown by:
    (A) Lichens (B) *Chlorella*
    (C) Yucca flower (D) *Zoochlorelle*
30. Nodules of a leguminous plant with nitrogen fixing bacteria is an example of:
    (A) Plant-animal Mutualism (B) Plant-plant mutualism
    (C) Animal-animal mutualism (D) None of the above

3 CHAPTER Population

3.1 INTRODUCTION

In environmental biology, the study of populations, especially animal populations, is considered very significant because biotic communities are made up of several or many population groups and cannot be understood without a study of their dynamics and interrelationships. A Population must adjust to its environment like the organisms, although its environmental relations are not so simple.

A *population* may be defined as a group of organisms of the same species occupying a given area at the same time. Thus, for the purpose of this text, it is generally taken to be as species population. A population may be subdivided into *demes* or local populations, which are groups of interbreeding organisms, or the smallest collective unit of a population. A population may consist of either unitary or modular organisms. Insects, fish, amphibians,birds and mammals are examples of *unitary organisms.* In all such organisms, each individual is produced from a zygote and the form and development of individuals is highly predictable. But in *modular organisms,* the zygote develops into a unit of construction or module, which produces further modules to form a branching structure; therefore, the form and timing of development of individuals are unpredictable. Sponges, corals and plants are examples of modular organisms. Plants grow by accumulating persistent modules. In grasses, different units or ramets are attached by rhizomes or stolons. However, some plants like duckweed, exhibit individual ramets.

3.2 POPULATION PARAMETERS

Each population is a separate entity showing several characteristics such as spacing, size, density, natality, mortality, age structure, growth, fluctuations and cycle.

3.2.1 Spacing

The pattern of spacing of organisms is an important feature of every population. Spacing may be uniform, random or clumped. *Uniform* spacing is shown by invertebrate populations; *random* spacing occurs among plants; clumped spacing may be shown by mating individuals. Besides such clumping for reproduction, animals form swarms, herds, flocks or colonies for feeding and protection. According to *Allee's principle,* each species has an optimum clumping value and overcrowding beyond this state or undercrowding below it tends to act as a limiting factor in population regulation.

3.2.2 Size and Density

The success of a population is reflected in its size and density. It is easy to define the population size when population consists of unitary organisms; it is simply the number of individuals present in a given area. *Population density* is the number of individuals per unit area(or volume). The unit of area in measuring the density varies. This may be expressed, for example, the number of microorganisms per litre of water. For larger animals or plants, square metre, square km or acre may be the unit. The density of a population may be studied in several ways. The abundance of a population in a given geographic region is termed the *average* or *regional density.* A region, however, usually includes unfavourable parts in which the individuals may be rare or absent, and more suitable niches in which the individuals occur in large number. The abundance of a population within its niche is called the *economic* or *niche density,* which is always higher than the regional density. However, in many cases, it is impractical to determine the *absolute density* of a population and it may be easy to know the *relative density* of the population (*i.e.*, the area *x* has more individuals than area *y*).

The density of a population may be almost constant, increasing or decreasing at a given time. If the population density is increasing, this means that the natality (birth rate) is higher than the mortality(death rate). If it is decreasing, then obviously the number of individuals in the population who are dying per unit time exceeds the birth rate, the number of individuals produced per unit time. When birth rate and death rate are almost equal, the population density will appear to be constant for a given time. Thus, changes in population size overtime can be calculated by adding new born individuals (*B*) and the number of immigrants (*I*) to the original population at time t (Nt) and subtracting the number of deaths (*D*) and emigrants (*E*) to give a new population size at the time t+1 (Nt+1). This may be represented by the equation:

$$Nt + 1 = Nt + B + I - D - E$$

According to krebs (1994), the four processes – natality, mortality, immigration, and emigration– are the primary population parametres and affect the density as shown below :

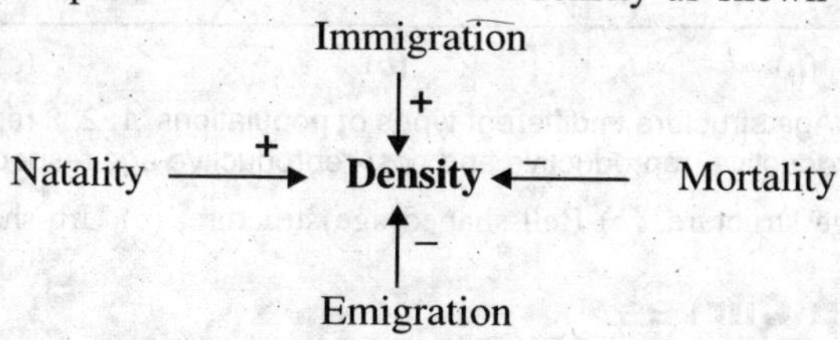

The estimation of population density may be made by several methods. The use of *quadrat* is the simplest method. A quadrat is a sampling area of any shape, but a quadrat of square or rectangular shape is preferred. The size or area of the quadrat depends on the sampling need and may be 0.5 m^2 or $1m^2$. The general procedure is to count all the individuals on several quadrates of known size and to extrapolate the average to the whole area. The use of quadrates is the most common method for sampling plants. The method has also been used for sampling animals such as ants, earthworms and mollusks. One of the recent and more widely used methods is *CMR technique, i.e.* capture a number of animals, tag or mark them in one fashion or another (*e.g.,* by clipping their tails, piercing their ears, or attaching some marker), release them, and then at a subsequent time obtain another sampling of animals. The proportion of the originally marked animals in the second sampling within the same area provides the basis for estimating the population density.

$$\frac{\text{Marked animals in second sample}}{\text{Total caught in second sample}} = \frac{\text{Marked animals in first sample}}{\text{Total population of animals}}$$

For example, $20/100 = 100/N$

or $N = 500$

This method is known as the *Petersen method.* It is also called the *Lincoln method* by wild life ecologists. Dahl (1919) used this method for estimating trout density in lakes.

3.2.3 Age Composition

There are three major *age groups* in a population. These are *prereproductive, reproductive,* and *postreproductive.* These categories, proposed by Bodenheimer(1985), have considerable value in visualising the status of a population. In natural population, the age structure is almost constantly changing. In a rapidly growing or *expanding population* as in the case of bacteria, yeast and housefly, birth rate is high and population growth may be exponential. Under these conditions each successive generation will be more numerous than the preceding one and, as a result, a *pyramid* age structure results. In a *stable* population, a bell-shaped age structure is obvious as the prereproductive and reproductive age groups become more or less equal in size and the post reproductive group remains the smallest. In a declining population, an urn-shaped age structure is shown (Fig. 3.1) indicating the smallest size of prereproductive category or young individuals. However, we rarely find a natural population having a stable age structure or a stationary age distribution. In some long-lived species such as fish and trees, *dominant year-classes* may be found. In human population, differences in mortality rates have little effect on age composition, while changes in birth rates have a large effect on age structure (Barclay, 1958).

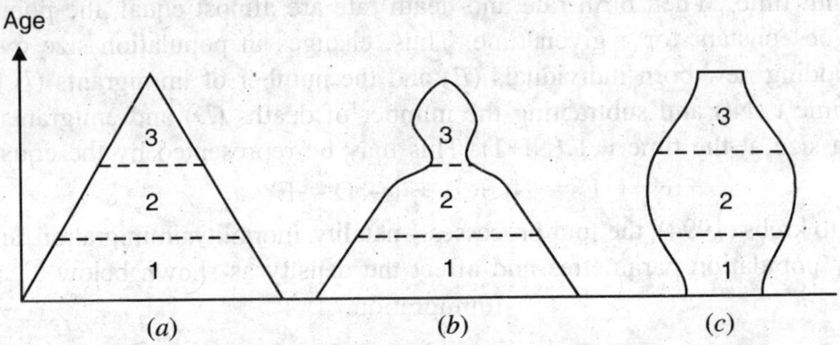

Fig. 3.1. Age structure in different types of populations. 1, 2, 3 represent pre-reproductive, reproductive and post-reproductive age respectively.
(*a*) Pyramid age structure, (*b*) Bell-shaped age structure, (*c*) Urn-shaped age structure

3.2.4 Survivorship Curves

In order to understand a survivorship curve we should first know what is a cohort. A *cohort* may be defined as a group of individuals of the same age recruited into a population at the same time. A cohort life table is also called the *dynamic life table,* which may be defined as the age-specific survival and fecundity of a cohort followed from birth to death of the last individual.

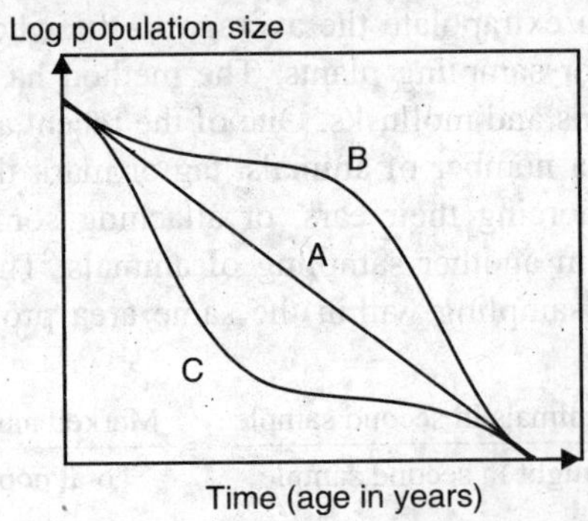

Fig. 3.2. The three general forms of survivorship curves (see text for details)

A survivorship curve is a plot of the declining size of a cohort as the individuals die usually with time on the horizontal axis and log population size ($\log_{10}$ lx) on the vertical axis where lx is the proportion of the original cohort still alive. A survivorship curve is a graph to show the number of individuals, which survive per thousand of population through each phase of life represented by age in years. Another method of showing a survivorship curve is to use a log scale for the number of individuals. This type of semi-log plot indicates the population dynamics of the ogranism and is therefore quite useful. Fig. 3.2 shows three general shapes of survivorship curves on a semi-log plot. These curves illustrate different patterns of population dynamics. (*A*) shows a steady mortality rate during each phase of life; (*B*) shows that the highest mortality rate occurs in older individuals; (*C*) shows high mortality in the young with very little in adults.

However, natural populations rarely show extreme cases like (*B*) or (*C*) and they often show a wavy curve. The shape of the survivorship curves may be used to show the critical periods in the life of an organism. As the curve becomes steeper there is an increase in mortality, which indicates some environmental or other effect on the population. The normal plot curve shows most clearly the early pattern of mortality in the young individuals, whereas the semi-log curve shows the pattern of mortality in the older individuals more clearly.

3.2.5 Population Growth

Of the several dynamic features of a population, the most important is the *growth,* the capacity for increase in individual numbers. By measuring the size or density of a given population from time to time we can obtain important information regarding its rate of increase, and can also predit future changes in its size. To measure the growth of a population of a particular species in the laboratory, say bacteria or yeast cells, first we transfer a relatively small number of the given individuals to a flask containing appropriate nutrient solution, and then proceed to measure the population increase over a desirable period of few days.

3.2.5.1 Logistic Growth

When a population is allowed to grow in a limited space (environment) it shows logistic growth. If we plot a graph of the number of bacteria or yeast cells against time, we find a typical S-shaped sigmoid curve called the *population growth curve* (Fig. 3.3), which usually consists of four phases. The first phase initially shows a slow rate of growth in population. This phase is often called the *lag period* , and is followed by a rapidly accelerating stage of multiplication. These together constitute the first phase of population growth curve ,which is called the *phase of accelerating multiplication.* It is followed by a phase of extremely rapid population increase known as the *logarithmic phase* in which the number of individuals increase at an exponential rate. During the logarithmic phase, which represents the period of maximum increase in population, the number of organisms increases in a geometric fashion (2,4,8,16,32,64 and so on) instead of in arithmetic manner (2,4,6,8,10,12 and so on).

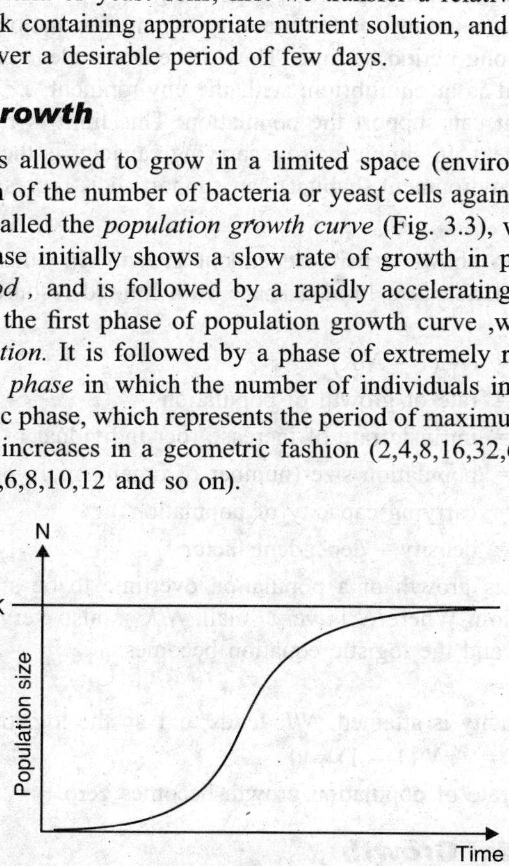

Fig. 3.3 Population growth curve, where the population growth rate slows down as the population approaches the environmental carrying capacity, K.

The logarithmic phase is succeeded by the *phase of decelerating multiplication,* during which the rate of population increase begins to slow down. The final phase, called the *equilibrium phase,* is one in which there is essentially no net change in the population since the birth rate is more or less balanced by the death rate.

The S-shaped logistic growth curve is an adequate description for the laboratory population growth of organisms having simple life cycles such as *Paramecium,* yeast and the like. However, the population growth in organisms with more complex life cycles often does not follow the logistic growth curve. In particular, the stable asymptote of the logistic growth curve is not achieved in natural populations, but numbers fluctuate.

Although the total number of individuals and the time intervals are considerably different for organisms, the S-shaped curve is typical under optimal conditions of the growth of all populations of living organisms ranging from bacteria to man. However, the population growth curves of all organisms fall off from the logarithmic phase to become ultimately the equilibrium phase instead of continuing unchecked. The reason is the environmental resistance comes to operate. In other words, the various physical, chemical and biotic factors of the environment are incapable of supporting a maximal rate of reproduction for any organism over a long period of time. Such limitations as the restricted supply of nutrients, the inhibitory effects of accumulated metabolic waste products, and the hindrances imposed under natural conditions by competition, parasitism and predation markedly increase the death rate besides restraining the rate of reproduction.

The environment, therefore, is capable of supporting only a certain number of organisms of any given species, thus accounting for the inability of any population to maintain itself in the logarithmic phase for a long period of time. Thus, it becomes clear that there is no net change in the population when it is at equilibrium with the environment, *i.e.,* it has reached the limit at which the environment can support the population. This limit is referred to as the *carrying capacity* of the environment. In simple words, carrying capacity is the number of individuals in a population which the environment (habitat) can support. It is represented by "K".

Logistic Equation

The logistic equation shows density-dependent growth. In other words, it describes the growth of a simple population in a limited space with limited resources. The logistic equation may be written as:

$$dN/dt = rN\,(1 - N/K)$$

where dN/dt = rate of growth of population

r = intrinsic rate of increase (per individual of population)

N = population size (number of organisms in population at time t)

K = carrying capacity of population

$(1 - N/K)$ = density – dependent factor

This equation predicts growth of a population overtime to be sigmoidal, as is commonly observed in real population. Where N is very small, N/K is also very small and therefore $(1 - N/K)$ is approximately 1 and the logistic equation becomes:

$$dN/dt = rN$$

As the carrying capacity is attained, N/K tends to 1 so the logistic equation becomes:

$$dN/dt = rN\,(1 - 1) = 0$$

In other words, the rate of population growth becomes zero.

3.2.5.2 Exponential Growth

When the curve of population growth quickly begins to rise very steeply, the population shows exponential growth (Fig. 3.4), A population growing exponentially increases according to the equation :

$$N_t = N_0 e^{rt}$$

Where Nt = the number of individuals in the population after t units of time

N_0 = initial population size ($t = 0$)

r = exponential growth rate

e = the base of the natural logarithm (about 2.72)

Exponential growth results in a continuous curve of increase (or decrease, where the rt term is negative) whose slope varies in direct relation to the size of the population. The term e^r is the factor by which the population increases during each unit of time, often written as the lower case Greek lambda (λ). That is, when $t = 1$,

$$Nt = N_0 e^r, \quad \text{or}$$

$$Nt = N_0 \lambda$$

The rate of exponential population increase may be written as:

$$dN/dt = rN$$

This equation encompasses two principles. First, the exponential growth rate (r) expresses population increase on a per individual basis. Second, the rate of increase (dN/dt)varies in direct proportion to the size of the population (N). Thus the equation is:

the rate of change in population size = the contribution of each individual to population growth × the number of individuals in the population

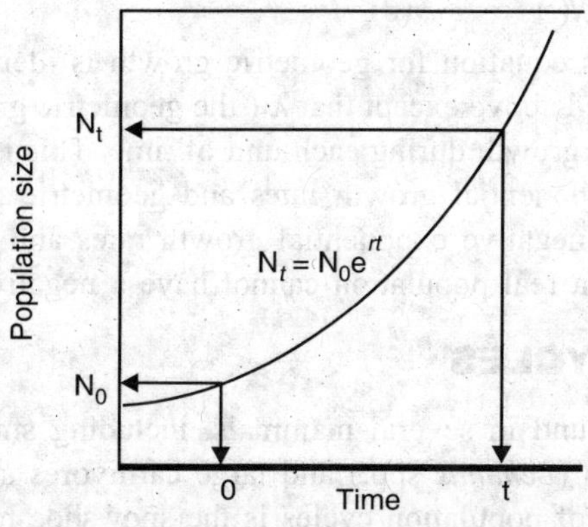

Fig. 3.4. The exponential growth curve for a population growing at rate r between time 0 and t. During this period the number of individuals increases from N_0 to N_t. (From Ricklefs and Miller, 2000)

Difference between Logistic and Exponential Growth

Logistic growth involves the growth of a simple population in a limited space with limited resources. On the other hand, exponential growth starts slowly and then proceeds faster as the population increases. If number of individuals is plotted against time, it yields a J-shaped exponential growth curve.

In logistic growth, the role of *environmental resistance* is important. After levelling off, a population with this type of growth typically fluctuates slightly above and below the carrying capacity. A plot of the number of individuals against time yields a sigmoid or S-shaped logistic growth curve.

3.2.5.3 Geometric Growth

Geometric growth (Fig. 3.5) may be defined as the population growth in which the rate of increase is proportional to the number of individuals in the population at the beginning of the breeding season. When young ones are added to the population only at specific times of the year, during well defined reproductive periods, the population is said to have geometric growth.

When populations increase periodically, the growth rate is most conveniently expressed as the ratio of the population in one year, $N(1)$, to that in the preceding year, $N(0)$. This ratio is expressed by the symbol λ, called the geometric growth rate. Therefore, $\lambda = N(1)/N(0)$. This may be compared to $\lambda = e^r$ as used in exponential population growth. In both the cases, λ represents a factor of population increase. To project the growth of a population, the initial number, $N(0)$, is multiplied by the geometric growth rate λ once for each unit of time passed. Hence,

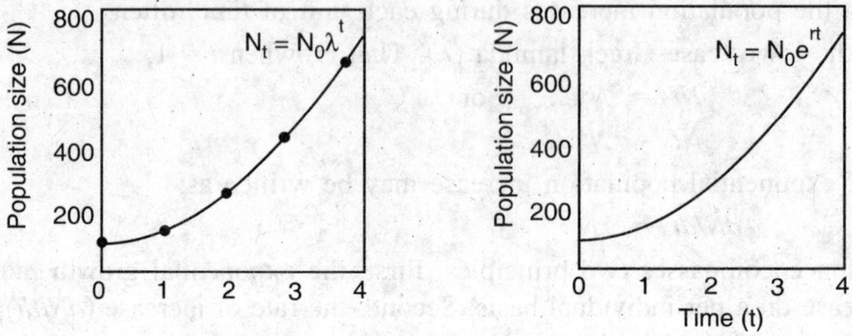

(a) (b)

Fig. 3.5 Populations undergoing (a) geometric growth and (b) exponential growth at equivalent rates ($\lambda = 1.6$, $r = 0.47$) (after Ricklefs and Miller, 2000)

$$N(t) = N(0)\,\lambda^t$$

It may be noted that this equation for geometric growth is identical to that for exponential growth ($Nt = N_0e^{rt}$) mentioned above except that λ (the geometric growth rate) takes the place of e^r, the amount of exponential growth during each unit of time. This is well known that increasing populations have positive exponential growth rates and geometric growth rates greater than 1; decreasing populations have negative exponential growth rates and geometric growth rates less than 1 but greater than 0 as a real population cannot have a negative number of individuals.

3.3 POPULATION CYCLES

Population cycles are found in several mammals, including small mammals such as voles (*Microtus* spp.)and lemmings (*Lemmus* spp.) and large carnivores such as lynx and foxes. One of the best known examples of population cycles is the snowshoe hare (*Lepus americanus*) and lynx (*Lynx canadensis*) cycle. The hare in the boreal forests of North America is a herbivore, and follows a 10-yr cycle. The hare population regularly displays a peak population every 10 years, which is then followed by a period of scarcity. The lynx also shows a somewhat similar cycle which reaches peak of abundance about a year after that achieved by the hare. These two cycles are apparently interrelated, since the lynx is the predator of the hare. Therefore, there is a close parallelism in cyclic abundance of these two species.

Other examples of cyclic populations are the lemmings (*Lemmus sibericus*) in the coastal tundra at Barrow, Alaska where it is the dominant herbivore every 3 or 4 years. Their population increases when they reproduce in nests of grass and sedges at the base of the snow pack, growing rapidly during the winter and attaining a peak in late spring. Their breeding ceases in May and during snow-melt many lemmings burrow to the surface and wander about to die. Large numbers of predators also attack the exposed lemmings. During the summer their survival is low and the population crashes to a low level where it remains for 1 to 3 years (Batzli *et al.*, 1980).

However, no single factor is made responsible for these population cycles. There are extrinsic theories of microtine abundance (Begon *et al.*, 1996). Predators play an important role in microtine cycles. Predators were absent when the lemming populations were low, but they multiplied fast during lemming peaks. Intrinsic theories of microtine abundance consider factors such as hormonal changes and dispersal.

3.4 POPULATION DYNAMICS AND MODELS OF POPULATION REGULATION

One of the central questions in population ecology is-what determines the size of a natural population? The answer is – different factors or mechanisms under different circumstances. In a modern synthesis on population regulation, Krebs (1994) emphasizes the role of the following factors:

(*a*) *Limiting factors* : A factor is defined to be a limiting factor if a change in the factor produces a change in the equilibrium or average density. For example, a disease may be a limiting factor for a fish population if fish population abundance is lower when the disease is present.

(*b*) *Regulating factors* : A factor is defined to be a regulating factor if the percentage mortality caused by the factor increases with population density; or, alternatively a factor is defined to be a regulating factor if the reproductive rate is reduced as the population rises. For example, a disease may be a regulating factor only if it causes a higher fraction of losses as fish density goes up.

Krebs (1994) has proposed two approaches to study population regulation. These are: key factor analysis and experimental analysis. *Key factor analysis,* proposed by Morris (1957), is a technique of analysing populations through the preparation of life tables (egg, larval, pupal and adult stage) and a retrospective analysis of year to year changes in reproduction and mortality. For each drop in numbers in the life table we define

$$k = \log (Ns) - \log (Ne)$$

where, k = instantaneous mortality coefficient

N_s = number of individuals starting the stage

N_e = number of individuals ending the stage

For example, 83 larvae entered the pupal stage, and of these, 54.6 were killed by predators which reduced the population to 28.4 m. Hence k (instantaneous mortality coefficient for pupal predation)

$$= \log (83.0) - \log (28.4)$$

$$= 0.47$$

However, key factor analysis assumes that all mortality factors are additive ($K = k_1 + k_2 + k_3 \ldots$) and it ignores compensatory mortality. Various mortality agents may cause additive compensatory losses in populations. Additive losses may limit or regulate population density, but compensatory losses may be irrelevant to both regulation and limitation.

In *Experimental analysis approach,* we try to identify limiting factors and study them experimentally. If we find that food is a limiting factor, we can increase the food supply and see if population size increases accordingly. But it should be noted that in many cases more than one factor may be involved (*e.g.,* food supply and parasite levels) and often it is not possible to manipulate a suspected factor (*e.g.,* weather). Krebs (1994) comments that experimental analysis is forward- looking and oriented toward hypothesis testing about mechanisms of regulation; key factor analysis is backward - looking and is confined to a descriptive analysis of a population; and theoretically both method should converge to provide an understanding of population regulation.

According to Krebs (1994), population regulation in plants must be discussed as regulation of biomass rather than numbers, because most plants are modular organisms. Crawley (1990) mentioned that plant ecologists have not usually addressed the problem of population regulation in the same way as animal ecologists, but the same general principles can be applied. As a plant population increases in biomass (and numbers), their reproduction or survival is reduced by shortage of factors like space, nutrients, water or light; or damage by herbivores, parasites and diseases. Since they have fixed location, competition for nutrients or light is considered an important factor in their population regulation. This competition has been described by the – 3/2 power rule, also called the yoda's rule or *self – thinning rule.* This rule describes the relationship between individual plant size and density in even – aged populations. Mortality or "thinning" from intrapopulation competition is postulated to fit a theoretical line with a slope of – 3/2, as mentioned below:

$$\text{Log } m = -3/2 (\log N) + K$$

where m = average plant weight (g)

N = plant density (individuals per square metre)

K = a constant

This line has been suggested as an ecological law (Westoby, 1984) that applies both within one plant species and between different plant species. Weller (1987) found that more shade tolerant gymnosperm trees had more shallow slopes than the predicated value. His results argue against a single, quantitative thinning rule for all plant species as they differ widely in competition for nutrients, water and light.

The questions as – what stops population growth beyond a limit, and what determines average abundance have been answered by Krebs (1994) on the basis of the following theories. These general theories focus on the interactions between the population and the environmental factors of food, weather, shelter and enemies like parasites, predators and diseases.

3.4.1 Biotic School

The biotic school suggests that density-dependent factors play a key role in preventing population increase and in determining average abundance. The two entomologists Howard and Fiske (1911) were the prototypes of biotic school of population regulation, which proposed that biotic agents like predators and parasitoids were the main agents of natural regulation. According to Nicholson (1933), who is considered as one of the pioneers of the biotic school, the only factor that can control populations is some kind of *competition* (competition for food, competition for space, or the competition of predators and parasites). Smith (1935) also favoured the main points of Nicholson and pointed out that density-dependent factors are mainly biotic in nature – competition, predation, disease; and that the density-independent factors are mainly abiotic or physical in nature, mainly the climate. But he also pointed out that climate might act as density-dependent factor under some circumstances. For example, in the case of protective refuges, if there are only so many of these to go around and all the unprotected individuals are killed by climate, the climate mortality would be density-dependent.

The biotic school emphasises that balance in nature is produced by density-dependent factors, which are usually biotic in nature such as predators and diseases.

3.4.2 Climatic School

Bodenheimer (1928) was the first to suggest that the population density of insects is regulated primarily by the effects of weather. According to him, weather effects both development and survival of insects. He was also impressed by the fact that weather was responsible for the largest part of the mortality of insects, as 80 to 90 per cent of the insects in their early stages of development are killed by weather factors. Uvarov (1931) in his paper "Insects and Climate" also pointed out the role of climatic factors on the growth, fertility and mortality of insects.

Thus, climatic school emphasises the role of weather factors and suggests that weather may act as a density-dependent control.

3.4.3 Self-Regulation School

According to this school, a population is regulated by *intrinsic factors.* From a study of rodent populations Chittay (1960) proposed that all species are capable of regulating their own population densities without destroying the renewable resources of their environment or requiring enemies or bad weather to keep them from doing so. The self-regulation school thus emphasises on events going on within a population and on individual differences in behaviour and physiology. The general premise of this school is that abundance may change because the quality of individuals changes. Population increase may be stopped by a deterioration in the quality of individuals as density rises rather than a change in environmental factors. Average abundance may be altered by genetic changes in populations. According to this school, both quality and quantity are important aspects of natural populations and therefore the most important environmental factor for such populations is – other organisms of the same species. Giving an example of territorial behaviour in birds, Wynne-Edwards (1962) suggested that food played an important role in the problem of self-regulation. In birds, population density is controlled by territoriality, which is a means to eliminate competition for food and further ensures that food supply will not be exhausted.

3.4.4 Comprehensive School

Thompson (1929) was the first to adopt the *comprehensive theory of natural control.* He included all factors from climate to parasites and suggested that the control was due to a complex of factors. Schwerdtfeger (1941) also favoured comprehensive theory and emphasised the role of joint action of many factors. According to him, a network of interactions are involved in population control. He termed this network the "*grandocoen*", which he defined as the totality of all the factors that affect the population changes of the species. This grandocoen varies from place to place and from time to time and differs for different species. Thus, he emphasised not only the connection between factors but also between species.

According to Andrewartha and Birch (1954), the numbers of animals in a natural population may be limited in three ways: (*a*) by shortage of food, or places to make nests; (*b*) by inaccessibility of these material resources relative to the animal's capacities for dispersal and searching; and (*c*) by shortage of time when the rate of increase (*r*) is positive. Of these three ways, they believe that the last is probably the mot important in nature. They also pointed out that in any population an animal's chance to survive and multiply depends on all the four components of the environment– weather, food, other animals and pathogens, and living place. Thus, there may be cases when population size is largely determined by weather, and other cases in which population size is mainly determined by other animals.

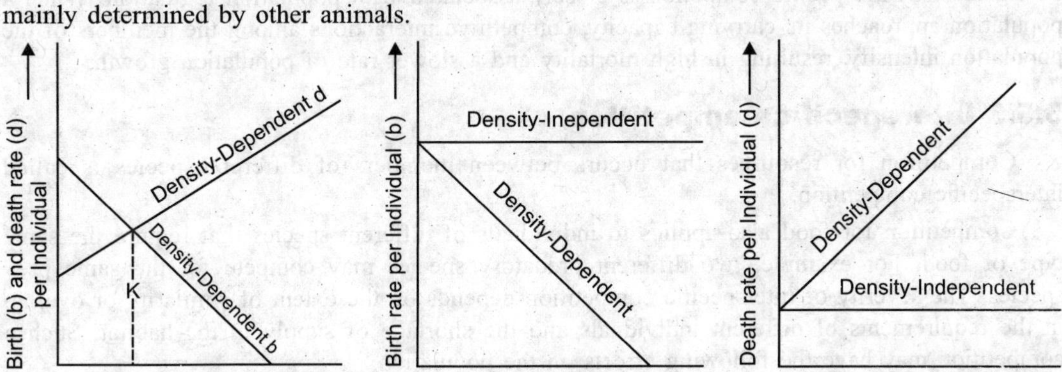

Fig. 3.6 Role of density-dependent and density-independent factors.

Population regulation implies that one or more factors are always involved in controlling population size whether they are constant or temporally variable in operation (Chapman and Reiss, 1995). The role of density-dependent and density independent factors is shown in Fig.3.5.

3.5 COMPETITION

Top-down control and bottom-up control of population size

These two hypotheses are not mutually exclusive (Krebs). According to biologists, genuine cases of *top-down control* systems by predators exist in a number of ecosystem; for example, large predatory fish controlling other fish populations in lakes.

According to *top-down control* hypothesis lynx preying on hares periodically reduce their population. The shortage of hares than reduces than lynx population, which allows the hare population to build up again. At some point the lynx population increases to take advantage of the increased supply of hares, starting the cycle again.

Researchers now hypothesize that periodic crashes in the hare population can also be influenced by the food supply of the hares. Large numbers of hares can die when they consume food plants faster than they can be replenished and have a decrease in the quality and quantity of their food. Once the hare population crashes, the plants recover and hare population begins rising again in a hare-plant cycle. If this *bottom-up control* hypothesis is correct, instead of lynx controlling hare population, the changing hare population may be causing fluctuations in the lynx population. With the help of these control mechanisms, the alternating abundance and scarcity every 10 years of the hare (*Lepus Americans*) and lynx (*Lynx Canadensis*) in Canada is explained.

3.5.1 Introduction

In nature, organisms compete for resources. Competition may be defined as an antagonistic relationship which is use or defence of a resource by one individual that reduces the availability of that resource to other individuals.

3.5.2 Intraspecific competition

Competition for resources that occurs between members of the same species is called intraspecific competition. It is very well known that members of the same species compete for food, shelter and mate.

Although there may be differences in the food requirements of the young and adult of a species, or the preferred pray size may differ for male and female, the general degree of overlap in resource use means that intraspecific competition is a major force in environmental biology of organisms. Intraspecific competion is closely associated with population regulation. When a population approaches its carrying capacity, competitive interactions among the members of the population intensify, resulting in high mortality and a slower rate of population growth.

3.5.3 Interspecific competition

Competition for resources that occurs between members of different species is called interspecific competition.

Competition for food also applies to individuals of different species that require the same type of food. For example, two different predatory species may compete for the same prey species. The severity of interspecific competition depends on the extent of similarity or overlap in the requirements of different individuals and the shortage of supply in the habitat. Such a competition may have the following effects on the population.

3.5.3.1 Regulation of population size

The role of competition in regulating population size is directly effective by causing mortalities through fighting, nest destruction and loss of food supplies. In also results into increased predation and dispersal into other regions.

3.5.3.2 Segregation of species into different niches

Interspecific competition is the most critical factor that confines a species to a particular niche. This becomes evident with the expansion of the species beyond the usual limits of its niche when this competition is over. Thus, interspecific competition results in the segregation of species into different niches.

3.5.3.3 Competitive exclusion principle

The principle has been variously named after its principal authors – Gause's principle (Lack, 1944), Volterra - Gause principle (Hutchinson, 1957) and Hardin's competitive exclusion principle. Gause's competitive exclusion principle states that no two species can coexist if they occupy the same niche. In other words, an ecological niche cannot be simultaneously and completely occupied by stabilised populations of more than one species.

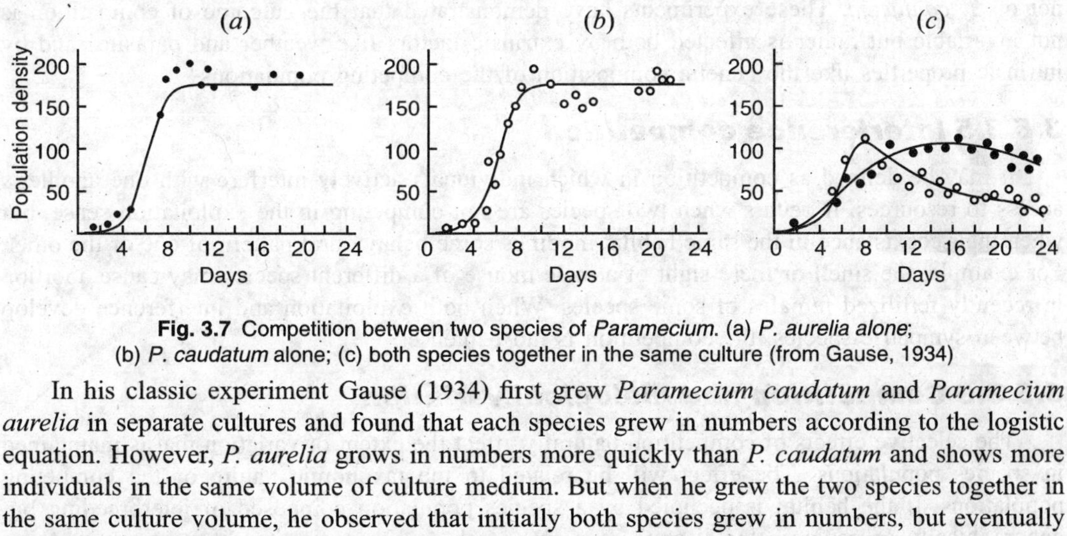

Fig. 3.7 Competition between two species of *Paramecium*. (a) *P. aurelia alone*; (b) *P. caudatum* alone; (c) both species together in the same culture (from Gause, 1934)

In his classic experiment Gause (1934) first grew *Paramecium caudatum* and *Paramecium aurelia* in separate cultures and found that each species grew in numbers according to the logistic equation. However, *P. aurelia* grows in numbers more quickly than *P. caudatum* and shows more individuals in the same volume of culture medium. But when he grew the two species together in the same culture volume, he observed that initially both species grew in numbers, but eventually *P. caudatum* declined and became extinct (Fig. 3.7). He repeated has experiment and found that *P.aurelia* always won the competition between the two species. Gause attributed the result to the existence of 'but a single niche in the conditions of the experiment'.

Table 3.1. Results of competition between two species of flour beetle under different laboratory conditions (after Park, 1954).

Temp.	*Relative Humidity (%)*	*% of times T. confusum wins*	*% of times T.castaneum wins*
24°C	30	100	0
24°C	70	71	29
29°C	30	87	13
29°C	70	14	86
34°C	30	90	10
34°C	70	0	100

There are other examples illustrating the principle of Gause. Park(1948,1954,1962) conducted laboratory experiments on two species of flour beetle, *Tribolium confusum* and *Tribolium castaneum.* He found that in mixed cultures one species always "wins" over another. But it depended on the environmental conditions. *T.castaneum* always wins (or *T. confusum* becomes extinct) under conditions of high temperature (34°C) and high humidity (70% R.H), while *T. confusum* always wins (*T. castaneum* becomes extinct) under cool-dry climate, *i.e.*, under conditions of low temperature (24°C) and low humidity (30% R.H.). Under some different environmental conditions (Table 3.1) sometimes one species survived,sometimes the other. This leads us to conclude that species can coexist even in the face of interspecific competition provided that their niches do not overlap too much.

3.5.3.4 Exploitative competition

Competitive interaction may cause exploitation, which develops when two species are competing for the same resource. It is mentioned earlier that when two flour beetles, *T. confusum* and *T. castaneum*, are kept together, one of them drives out the other in competition depending upon the laboratory conditions. However, when their cultures were infested with *Adelina*, a sporozoan parasite that may kill the beetles, it was found to have important effect on *T. castaneum* but not on *T. confusum*. These experiments have demonstrated that the outcome of competition is not invariable but rather is affected both by extrinsic factors like weather and parasites and by intrinsic properties like the genetic composition of the competing populations.

3.5.3.5 Interference competition

It may be defined as competition in which individuals actively interfere with one another's access to resources. It occurs when two species are not competing in the exploitation sense, but when their coexistence in the same habitat modifies some behavioural pattern of one or the other. For example, the smell or mere sight of a male mouse of a different species may cause abortion in recently fertilized females of some species. When both exploitation and interference develop between sympatric species, the competition is most intense.

3.5.3.6 Competition and biological variation

The selective effects of competition naturally affect the extent of variation that is maintained in species populations. The effect will be related to the taxonomic nature of the competing populations. If the habitat is occupied by a species population composed of interbreeding but geographically or ecologically discrete segments, such as subspecies or ecotypes, competition between them will tend to extend variation within the individual populations until it approximates to the variation within the habitat. In other words, the overall tolerance range of the species will widen or competition will tend to widen the biological variation already present in the segments. If, on the other hand, the habitat is occupied by competing independent species populations, gene flow will accordingly be very much restricted. The pressure of interspecific competition will in this case select from each species population those genomes that confer a special adaptation to a particular and limited range of the environment. In this case the competition will, therefore, tend to narrow the biological variation present in each of the competing species.

3.5.3.7 Competition and character displacement

Character displacement is defined as the divergence in the characteristics of two otherwise similar species where their ranges overlap, caused by the selective effects of competition between the species in the area of overlap. In the phenomenon of character displacement, the character traits of two closely related species differ more when they occur in *sympatry* (two species coexisting within the same geographic area) than in *allopatry* (when their distributions do not overlap, they are said to be allopatric). The ground finches (*Geospiza* spp.) of Galapagos Islands, first described by Darwin, furnish an example of character displacement. Character displacement occurs when interspecific competition results in natural selection causing changes in the morphology of two closely related species.

3.5.3.8 Logistic equation and interspecific competition

For a single population, *i*, we may write the logistic equation –

$$dNi/dt = ri\ Ni\ (1 - Ni/Ki)$$

In this equation all the parameters have been subscripted to show that they are specific to population *i*. Intraspecific competition appears as the term *Ni/Ki*. As *Ni* approaches *Ki*, *Ni/Ki* approaches 1 and the quantity (1 – *Ni/Ki*) approaches 0. A stable equilibrium (*dNi/dt* = 0) is reached when *Ni* = *Ki*.

If there is a population *j* that competes with population *i* for resources; and population *j* has a density *Nj* and a carrying capacity *Kj*. Naturally the competitive pressure of population *j* will have a negative effect on population *i* and will reduce the growth rate of population *i*. This idea may be incorporated into the logistic equation by subtracting a term that represents the effect of interspecific competition (with species *j* on species *i*)

$$dNi/dt = ri\ Ni\ (1 - Ni/Ki - [\text{Competitive effect of } j \text{ on } i]\ )$$

This equation now includes terms for both intraspecific competition (*Ni/Ki*) and interspecific competition. Both terms act as discounting terms (subtractions) and the equation may be written as:

$$dNi/dt = ri\ Ni\ (1 - [\text{intraspecific competition}] - [\text{competitive effect of } j \text{ on } i]$$

In must be appreciated that in this model the intraspecific and interspecific effects are independent of each other. In order to show an interaction between the two effects, we must include a term where the product of the two was subtracted from 1.

Volterra incorporated interspecific competition into the logistic equation by incorporating the term aij *Nj/Ki* into the quantity within the parantheses. Hence equation becomes –

$$DNi/dt = ri\ Ni\ (1 - Ni/Ki - aijNj/Ki)$$

Where *Nj* = the number of individuals of the competing species j

aij = coefficient of competition

The *coefficient of competition* may be defined as a measure of the degree to which one species, *j*, utilizes the resources of another, *i*, expressed in terms of the population consequences of the interaction. In other words, *aij* is the effect of an individual of species *j* on the exponential growth rate of the population of species *i*. The multiplication of aij by *Nj* (the size of the competing population *j*) converts the population of the competing species into units of species i on which it has an effect. Dividing the product *aij Nj* by *Ki*, as shown in the equation, gives the proportion of the carrying capacity of *i* that is taken up by equivalent units of individuals of *i*. However, population *i* will also have some competitive effect on population *j*. This may be shown by the following equation:

$$dNj/dt = rjNj\ (1 - Nj/Kj - ajiNi/Kj)$$

This equation shows how intraspecific competition in population j and the competitive effect of population *i* on population *j*, affect the growth rate of population j.

3.6 PREDATION

3.6.1 Introduction

Predation is an interaction between two different species and may be defined as the consumption of all or part of another individual (the prey). According to C.J.Krebs (1994), five specific types of predation may be distinguished. (*a*) *Herbivores* are animals that prey on green plants and often the plants are not killed but may be damaged; (*b*) *Carnivores* are typical predators, preying on herbivores or other carnivores; (*c*) *Parasites* are animals or plants that depend on the host for nutrition; (*d*) *Insect parasitoids* are another type of predator, which lay eggs on or near the host insect which is subsequently killed and eaten; and (*e*) *Cannibalism* is a special form of predation in which the predator and prey belong to the same species.

Predators which play a crucial role in maintaining the structure of the community, for example star fish on the sea coast and tiger in the forest, are called *keystone predators* (Paine, 1966; Hunter and Price, 1992). Removal of a keystone species allows spread of some other organisms which may in turn crowd over an area reducing the diversity and complexity of local food webs.

Predation is an important ecological process from the following three points of view. First, predation may restrict distribution or reduce abundance of the prey. If the prey is a pest, we may think the predation useful. However, if the effected animal is a valuable resource, we may think the predation undesirable. Second, like competition, predation influences species diversity in a community. Third, predation is a major selective force, and many adaptations, such as warning colouration, may be explained on the basis of predator- prey co-evolution. There are many cases of co-evolution of predator and prey, which is the ecological basis for biological control of undesirable species. The predators have evolved to become more efficient at searching and capturing its prey, and prey species in response have evolved to become more efficient at evading capture by the prey. The mechanisms underlying the dynamics of predator and prey populations are best understood by studying predator- prey interactions at the individual level.

3.6.2 Predator - Prey Interaction

The act of predation can be described into sequential stages of search, pursuit, attack, capture and ingestion, and each stage calls for decision making by the predator. Prey capture success depends on various parameters relating to the predator (body size, mouth size, searching efficiency), prey (body size, evasive behaviour, visibility) and the ambient medium (light intensity, turbidity). There are many examples of evolutionary compromise in body size of the prey species. For example, In case of *Daphnia* (prey) eaten by copepods (predator), large size of prey is helpful to avoid predation; but in case of *Daphnia* eaten by fish, small size of prey is helpful to avoid predation.

3.6.3 Prey Profitability

The profitability of a prey to the predator is a ratio of its energy content and the handling time. Given a choice between two potential prey types, a predator, which is optimizing its effort, should choose the most profitable prey. Suppose in one case it is 100/5 sec and in the other case it is 100/15 sec., the predator would like to capture the first prey for obvious reasons. Optimal foraging models try to determine in which patch a predator should forage, how long he should spend foraging in a given patch, and what prey species he should include in his diet, to maximize net energy intake.

According to *optimal foraging theory* (Stephens and Krebs, 1986), the ability of predators to aggregate in patches of high prey density is a critical element in determining how effective the predator can be at limiting prey populations. The *ideal free distribution theory* suggests that predators should move among sites until profitability is equal at all sites. If profitability was higher at one site, the theory predicts that predators would naturally redistribute themselves. This assumes that predators are *ideal* in their assessment of the quality of patches, and *free* to move from patch to patch. Many experimental demonstrations of this process have been performed. In one case six stickleback fish were placed in a tank and food added at both ends, with one end offering low food density and the other high food density. The distribution and movement of the fish reflected accurately the variation in the food density.

3.6.4 Predation-Deterrent Adaptations

Prey species have evolved many predation–deterrent adaptations, which are morphological (increased transparency to elude detection by visual predators, changes in body size , coloniality, spination), chemical (toxic chemical secretion, unpalatibilty) and behavioural (evasiveness, diel vertical migration). Many of these defences are readily inducible by the predator through infochemicals called *kairomones.* The prey also responds to predation pressures by adaptive changes in life history traits such as age at first reproduction, frequency of reproduction, clutch size and longevity.

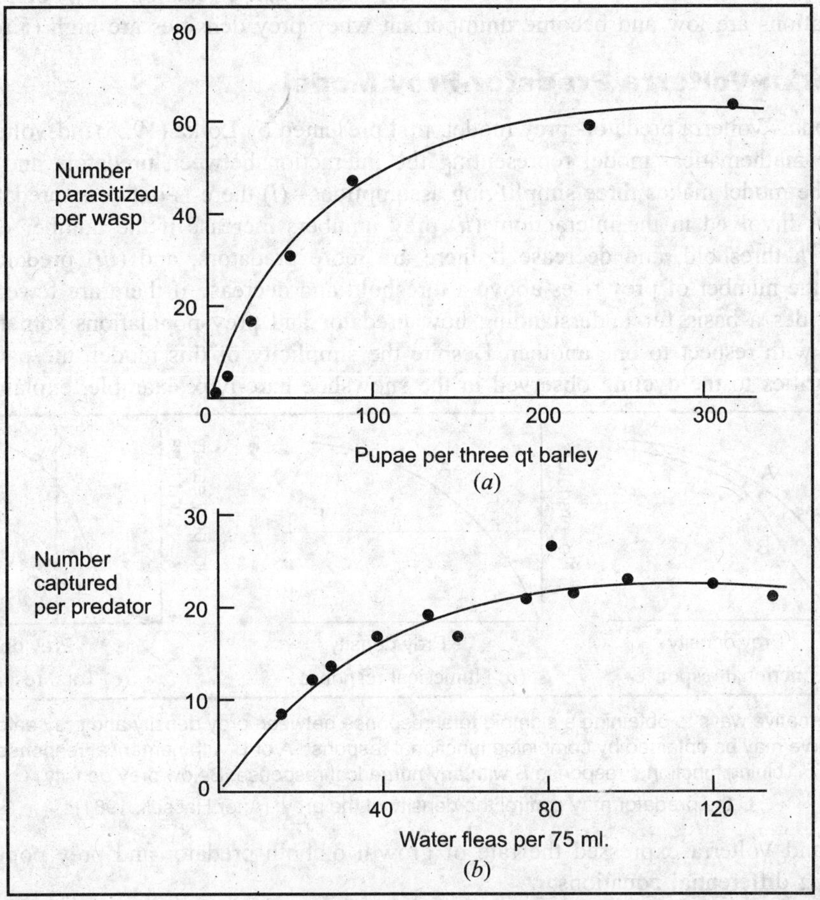

Fig. 3.8 Functional responses from an insect parasitoid and a predatory larva in the laboratory. (*a*) The parasitoid wasp parasitizing housefly pupae, (*b*) The larvae of damselfly feeding on the water flea (After Hassell, 1976)

3.6.5 Predator's Response to Changes in Prey Population Density

The response of a predator to changes in population density of its prey is of considerable relevance in evaluating the biological control potential of a predator. In a simple one predator— one prey system, there are four possible responses of a predator to an increase in prey population density. These are : (*a*) *a numerical response*, in which the density of predators increases by reproduction in a given area; (*b*) *a functional response*, (Holling, 1959), in which the number of prey eaten by individual predators changes (Fig. 3.8); (*c*) *an aggressive response*, in which individual predators move into and concentrate in certain places within the study area; and (*d*) a *developmental response*, in which individual predators eat more or less prey as predators gradually become mature (Murdoch, 1971). If we can measure the functional, numerical, developmental and aggressive responses for a predator-prey system, we can easily determine the total response of the predators (Hassell, 1981). If the total response is sigmoid (Fig. 3.9), the predator may limit the density of the prey at low prey densities. An important question is whether mortality caused by the predators increases with rise in prey density. The mortality rate increases with prey density only for the first part of the sigmoid curve (up to C in Fig 3.9). At higher prey densities, the predator will exert no controlling influence on the prey because it will be swamped by prey numbers. Thus, it may be concluded that predators may have important effects on prey abundance when prey populations are low and become unimportant whey prey densities are high (Krebs, 1994).

3.6.6 Lotka-Volterra Predator-Prey Model

The Lotka- Volterra predator- prey model, first presented by Lotka (1925)and Volterra (1926), is a simple mathematical model representing the interaction between predators and their prey. However, the model makes three simplifying assumptions– (*i*) there is only one predator and one prey species involved in the interaction; (*ii*) prey numbers increase if the number of predators falls below a threshold, and decrease if there are more predators; and (*iii*) predator numbers increase if the number of prey rises above a threshold and decrease if there are fewer prey. This model provides a basis for understanding how predator and prey populations sometimes come to oscillate with respect to one another. Despite the simplicity of this model, the pattern shows clear similarities to the cycling observed in the snowshoe hare-lynx example, explained earlier.

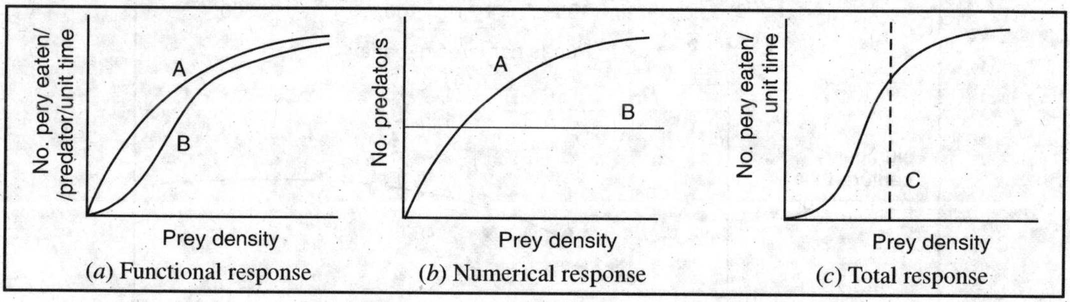

Fig. 3.9 Alternative ways to obtaining a sigmoid total response between prey density and prey eaten. The total response curve may be obtained by combining functional response A or B with numerical response A, or combining functional response B with any numerical response. Below prey density C the predator may control the density of the prey. (After Hassell, 1981)

Lotka and Volterra expressed the rate of growth oı both predator and prey populations by the following differential equations:

$$dP/dt = g\,(H, P) \text{ – – – – for the predator}$$

$$dH/dt = f\,(H, P) \text{ – – – – for the prey}$$

where

P = predator population size

H = prey population size

f and g = arbitrary functions of variables H and P

Thus, the rate of growth of prey population, dH/dt, is some joint function (f) of the size of prey population (H) and the size of predator population (P). It may be recalled that the general form of these models is the same as that of the population growth models in which the change in population size (dN/dt) was shown as some function (f) of N, $f\,(N)$. The Lotka - Volterra predator- prey model is a simple elaboration of this idea. What has been included in the functions $f\,(H, P)$ and $g\,(H, P)$ are terms that account for how consuming and being consumed add and remove individuals from the populations of predator and prey.

3.6.7 Nicholson - Bailey Model

Nicholson and Bailey (1935) have proposed a model for parasitoid- host population dynamics. The model indicates that at least *two* factors may lead to coexistence of parasitoid and host. First, if the birth rate (*b*) of the parasitoid is made to decrease as the density of the parasitoid increases, the expanding oscillations of the model will be changed to damped oscillations. Second, we recognize that as the parasitoid population density increases, the efficiency with which each parasitoid finds unattacked hosts, *a*, will decrease. If either of these enhancements is included in the model a stable interaction is predicted.

Thus, in these models the rate of prey removal was an asymptotic function of the predator density. The model predicts that parasitoid-host interactions will be unstable because they will undergo ever-increasing oscillations (Ricklefs and Miller, 2000)

3.6.8 Predation-Sensitive Food Hypothesis

A model of prey-predator interaction where the predator population is less strongly dependent on prey levels, prey density is held near its carrying capacity and predation is more likely to be influenced by risky behaviour on the part of the prey, is called predation- sensitive food hypothesis (Sih, 1982; Sih and More, 1990; Abrams, 1991). The hypothesis recognizes the trade-off between the need to avoid the action of the predator and the need to obtain sufficient nourishment to survive in a limited-food environment (Ricklefs and Miller, 2000). The hypothesis emphasizes the idea that predation may not be the only factor operating to limit prey population size; intraspecific competition during times of food limitation may also play an important role (Sinclair and Arcese, 1995).

3.6.9 Plant Defense from Predation

Plants defend themselves from predation in two ways: (*i*) defensive structures, and (*ii*) toxicity and unpalatability. Defensive structures exist in a variety of forms from simple small hairs on the leaf surface, which may trap insects and other invertebrates, to large hooks, barbs and spines (*e.g., Utrica dioca, Rosa canina* and *Acacia* spp.) which deter mammalian herbivores. The size and prevalence of such defensive structures may also be induced in defoliated plants.

In plants, there is a vast variety of chemical compounds used to defend plants against predators and parasites. These secondary compounds may either be directly toxic (*e.g.,* mustard oils in cabbages, nicotine in tobacco, cyanide in white clover, cardiac glycosides in milkweeds) or they may reduce the food value of the plant, for example, by reducing the availability of the leaf tissue protein to the animal gut. The tannins in the mature leaves of many woody plants bind to proteins, thus making them inaccessible to the predator's gut. Similarly, tomato plants produce protease inhibitors, which inhibit the protease enzymes in the herbivore's gut.

3.7 BIOLOGICAL CONTROL

3.7.1 Introduction

Biological control may be defined as the utilization of a pest's natural enemies in order to control that pest. It is the control of pests and parasites by the use of other organisms, *e.g.,* of mosquitoes by fishes which feed on their larvae. In other words it is a practice in which an organism is used against another organism. Under this practice, there are four types of pest control: (*i*) *classical biological control* or *importation,* in which a natural enemy from another geographical area, often the area in which the pest originated from, is introduced to contain the pest below the *economic injury level,* EIL, the definition of EIL is the pest density at which the difference between the curve showing value of the crop and the curve showing cost of achieving this pest density is greatest; (*ii*) *inoculation,* in which the periodic release of a control agent is required so that it is available throughout the year. Inoculation is widely practiced in the control of arthropod pests in glasshouses, where crops are removed, along with their pests and their natural enemies at the end of the growing season; (*iii*) *augmentation,* which involves the release of an indigenous natural enemy in order to supplement an existing population, and is therefore carried out repeatedly usually to coincide with a period of rapid growth of pest population; and (*iv*) *inundation,* which is the release of large numbers of natural enemy, with the aim of killing those pests present at the time. These are usually termed *biological pesticides.* However, insects have been main agents of biological control against both insect pests and weeds.

3.7.2 General Theory of Biological Control

The classical theory of biological control based on the Nicholson-Bailey model is an equilibrium theory (Huffaker and Messenger, 1976). According to this theory, a successful biological control is produced by the predator imposing a low, stable host equilibrium (Fig. 3.10). But a successful biocontrol agent should be host-specific, synchronous with the pest, should have high intrinsic rate of increase (*r*), should be able to survive with few prey available, and should have high searching ability. All these properties are shown by insect parasitoids than predators. Successful biocontrol agents cause density-dependent losses in the host population. Spatial density dependence occurs when parasitoids or predators cause a higher fraction of losses in dense host patches than in sparse host patches (Hossell, 1977). If predators can aggregate in patches of high host density, then, according to this theory, biological control of the pest is much more likely. The theory has been challenged recently by Murdoch *et al.* (1985). They have based their view on a nonequilibrium model of predator-prey interaction. The model assumes that a stable equilibrium of predator and prey is not necessary for satisfactory biological control. Pest populations may fluctuate wildly without pest densities exceeding the economic threshold. According to Krebs (1994), the nonequilibrium model is a metapopulation model and, as such, emphasizes that population in different patches may fluctuate independently.

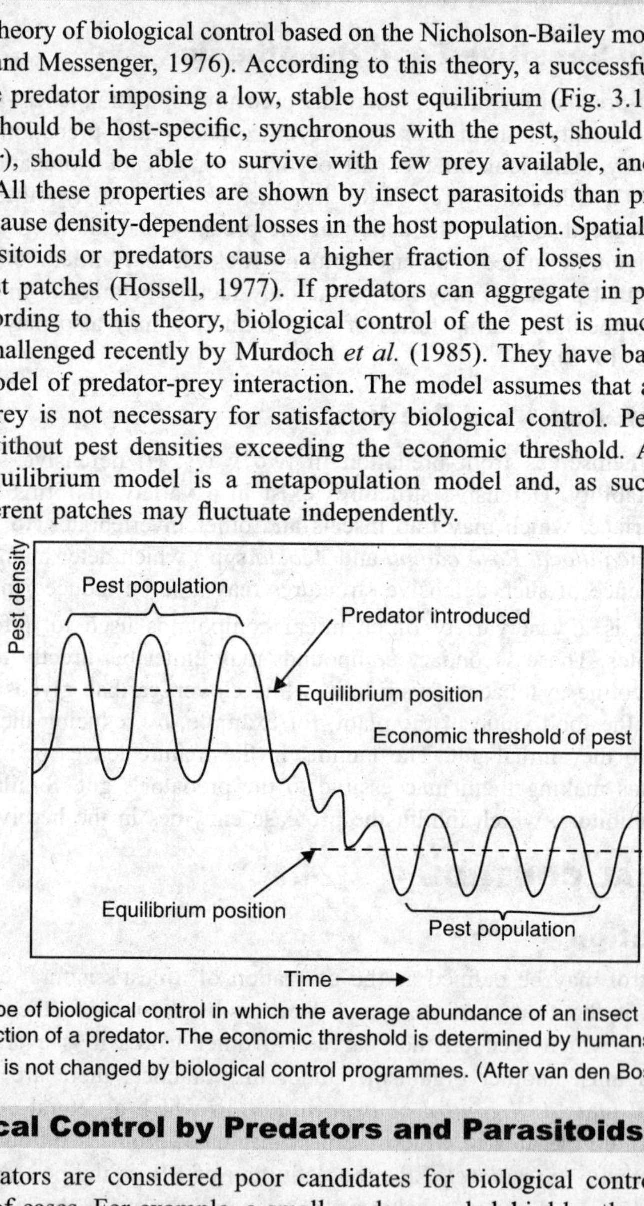

Fig. 3.10 Classic type of biological control in which the average abundance of an insect pest is reduced after the introduction of a predator. The economic threshold is determined by humans' activities, and its position is not changed by biological control programmes. (After van den Bosch *et al.* 1982)

3.7.3 Biological Control by Predators and Parasitoids

Although predators are considered poor candidates for biological control, they have been used in a number of cases. For example, a small predaceous ladybird beetle, *Rodolia cardinalis,* commonly called vedalia, has been used to control the cottony- cushion scale insect (*Icerya purchasi*), a pest of citrus trees. Adult Parasitoids (Hymenoptera) lay their eggs in or near other insects. The larval parasitoid then develops inside its host and kills it before or during the pupal stage.

3.7.4 Biological Control by Parasites

Some calcid wasps control a number of major pests. The oriental fruit fly, *Dacus dorsalis,* a pest of ripe fruits in Hawaii has been controlled by three species of parasitic wasps of the genus *Opius* (*O. vandenboschi, O. longicaudatus* and *O. oophilus*). This example also illustrates that several parasites of the same pest can be released without having any adverse effect on the overall control. Although the three control agents competed for the same host, the one with superior qualities displaced the others and became dominant. In this case *O. vadenboschi* derived the advantage from attacking first instar larvae and thereby inhibiting the development of the eggs and larvae of *O. longicaudatus,* which favoured older host larvae for oviposition. Likewise, *O. oophilus,* which oviposits in the eggs of the host, are already present as larvae by the time hosts are suitable for attack by *O. vadenboschi.*

The geometrid moth *Operophetera brumata* or winter moth, a pest (defoliator) of hardwood forest and ornamental trees in Canada and Europe, has been controlled by a tachinid fly, *Cyzenis albicans,* and a wasp *Agrypon flaveolatum.* However, in this case there was no displacement. Instead, the two species that are compatible and complimentary to each other, were able to bring about control. *C. albicans* was very effective at high host densities, whereas the superior searching ability of *A. flaveolatum* made it effective at low host densities.

3.7.5 Biological Control by Pathogens

(*a*) **Bacteria :** The use of spore-forming bacteria as a means of controlling the larvae of the Japanese beetle (*Popillia japonica*), a serious pest of fruits and vegetables, provided the first encouragement for the application of bacteria in insect control. *Bacillus popilliae* and *Bacillus lentimorbus* that cause types A and B milky disease of Japanese beetle can both be mass- produced and are sold as a spore dust for injection into the soil. Infected larvae that die in the soil become a source of contamination for other larvae feeding in the vicinity. Larval population can be substantially reduced in this way and the *Bacillus* spores persist in the soil to infect larvae from generation to generation.

Another spore- forming bacteria *Bacillus thuringiensis* is a facultative pathogen that infects a variety of insects, including the larvae of lepidopterans, flies, and beetles. The bacteria can be cultured on artificial media and is therefore quite economical to produce. Commercial preparations of *Bacillus thuringiensis* (Biotsol, Dipel, Thuricide) containing both spores and crystals are used as a biological insecticide on a variety of crops.The rather specific nature of *Bacillus thuringiensis* to kill a few groups of foliage feeders and not to harm beneficial species is of great value in management programmes.

(*b*) **Fungi :** Most entomogenous fungi are internal pathogens. They belong to all the four major taxonomic groups of true fungi, but only a few are frequently associated with insect disease outbreaks. The most commonly used in insect control are *Beauvaria bassiane* (white muscardine disease) and *Metarrhizium anisopliae* (green muscardine disease), both of which are fungi imperfect. The infective unit of an entomogenous fungi is usually a spore which germinates on the surface of the host's integument. Once the host tissue is invaded, the fungus can complete its life cycle, but the survival and germination of spores is critical to the development of an epidemic.

Facultative fungi such as *Beauvaria* and *Metarrhizium* can be cultured on artificial media, thereby facilitating the production of spore preparations which may be used in biological control.

As with most biological control agents, fungi can be used for either persistent or short-term control. A fungus can be introduced into an area where it becomes established and kills the host year after year. Alternatively, fungal spore preparations can be used as microbial insecticide similar to the way *Bacillus thuringiensis* is used.

However, few attempts have been made to colonize entomogenous fungi. Most projects have involved the redistribution of indigenous fungi or those associated with introduced pests, rather than the importation of foreign species. The best example of attempts to establish new fungal pathogens in disease free areas involves the introduction of *Coelomomyces* against mosquito larvae, but so far the success has been limited. The successful use of repeated applications of fungal spores as microbial insecticides have been reported for achieving short-term reductions of pest populations. The major limiting factor in initiation of fungal disease in insect populations is the effect of the microclimate on spore survival and germination. The optimal temperature range for the growth of entomogenous fungi is fairly narrow, and relatively high humidity is needed by most fungi to germinate and successfully prenetrate their host before they can produce the new spores required to spread the disease. Sunlight also kills the spores. Consequently, the application of a spore preparation must coincide with both the presence of susceptible hosts and suitable environmental conditions. Best success can usually be obtained by applying the spores in the absence of sunlight such as on a warm evening after either rain or irrigation which provides the needed humidity.

(*c*) **Viruses :** The insect pathogenic viruses are called *inclusion viruses,* as opposed to *noninclusion viruses* in which the virus particles or *virions* are free within the cells of the host. The virus particles first multiply in the nuclei, but later continue to replicate in the cytoplasm. The disease eventually kills the insect, leaving it hanging as a fragile sac of virus like the one which results from nuclear polyhedroses infection.

A few non-inclusion viruses also attack insects. But with the exception of Tipula Iridescent Virus (TIV) and Mosquito Iridescent Virus (MIV) that might prove useful in mosquito control, most attention has been given to the inclusion viruses. The very fact that the virus particles enclosed in a protein matrix maintain their infectivity for many years means that the inclusion viruses can be stored as concentrated preparations for later application with conventional pesticide spray equipment.

It has been shown that a nuclear polyhedroses virus is highly effective against a variety of forest sawflies and, as it persists in the environment, it provides continuous regulation of the pest in some areas. Several nuclear polyhedroses virus are being mass produced for possible use against a variety of pests, including cotton bollworm, tobacco budworm, corn earworm, cabbage looper, forest tent caterpillar, and alfalfa butterfly. However, one of the problems with viruses is that there are periods when they have little effect on the pest populations. A virus may remain latent in a pest population for several generations and then develop epizootics when the pest population comes under stress.

Generally, short-term control can be achieved by frequent applications of virus preparations so that there is an active innoculum in the pest environment for a long period.

Role of larvicidal fish in biological control

There are a number of larvivorous fish (indigenous species– *Aplochoilus panchax, Oxygaster, Danio, Rasbora,* etc; *exotic species–Gambusia affinis, Lebistes reticulatus,* etc.) Which feed at one stage or other on mosquito larvae. For mosquito control, larvivorous fish are introduced into a water body preferably after cleaning any dense vegetation and removal of predatory fish so as to allow the larvicidal fish to propagate.

3.7.6 Genetic Control

Genetic control is a type of biological control that uses two strategies to reduce pest problems. First, crop plants can be manipulated to increase their resistance to pests. Second, we can attempt to alter the genome of the pest species so that they become sterile or less harmful.

Resistant varieties of many crop plants have been developed by selective breeding (Maxwell and Jennings, 1980). However, resistant plants do not necessarily have chemical defenses. Strains of cotton plant produced with low gossypol (a chemical that occurs in green parts and seeds of cotton plant and is toxic to chickens and pigs) content are quite low in resistance to insect pests. Resistant crop plants have also been developed by genetic engineering. Genes that produce resistance in one species can be transferred into a crop plant to make the crop genetically resistant to specific pests. Bacteria may also be used as vehicles to carry biopesticide genes. For example, in 1987 the first success was reported of inserting a gene (the δ – toxin gene of *Bacillus thurengiensis*) into tobacco plants, conferring resistance against Lepidoptera. *Bacillus thurengiensis* (Bt) is the main focus at present for developing insect – resistant crops (Lambert and Peferoen 1992). This bacteria normally lives in the soil and carries a gene for a toxic protein that kills the larvae of moths and butterflies. By splicing this gene into bacteria that normally live on crop plants, genetic engineers have produced insect– resistant crops. Insect pests would inject the bacteria while feeding on the plant and thereby be poisoned. Alternatively, the Bt genes that produce the toxins can be transferred directly into the plant's genome, so that the plant would protect itself. As of 1992 tobacco, potato, cotton and tomato plants have been genetically engineered with Bt genes (Lambert and Peferoen, 1992). The development and use of such transgenic plants has immense potential. However, one major problem is that pest insects will become resistant to the biopesticide, just as they become resistant to chemical pesticides (Pimentel 1991).

The simplest genetic manipulation that can be carried out on a pest species is sterilization. A large number of pests are sterilized by radiation or by chemicals and released into the wild where they can mate with normal individuals. This technique leads to a decrease in birth rate of the pest and control can be achieved. The most notable success of this technique is the near extinction of the screw-worm fly, *Cochliomyia hominivorax,* which lays its eggs on fresh wounds of livestock and wild animals. Another example of sucessful use of sterile-insect method was the suppression of mosquito *Culex pipiens quinquefasciatus* on a small island off Florida (Patterson, *etal* 1970). However, the sterile insect method cannot be used for all pest populations because it requires the rearing and sterilizing of a large number of individuals and isolation of target area so that natural males from outside the area may not be able to reach there to undrmine the programme.

3.8 INTEGRATED PEST MANAGEMENT

Integrated control (Stern *etal* 1959) or integrated pest management (IPM) uses biological, chemical and cultural methods of control. This system is ecologically sound because it relies on natural biological control as much as possible and depends on chemical treatments only when absolutely necessary. However, to achieve integrated control, we need to understand the population dynamics of the pest species and also of the crop system.

Gonzaliz (1970) has compared developing an IPM programme to building a house. Sampling, establishment of economic thresholds, and knowledge of natural mortality in agro-ecosystems form the foundation of the house. The variety of control methods available, such as, chemical, biological, host resistance and autocidal methods form the studs. The integrated programme forms the roof. How well the roof holds up depends upon the strength of the foundation and the number of studs that support it (Fig. 3.11).

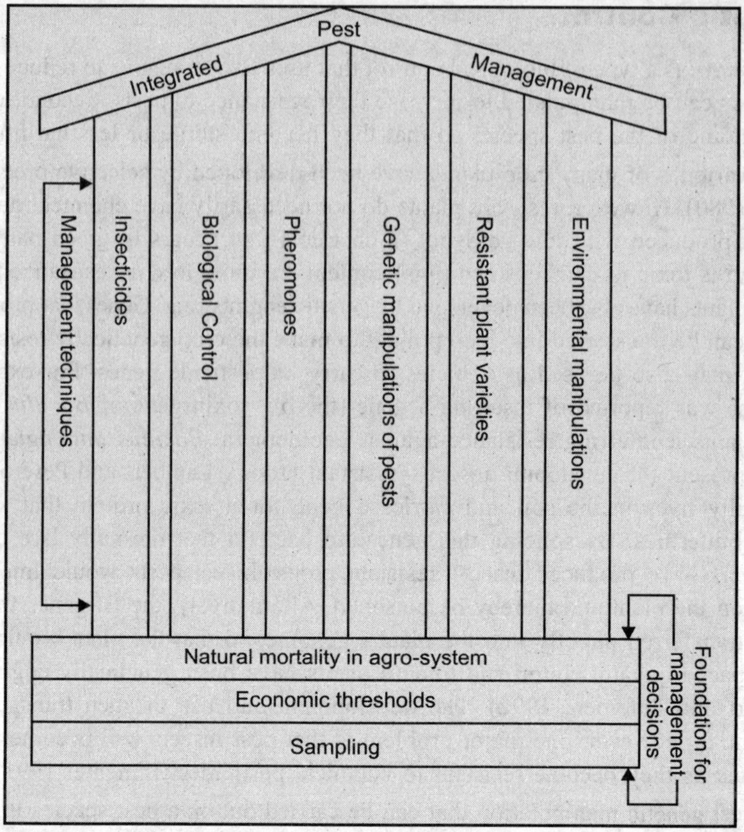

Fig. 3.11 Diagrammatic representation of an integrated pest management programme. (After Gonzaliz, 1970)

Based on the above consideration, the integrated pest management prescribes the use of the following:

(*a*) *Predators:* for example, the use of ladybeetles and lacewings against agricultural pests, or beetles to control weeds, or herbivorous fishes to control weeds.

(*b*) *Parasites:* For example, the use of calcid wasps against pests.

(*c*) *Pathogens:* For example, the use of viruses and bacterial infections that are specific for a pest.

(*d*) *Decoy plant:* Cultivation of low-value crops to attract pests away from high-value crops.

(*e*) *Hormonal stimulants:* Such as juvenile hormones which prevent insects from completing their cycle.

(*f*) *Pheromones:* Sex lures and other biochemicals that regulate pest behaviour.

(*g*) *Degradable chemical insecticides:* Organic phosphates and others.

(*h*) *Biological insecticides:* Such as *Bacillus thuringiensis* because their action is specific.

(*i*) *Slow-release herbicides:* Such as copper sulphate is used for algae control.

(*j*) Artificial selection for disease and pest resistance rather than for short term yield as such.

(*k*) Rotation and diversification of crops; strip-cropping; the use of trap crops; planting pest-resistant strains of crops; timing crop-planting so as to avoid pest outbreaks; altering the distance between crop rows.

(*l*) Chemical or radiation sterilization.

(*m*) Modifications in the use of water and fertilizers.

Thus, the strategies employed in integrated pest management are multiple and diverse. The various strategies may be used singly or in combination, simultaneously or in sequence, depending upon the particular pest problem.

As the programme involves lot of time, effort and money, actual examples of IPM are limited. As example we can take the IPM programme developed to control cotton pests in California. The main pest of cotton was *Lygus hesperus,* which feeds on cotton buds. In addition, the cotton bollworm (*Heliothis zea*) was a secondary pest. The IPM programme required minimum insecticide use to prevent secondary outbreaks. Subsequent studies showed that the lygus bug could only inflict serious damage during the budding season (June-July). Therefore, insecticide applications were made during this time only. Cultural control was made by introducing thin strips of alfalfa into the cotton field which diverted the lygus out of the cotton. Each field was monitored twice a week from the beginning of budding to the end of August. Plant development, pest and natural enemy date were all collected, leading to a successful pest control programme. Overall, the programme represents a proper integration of biological controls and carefully timed insecticide treatments. The success of the programme is illustrated in Table 3.2:

Giese *etal* (1975) have given an example of integrated control in respect of alfalfa pest management. The alfalfa weevil (*Hypera postica*) is the most important alfalfa pest affecting the alfalfa crop and an integrated control programme was designed for this weevil (Armbrust and Gyrisco, 1982).

Table 3.2. The success of IPM in California, 1970-71 (after Hall *etal* 1975; Flint and van den Bosch, 1981)

	IPM Users	*Non-Users*
Average insecticide per acre of cotton (US $)	4.9 (± 3.9)	12.0 (± 7.4)
Average yield per acre of cotton (US $)	270.2 (± 27.5)	247.8 (± 55.1)
Average insecticide cost per acre of citrus (US $)	20.5 (± 15.0)	42.4 (± 18.3)
Average yield per acre of citrus (US $)	515.8 (± 260.6)	502.9 (± 157.0)

MODEL QUESTIONS

1. Define 'ecological population'. Describe the phases of population growth curve. How growth is affected by the carrying capacity of the environment?
2. What is logistic equation of population growth? How it differs from exponential growth?
3. How a natural population regulates itself? Describe the theories of population regulation?
4. What is interspecific competition? How it affects a population?
5. Define predation. How it differs from parasitism? Write a note on predator-prey interaction.
6. Define predation and review the models of predator- prey interaction.
7. Define 'biological control'. Describe the methods of biological control.
8. Define integrated pest management (IPM). How this approch is better than other methods of pest control?
9. Comment on the statement " IPM is a philosophy of pest management rather than a specific and well defined strategy".
10. Write note on the following:
    (*a*) Population density, (*b*) Population cycles,
    (*c*) Competitive exclusion principle, (*d*) Predation,
    (*e*) Models of predator- prey interaction,
    (*f*) Biological control, (*g*) IPM.

MULTIPLE CHOICE QUESTIONS

11. Niche density of a population is :
    (A) Equal to average density, (B) Equal to regional density,
    (C) Higher than regional density, (D) Lower than regional density
12. Intraspecific competition for a limited resource in population creates :
    (A) Displacement of some individuals,(B) Death of some individuals,
    (C) Nonavailability of food, (D) All the above
13. Ecological niche parameters are affected by :
    (A) Competition, (B) Predation,
    (C) Available resources, (D) All of them
14. In logistic growth of a population :
    (A) The percentage rate of increase decrease in a linear fashion as population size increases,
    (B) There is unlimited growth, (C) Growth is not affected by resources,
    (D) Growth is not affected by the carrying capacity of the environment
15. Which equation is correct for exponential growth ?
    (A) $Nt = N_0$, (B) $Nt = N_0 e^{rt}$,
    (C) $Nt = Nrt$, (D) $Nt = Nert$
16. Population growth in which the rate of increase is proportional to the number of individuals in the population at the beginning of the breeding season is called :
    (A) Exponential growth, (B) Logistic growth,
    (C) Geometric growth, (D) Unregulated growth
17. Plants defend themselves from predation by :
    (A) Toxicity and unpalatability, (B) Toxic secondary compounds,
    (C) Defensive structure, (D) All the above
18. Biological control includes :
    (A) Introduction of a natural enemy from another geographical area (importation),
    (B) Inoculation, (C) Augmentation and Inundation
    (D) All the above
19. IPM comprises the following:
    (A) Physical and cultural control, (B) Biological and chemical control,
    (C) Biological control and use of resistant varieties,
    (D) Physical, cultural, biological and chemical control and use of resistant varieties
20. Which of the following is a microbial pest control agent?
    (A) *Puccinia chondrillena*, (B) *Bacillus thuringiensis*,
    (C) *Beauveria bassianca,* (D) All of them
21. Which of the following fish is used for biological control of mosquto?
    (A) *Tor putitora*, (B) *Labeo rohita*,
    (C) *Gambusia affinis*, (D) *Clarias*
22. Which of the following is voted as an example of population cycle?
    (A) *Hare-lynx*, (B) *Tor-Labeo*,
    (C) *Gambusia-Clarias*, (D) *P. aurelia-P. caudalum*

23. Who wrote the book, for the first time telling the adverse affect of pesticides?
    (A) Rachael Carson, (B) Fredric Clemets,
    (C) E.P. Odum, (D) Charles Elton
24. The impact of industrial effluents on fish populations of our rivers is such that these are:
    (A) Decreasing, (B) Increasing,
    (C) Showing no change, (D) Can't say
25. At Galapogos islands Darwin studied the population of:
    (A) Turtles, (B) Finches,
    (C) Snakes, (D) Earthworms
26. The niche parameters include niche–
    (A) Form, (B) Position,
    (C) Width, (D) All of them
27. Gause used the following in his experiment on competition.
    (A) Two species of *paramecium,* (B) Two species of flour beetle,
    (C) Two species of earthworms, (D) Two species of bacteria

CHAPTER 4

Biotic Community

4.1 INTRODUCTION

Ecologists have given a variety of meanings to the term community (Schoener, 1986; Fauth *et al.*, 1996). The term has been used by many of them for associations of plants and animals occurring in a particular locality and dominated by one or more prominent species or by some physical characteristic (Daubenmire, 1968). A community, technically often referred to as biota or biotic community, is a local association of several species populations. According to Krebs (1994), a community is an assemblage of the populations of living organisms in a prescribed area or habitat.

According to Clarke (1954), a group of mutually adjusted plants and animals inhabiting a natural area is known as a community. He prefers to use the term *biocenose* (a shortened form of the word "biocoenosis" coined by Mobius in 1880) for the community.

When territorial ranges of several species populations overlap, they may coexist within a given limited area. The organisms in such an area represent a community (Fig. 4.1). A community always contains plants as well as animals, because both are very necessary for the survival of the community. Just as populations possess characteristics above and beyond those of their component organisms, the community also exhibits characteristics above and beyond those of its constituent populations.

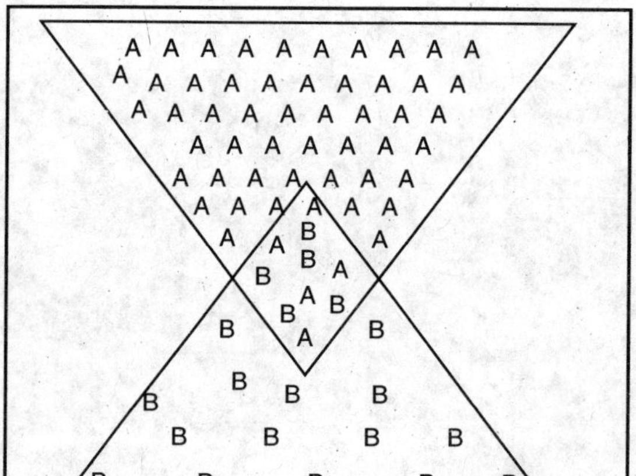

Fig. 4.1 Populations of species A and B coexisting within a given limited area as a community. There are more individuals in population A than population B.

A *major community* is the smallest ecological unit that is self-sustaining and self-regulating. It is made up of a large number of smaller minor communities that are not altogether self-sustaining. For example, forests and ponds are major communities; decaying logs and ant hills are minor communities. Members of a major community are relatively independent of other communities, provided they receive radiant energy from the sun. These members normally show a similarity in their physiological make-up, behaviour, and mode of life.

A number of parameters can be estimated in a community, but their interpretation depends on the nature of the community. In plant ecology, two schools have developed over the question of the nature of the community. The *organismic school* holds that communities are integrated units with discrete boundaries. The *individualistic school* holds that communities are not integrated units but collections of populations that require the same environmental conditions. However, the individualistic interpretation of the community is favoured more.

4.2 STRUCTURE AND CHARACTERISTICS OF A COMMUNITY

All communities have certain general characteristics in common, and their maintenance is governed by similar forces. Communities do not have exact limits but tend to overlap each other. Animals frequently shift from one community to another because of seasonal or other variations. Communities may be widely separated, but if the environmental factors are the same, similar kinds of animals may be found in them. Some characteristics of the community are as follows:

(*a*) Stratification : Communities that possess a large biomass usually exhibit stratification, that is, the populations they contain are spaced out or distributed into definite horizontal or vertical strata. Five vertical subdivisions of the forest community are (*a*) subterranean, (*b*) forest floor, (*c*) herbaceous, (*d*) shrub, and (*e*) tree strata. Similarly, there are animals that live on the forest floor, others on shrubbery and low vegetation, and still others in the treetops. Many forms shift from one substratum to another, especially in a diurnal manner. Many of the adjustments and requirements of a particular stratum are very similar in forests widely separated from each other in many parts of the world. The animals that occupy such similar strata, or ecological niches, although geographically separated, are called *Ecological equivalents.*

(*b*) Ecotone : There may be an intermediate transitional zone between two distinct communities. This is called an *Ecotone,* or a tensionl zone. An ecotone can be defined as a zone of transition between adjacent ecological systems having a set of characteristics uniquely defined by space and time scales and by the strength of interactions between adjacent ecological systems (*Holland,* 1988). An example may be the marginal region between a forest and a pasture or open land. Several examples of aquatic-terrestrial ecotones may be given, such as wetland ecotones (ecotones between wetlands and other types of ecosystems), lotic ecotones (fluvial boundaries of rivers and stream ecotones). As a rule, the ecotone contains more species and often a denser population than either of the neighbouring communities and this is known as the *Principle of edges.*

Recently is has been recognised that ecotones are not simply static zones where two communities join but are dynamic and have unique properties (Risser, 1990). They provide unique habitats for biota; they regulate interpatch dynamics; they may serve as early indicators of hydroclimatic change; and they have strong visual quality. These characteristics of ecotones are significant for aquaitc landscape management (*Petts,* 1990).

(*c*) Ecological dominance : In every community some plants and animals exert a dominant influence, in determining the nature and function of the whole community, because of their numbers, activities, or other important reasons. They are called *Ecological dominants.* An ecological dominant may be defined as a species population that exercises a major controlling effect on the nature of the community.

Thus, each community usually contains one or two species populations at the producer, herbivore, carnivore, and reducer levels that are recognizable by the controlling influence they exert on the community. These dominant species populations are the members of the community through which a major portion of the energy transfer is affected.

In land communities, plants are usually the dominants and some communities are named according to their dominant vegetation. However, in plant communities *abundance* is rarely correlated with *dominance.* For example, in a forest, a herbaceous species may be very abundant while a tree species is hardly so, in terms of the numbers of individuals present; but it is the tree species that determines the nature of such a community and is therefore the dominant species. In plant communities, the dominants are simply those plants that overtop all others in the community. In doing this, they modify the amount of light the subordinate species receive, the humidity, the amount of precipitation, the extent of air movement, and the composition and temperature of the air. Some plant communities have only a single dominant; for example, pine trees in a pine forest, or saal trees in a saal forest. But in a mixed forest there may be several *codominants.* In the ecological cycle of a community, the removal of a dominant species causes distrubances and changes in the character of the community.

(*d*) Seasonal and diurnal fluctuation : The component populations of a community may succeed one another in time as well as in space. The most conspicuous temporal variation of communities is seasonal. *E.J. Salisbury* (1925) was actually the first to demonstrate such temporal separation of plant communities in a temperate woodland. Local fluctuation of populations in communities, similar to seasonal migrations but over much more restricted distance, result from diurnal responses. The active day time mammalian population of a wooded area may include squirrels, but by night there may appear reccoons, opoossums and wood rats. These temporal relations between populations in a community, like the spatial ones, must be considered in terms of the niche concept.

(*e*) Pattern diversity : *Pattern diversity* is the basic characterstic of community organization. There are several patterns. Such as stratification patterns (vertical layering), zonation pattern (horizontal segregation), activity patterns (periodicity), food-web patterns (network organization in food chains), reproductive patterns (parent-offspring association and plant clones), social patterns (flocks and herds), coactive patterns (resulting from competition, antibiosis, mutualism), and stochastic patterns (resulting from random forces).

(*f*) Periodicity : *Periodicity* (activity patterns) refers to the rhythmic patterns of organisms in search for mates, food, and shelter. Some community periodicities are correlated with the daily rhythms of day and night, some are seasonal, and others represent tidal or lunar events.

(*g*) Turnover : In a community, as on all other levels of living organization, *turnover* occurs continuously. Individuals of the various populations emigrate or die out and are replaced by others. The important point is that this flux is automatically self-adjusting. As a result of this, the community remains internally balanced and exhibits a numerical steady-state, *i.e.*, in all populations making the community, the numbers of individuals remain relatively constant. For example, in a large permanent pond the number of algae, frogs, fish and any other plants or animals, will be more or less the same from decade to decade. Annual fluctuations are common; but over longer periods of time constancies of numbers are characteristic to most natural communities.

(*h*) Interdependence : Various kinds of interactions bring about a certain degree of integration within the members of a community. Food, reproduction and protection are the principal links which make the members of a community interdependent (as discussed ahead in detail).

According to Krebs (1994), there are five characteristics of communities that can be studied.

(*i*) Species diversity : This means what species of animals and plants live in a particular community. An important characteristic of the community is its diversity, which is a function of the number of different species it contains and their abundance. Diversity actually depends on *Species richness* and on the evenness (equitability) of species abundance. Two hypothetical communities comprising the same species may differ in their structure and diversity, depending on their relative abundance distributions. Communities in which the species are all more or less equal in abundance exhibit evenness, whereas communities with one or a few abundant species and many rare ones show *dominance.* However, communities possess both *taxonomic diversity* (diversity of species) and *ecological diversity,* which reflects the variation in ecological roles of species in the community.

(*ii*) Growth form and structure : A community may be described by major categories of growth forms such as trees, shrubs, herbs and mosses. These different growth forms determine the stratification of the community.

(*iii*) Dominance : This refers to the fact that all species in the community are not equally important. Of the hundreds of species, relatively few exert a major controlling influence by virtue of their size, numbers, or activities. Thus, dominant species are those which are highly successful ecologically and largely determine the conditions under which the associated species have to grow or live.

(*iv*) Relative abundance : This emphasizes the relative proportions of different species in the community.

(*v*) Trophic structure : The feeding relations of the species in the community determine the flow of energy and materials from plants to herbivores to carnivores. Thus, "who eats whom" decides many things in the community. These characteristics can be studied in all the communities that are in equilibrium and also in communities that are changing. The changes may be spatial along environmental gradients such as altitude, temperature and moisture. The changes may also be temporal, which are called *succession* and may lead to a stable community or the so-called *climax community.*

4.3 INTERDEPENDENCE IN A COMMUNITY

Each species of organism—microbe, animal, or plant making up the community is obviously influenced in its existence and activities not only by the inanimate physical and chemical environment but by other species of the community as well. This *interdependence* among different species is usually for *food, reproduction,* and *protection.*

(*a*) Interdependence for food : Charles Elton described communities in terms of their feeding relationships. In every stable community, green plants produce their own food; herbivorous animals eat the plants and carnivorous animals eat each other or herbivores. The elimination products and the dead bodies of plants and animals replenish the soil or the ocean; and this, plus solar energy and raw materials from the environment, then make new plant growth possible. There are, in fact, chains of animals linked together by food, and all dependent in the long run upon plants. We refer to these as *"food chains",* and to all the food chains in a community as the *"food cycle"* (Elton, 1927). Now the term "food cycle" has been replaced by the term "food web".

(*b*) Reproductive interdependence : A second link between the members of a community is *reproductive interdependence.* A familiar and important example of this is the pollinating activity of insects. In some well-known cases of remarkable specialization a given insect visits only one or a few specific flower types for pollen and nectar. The flowers (*e.g., Snapdragons*) in turn are structurally adapted to facilitate entry of the insect. Such intimate reciprocity testifies to a closely co-related evolutionary development of animal and plant. It is fairly obvious how such interdependence contributes to population balance: reduction of the insect population entails reproductive restriction of the plant, and vice-versa. Similarly significant in balancing the reproductive growth of plant population is the seed-dispersing activity of birds and mammals, and man in particular.

Other examples of reproductive interdependence are many. Birds such as *cuckoos* lay their eggs in nests of other birds. Insects such as gall wasps embed their eggs deep in the tissues of particular plants, where the hatching larvae find food and protection. Other insects deposit eggs on or under the skin of various animals. Certain wasps kill tarantulas and lay their eggs in them. Reproductive growth and geographical expansion of a community are intimately correlated with nutritional balances.

(*c*) Protective interdependence : The third main link among the members of a community is *protective interdependence.* Plants of forest and grassland usually protect animals by providing shelter against enemies and adverse weather. If the opportunities for *protection* are reduced, both the animal and the plant population may suffer. For example, if an overpopulation of insects makes available plant shelter inadequate, the insects will become easier prey for birds and bats. But this condition may also reduce the plant populations, for their major pollinating agents may no longer be sufficiently effective.

Animals in many cases are protected from other animals by *camouflage.* Such a protective device may involve body colour, or body shapes or both. Probably the most remarkable instance of colour camouflage is the phenomenon of *mimicry,* widespread particularly among butterflies and moths. In certain of these animals, pigmentation patterns exist which are virtually indistinguishable from those of other unrelated species. Usually those species are mimicked which are strong and have a few natural enemies. The advantage is that an animal resembling even superficially another more powerful one will be protected too, by scaring off potential predators. Insects also display a variety of camouflages. For example, the individuals of certain species possess the detailed shape of leaves, or branches, or of thorns. A praying mantis, coloured green, resembles a thin stem and a leaf insect and leaf butterfly have a striking resemblance to a leaf on which they habitually live.

Other protective devices vary widely in type. Various birds and some mammals mimic the song and voice of other species, either defensively or as an aggressive lure. The hermit carb protects its soft abdomen in empty shells of appropriate size. Schools of small pilot fish scout ahead of large sharks, leading their protectors to likely prey. Significant protection to various animals is also afforded by man, through domestication, game laws, parks, and sanctuaries.

4.4 ENERGY FLOW IN THE COMMUNITY

For the survival of a community a continuous flow of energy should be maintained from producers to herbivores to carnivores. Charles Elton (1927) wrote "there are, in fact, chains of animals linked together by food, and all dependent in the long run upon plants. We refer to these as "food chain", and to all the food-chains in a community as the "food-cycle" (now called "food web").

4.4.1 Food Chain

This is defined *as a series of orgnisms through which the food energy is transferred.* In other words, *the transfer of food from the plants through herbivores to carnivores is referred to as the food chain.* Each organism in the series feeds and dervies energy from the preceding one, as shown in the following simple linear food chain on land.

Level	*Type*	*Example*
Producer	Herbage	Grass
↓	↓	↓
Primary consumer	Herbivores	Grasshopper
↓	↓	↓
Secondary consumer	Small carnivores	Thrush
↓	↓	↓
Tertiary consumer	Large carnivores	Hawk

A freshwater ecosystem has different food chains than the one described above. A simple linear food chain in freshwater may be shown as follows :

Level	*Type*	*Example*
Producer	Phytoplankton	Algae, diatoms
↓	↓	↓
Primary consumer	Zooplankton	Copepods
↓	↓	↓
Secondary consumer	Fish	Minnow
↓	↓	↓
Tertiary consumer	Bird	Kingfisher

In a freshwater ecosystem floating microscopic phytoplankton are an example of the producer class, which provide food for the primary consumers or zooplankton. The latter consist of microscopic animals and larval forms, which in turn provide food for the secondary consumers such as fish. The tertiary consumers feed on fish and they may be predatory fish, or birds, or mammals.

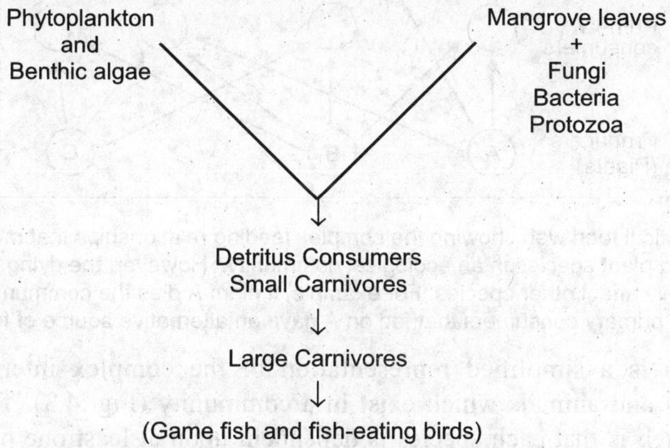

Fig. 4.2 A detritus food chain (simplified from ***W.E. Odum,*** 1970)

Food chains may be *simple or complex.* A simplest food chain involves essentially two links consisting of photosynthetic plants and microbes that obtain their nutrients from these plants by degrading or decaying them after their death. Complex food chains consist of many links. For example, in the ocean, the microscopic plants of the photic zone are primarily responsible for the production of organic matter by photosynthesis. The plants and their products may be consumed by certain unicellular or multicelluar herbivorous zooplankton. These are then ingested by other multicelluar carnivorous zooplankton, which in turn may be eaten by certain worms and crustaceans. The smaller crustaceans may be devoured by small fish, which are eaten by larger fish. The latter in turn may serve as food for various birds and mammals whose eventual death and subsequent decomposition by microbes of decay terminates the food chain.

According to Odum, food chains are of two basic types: grazing food chain and detritus food chain. A grazing food chain starts from a green plant base and goes through grazing herbivores to carnivores. The food relations among lemmings and their predators in northern eastern Alaska present an interesting example of grazing food chain. A detritus food chain starts from dead organic matter and goes through detritus-feeders to carnivores. A good example of detritus food chain is given by Odum (1970). The chain is based on mangrove (*Rhizophore mangla*) leaves, which fall into shallow estuarine waters of south Florida. Leaf fragments acted on by saprotrophs (or detritivores) and colonized by algae are eaten and re-eaten (coprophagy) by a group of small detritus consumers, which in turn provide the main food for game fish and fish-eating birds (Fig. 4.2)

4.4.2 Food Web

Food chains are not always simple and isolated but are interconnected with one another. The interlocking pattern of food chains or a matrix of food chains, with all sorts of short circuits and connections is often called the *food web* or *food net.* However, in any food web all the species are not equally important and many could be removed without seriously affecting the more important species.

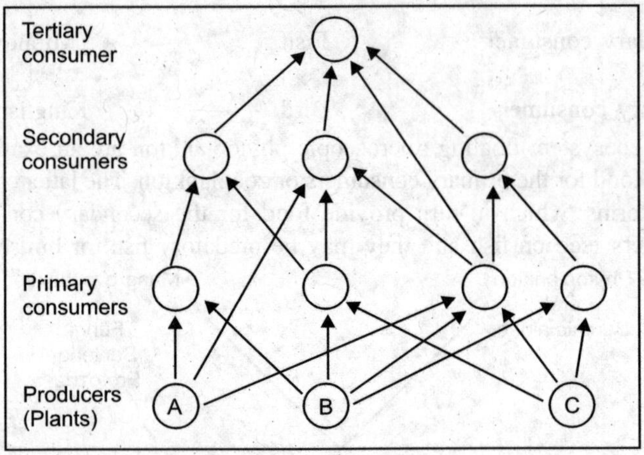

Fig. 4.3 Hypothetical food web showing the complex feeding relationships that may exist between different animal and plant species in an ecological community. However, the dying out of one species does not necessarily affect other species. For example, if plant A dies the community can still remain stable since the primary consumers that fed on A have an alternative source of food in B and C.

Thus, food web is a simplified representation of the complex interrelationships of the populations of plants and animals which exist in a community (Fig. 4.3). The basic operational principle in a food web is that each species is dependent upon at least one other species, and the numbers of each link species must be sufficient for their continued existence. If these conditions are maintained, the web will exist in an ecological nutritional equilibrium.

However, food webs are affected by changes in the environment, which may affect the food supply and reproductive rate. A food web on land is shown in (Fig. 4.4). In an aquatic ecosystem a more complex food web will exist, as shown in (Fig. 4.5). This is also affected by changes in the environment, such as variations in water temperature and oxygen content as well as nutrient supply, which can reduce or increase the population of link species.

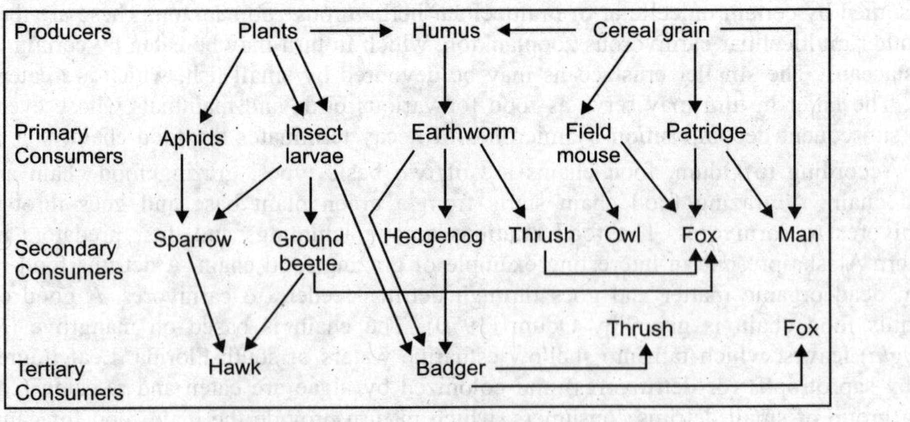

Fig. 4.4 A simplified food web on land.

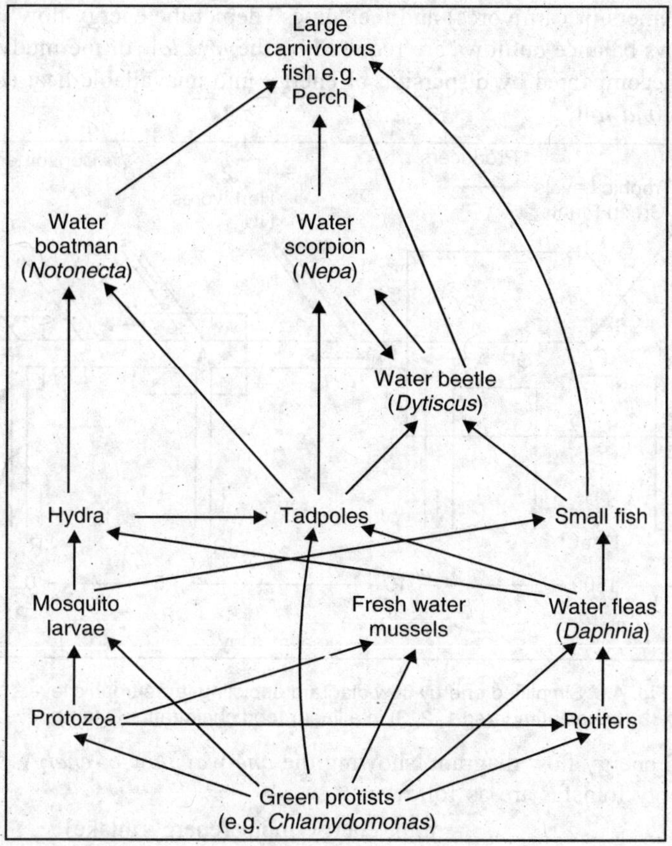

Fig. 4.5 Part of a food web in a typical fresh water pond.

4.4.3 Energy Flow in Food Chains

The transfer of the various forms of energy in a food chain is governed by the following laws of thermodynamics.

The *first law of thermodynamics,* also known as the law of conservation of energy, states that energy may be transformed from one form (type) to another, but is never created nor destroyed. Thus *radiant energy* in the form of light falling on a green plant is transformed partly into chemical energy or *potential energy* of food. This part of the transformation is initiated by the metabolic process called *photosynthesis.*

The *second law of thermodynamics* states that non-random energy (mechanical, chemical, radiant) cannot be converted into random form (heat energy) without some degradation. Because some energy is always dispersed into unavailable heat energy, no spontaneous transformation (for example light to food) can be 100% efficient.

Therefore, when the chemical energy accumulated by the plants is converted into kinetic energy by herbivores when they consume the plants, some degradation of energy occurs through its conversion into heat. Similarly, further energy conversion and degradation will occur when the herbivore is consumed by a primary carnivore, and when the primary carnivore in turn is eaten up by a secondary carnivore.

The organisation of food chain and the working of the two laws of thermodynamics can be understood by means of a simplified energy-flow diagram (Fig. 4.6). In this diagram the "boxes" 1, 2, 3 represent the trophic levels (1 : producers or autotrophs, 2 : primary consumers or herbivores,

3 : secondary consumers or carnivores) and the "pipes" depict the energy flow in and out of each level. Energy inflows balance outflows are required by the *first law* of thermodynamics, and each energy transfer is accompanied by dispersion of energy into unavailable heat (*i.e.* respiration) as required by the *second law*.

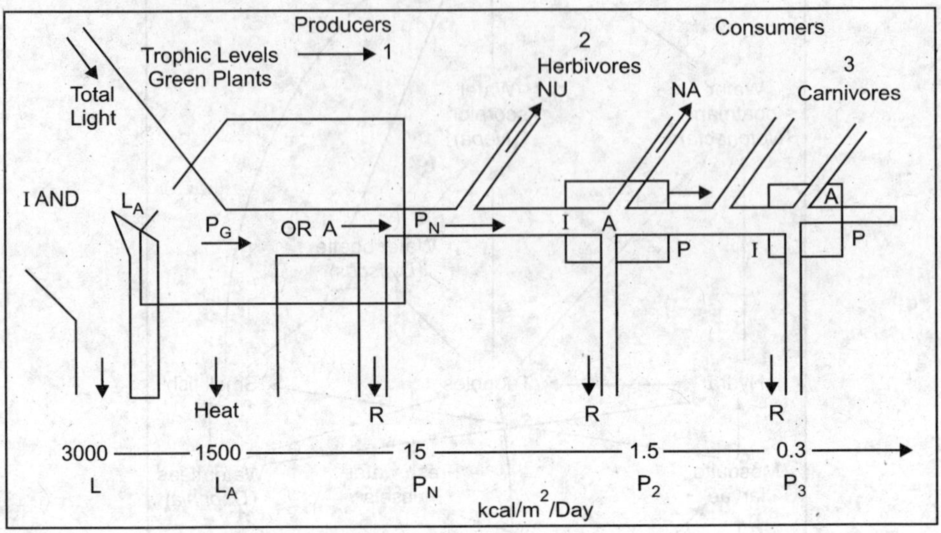

Fig. 4.6 Simplified energy flow diagram depicting three trophic levels (boxes numbered 1, 2, 3) in a linear food chain (after Odum).

In a simplified energy flow diagram, showing the *one-way flow of energy,* standard notations for successive energy transfer are as follows:

I = total energy input (energy intake)
L = total light
L_A = light absorbed by plant cover
P_G = gross primary production
A = total assimilation (assimilated energy)
P_N = net primary production
P = secondary production
N_U = energy not used (stored or exported)
N_A = energy not assimilated by consumers (egested), and
R = respiration (respiratory energy loss).

Bottom line in the diagram shows the order of magnitude of energy losses expected at major transfer points, starting with a solar input of 3000 kcl per square metre per day. (Fig. 4.6)

Thus, it is clear that energy flow is greatly reduced at each successive level regardless of whether we consider the total flow(I and A) or the components P and R. The diagram also shows the double metabolism of producers (*i.e.*, gross and net production) and the approximately 50 per cent absorption by the plant cover and one per cent conversion of light at the first trophic level. Secondary productivity (P_2, P_3, in the diagram), tends to be about 10 per cent at successive consumer trophic levels, although efficiency may be higher (say 20 per cent), at the carnivore levels as shown in the diagram.

The energy flow diagram may, in principle, be applied to any ecosystem or community. As the initial amount of energy introduced into a food chain in a given habitat is limited by the duration and intensity of sunlight, there will be some point along the food chain of the habitat when this initial load of radiant energy has been transformed and finally all converted to heat energy. At this point the food chain will have reached the *steady state,* in which the amount of energy taken into its biological system in unit time as radiant energy exactly balances that returned as heat energy to the habitat.

4.4.4 Relationship Between Standing Crop and Energy Flow

The relationship between the *standing crop* and the *energy flow* is very important. In the energy flow diagram (Fig. 4.6), the "boxes" represent the biomass of the standing crop of organisms functioning at the trophic level indicated, and the "pipes" represent the energy flow in and out of each level. The energy flows are greatly reduced with each successive trophic level, whether we consider the total flow (*A*), or the production (*P*), or respiration (*R*) components. Similarly, in many cases, the standing crop also decreases.

The standing crop biomass is much influenced by the size of the individual organisms forming the trophic level in question. According to the *inverse size-metabolic rate* principle, the smaller the organism the greater would be the rate of metabolism per gram of weight. For example, one gram of small algae may be equal in metabolism to many grams of forest tree leaves. Thus, if the producers of an ecosystem are formed of very small organisms, and the consumers are large, the standing crop of consumers may be greater than that of the producers, though the energy flow of the producers must be greater.

The sea ecosystem is a typical example of such a situation in which the consumers (clams, crustaceans, echinoderms, fish and the like) often outweigh the producers (microscopic phytoplankton) on which they depend. By harvesting at frequent intervals, man(as well as the clam) may obtain as much food (net production) from mass cultures of small algae as he obtains from a grain crop harvested after a long interval of time. However, the standing crop of algae at any one time would be much less than that of a mature grains crop. The standing crop biomass is usually expressed in terms of grams of organic matter, grams of carbon, or calories per unit area; but productivity is expressed as grams or calories per unit time, and the relationship between the two depends on the kind of organisms involved.

4.4.5 General Model and Patterns of Energy Flow in Communities

A generalized model of energy flow (Heal and Maclean, 1975) provides a complete description of the trophic structure of a community (Fig. 4.7). Besides the *grazer system* pyramid of productivity, it has two additional elements—a decomposer system and subcomponents of each trophic level. Thus, a distinction is made between vertebrate and invertebrate types, between detritivores and microbes which belong to the same trophic level and utilize dead organic matter, and also between consumers of microbes (microbivores) and of detritivores.

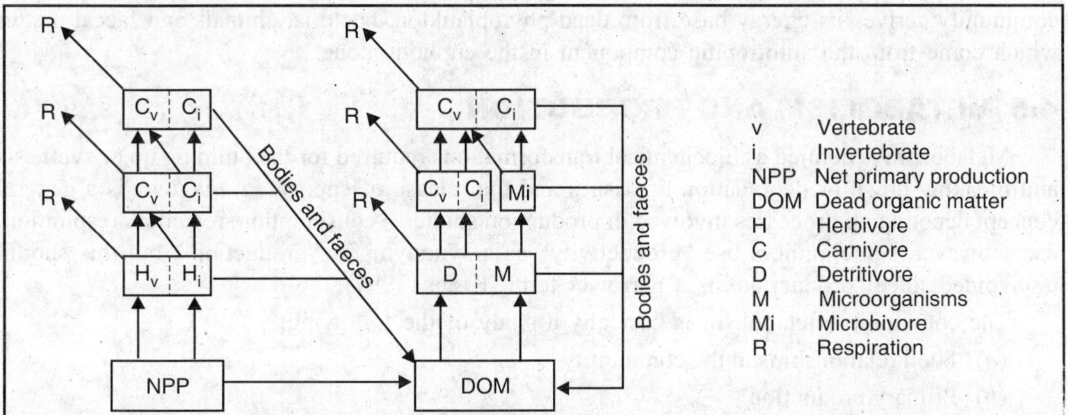

Fig. 4.7 A generalized model of trophic structure and energy flow for a terrestrial community. (After Heal and Maclean, 1975)

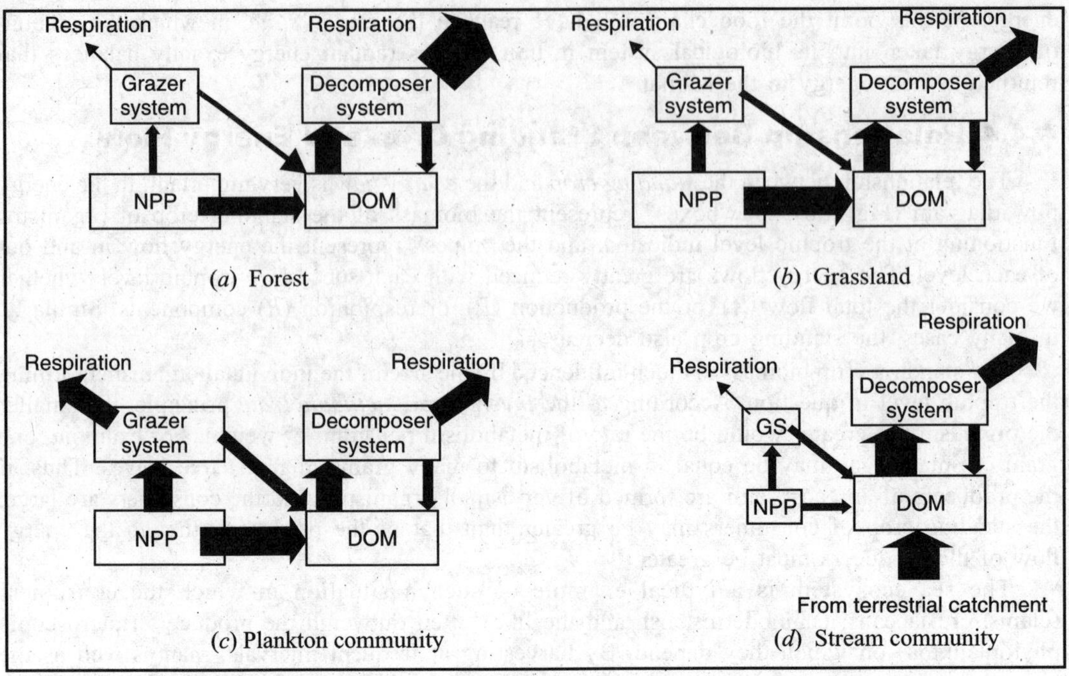

Fig. 4.8 General patterns of energy flow for (*a*) a forest, (*b*) a grassland, (*c*) a plankton community in a sea and (*d*) the community of a stream or small pond. Relative sizes of boxes and arrows are proportional to relative magnitudes of compartments and flows. DOM, dead organic matter; NPP, net primary productivity. (After Begon *et al*, 1996)

However, patterns of energy flow differ in different communities when we compare the patterns of energy flow in a forest, a grassland, a small stream, a large lake or ocean, we find interesting differences not only in their net primary production but also in their grazer system (Fig. 4.8). In every community, the decomposer system is responsible for the major part of secondary production and therefore heat loss in respiration. The *grazer system* has its greatest role in plankton community of a large lake or of the ocean, where a large part of net primary production is consumed alive and assimilated at a high efficiency. However, the grazer system has little influence in terrestrial communities because of low herbivore consumption and assimilation, and it has very little or no role in small streams and ponds as their primary production is quite low. Streams and ponds depend for their energy on dead community. In the deep ocean also the community derives its energy base from dead phytoplankton, bacteria, animals and faecal matter which come from the autotrophic component in the euphotic zone.

4.5 METABOLISM AND PRODUCTION

Metabolism is defined as biochemical transformations required for the building up or synthesis and breaking down or degradation of tissues and the release of energy. *Productivity* is a general concept denoting all processes involved in production studies—consumption, rejection, respiration, etc. However, some authors use "productivity" as a synonym for "production", but this should be avoided since "production" is a narrower term (Krebs, 1997).

The community metabolism is basically a study of the following:

(*a*) Food relationships in the community
(*b*) Primary production
(*c*) Secondary production

The feeding relationships in a community are considered under the headings food chain, food web, and trophic levels.

4.5.1 Primary Production

A very important factor in community metabolism is the rate of energy-trapping by green plants, which in turn governs the rate of production of organic material from simple inorganic substances. The solar energy is transformed into chemical energy in photosynthesis.

$$6CO_2 + 6H_2O + \text{solar energy, Chlorophyll and enzymes} \rightarrow C_6H_{12}O_6 + 6O_2$$

Gross primary production (assimilation) means the total energy fixed in photosynthesis or the total production as a result of photosynthesis per unit time. It is a measure of the total amount of dry matter made by a plant in photosynthesis. *Net primary production* means the energy fixed in photosynthesis minus the energy lost by respiration per unit time. Gross primary production (GPP), net primary production (NPP) and respiration (R) are therefore related to one another by the equation: NPP = GPP – R. But a considerable part of fixed energy is always lost in respiration.

$$C_6H_{12}O_6 + 6O_2 \xrightarrow[\text{enzymes}]{\text{metabolic}} 6CO_2 + 6H_2O + \text{energy}$$

These two aspects of primary production can be measured. The simplest method of measuring primary production is the *harvest method.* The amount of plant biomass produced in a unit of time can be determined from the difference between the amount present at the two times :

$\Delta B = B_2 – B_1$, where

ΔB = biomass change in the community between time t_1 and t_2

B_1 = biomass at t_1

B_2 = biomass at t_2

However, two possible losses must also be considered. There are biomass losses due to the death of whole plants or their parts (L) and biomass losses due to consumption by other organisms (C). If we know these values, the production can be determined as follows:

$$\text{Net primary production} = \Delta B + L + C$$

Net primary production in some community types is given in Table 4.1 from Whittaker and Likens (1973). It is clear from the Table that certain communities are more productive. The following factors affect the primary production in all natural communities.

Table 4.1 Primary production in some community types (after Whittaker and Lichens, 1973)

S.No.	*Community*	*Net primary production (g m^{-2} yr^{-1})*
	Terrestrial	
1.	Tropical rain forest	2000
2.	Temperate deciduous forest	1200
3.	Temperate grassland	500
4.	Desert scrub	70
5.	Cultivated land	650
	Aquatic	
6.	Lake and stream	500
7.	Estuaries	1800
8.	Open ocean	125
9.	Algal beds and reefs	2000
10.	Swamp and marsh	2500

***(i)* Light :** Light affects primary production to a great extent. The depth to which light penetrates in a water body is critical in defining the zone of primary production. The decrease in radiation with depth is shown by the following equation:

$$dI/dt = kI, \text{ where}$$

I = amount of solar radiation

t = depth

k = extinction coefficient (a constant).

The value of k increases with the loss of light. An extinction coefficient of 0.02 would occur in clear water whereas the value of k in seawater would be 0.10. Thus, light is an important factor in limiting primary production in the ocean (Platt and Sathyendranath, 1988).

***(ii)* Nutrients :** Nutrients also appear to limit primary production in subtropical and tropical oceans. Two elements, phosphorus and nitrogen, often limit primary production in the surface layers of the oceans.

***(iii)* Temperature :** Temperature is also a major controlling factor in freshwater communities. Low temperatures in lakes, streams and oceans tend to reduce primary production.

***(iv)* Evapotranspiration :** It is a measure of solar radiation, temperature and rainfall. In other words, it is the amount of water pumped into the atmosphere by evaporation from the ground and by transpiration from the vegetation. Rosenzweig (1968) showed that actual evapotranspiration could predict above ground production of terrestrial communities. Lieth (1975) showed that in addition to evapotranspiration, net primary production can be predicated from the length of the growing season, precipitation or temperature. Primary production varies in different types of forests. Evergreen broadleaf tropical forests are most productive. The differences in primary production among forests are mainly due to differences in the length of the growing season and leaf area index. Primary production in grasslands seems to be largely limited by rainfall (Sala *et al.* 1988).

According to *Krebs* (*1994*), primary production varies greatly over the globe. It is highest in the tropical rain forests and lowest in arctic, alpine, and desert habitats. The production in the sea is less than on land due to the limitations imposed by light and nutrients. Measurements of the amount of respiration by primary producers in terrestrial communities show that GPP is approximately equal to 2.7 times NPP. In the oceans, GPP is only about 1.5 times NPP (Whittaker, 1975). Remote sensing with satellites is now used for measuring the spatial and temporal variations of primary production on land and in the oceans of the world.

4.5.2 Secondary Production

This means the production at the level of consumers. In other words, secondary production is the rate of production of biomass by heterotrophic organisms (bacteria, fungi and animals). This occurs by using metabolizable energy (A) for growth and reproduction (Fig. 4.9).

There are several techniques to measure the components of secondary production in animal community. According to the general procedure each animal species is considered separately. To determine the *gross energy intake,* the feeding rate can be measured by confining an animal to a feeding plot and measuring herbage biomass before and after feeding. In some species, like birds the number of food items consumed can be counted directly. The *metabolizable energy* (assmilation) can also be measured in the laboratory by regulating the gross intake and collecting the faeces and urine. But in the field it is measured indirectly as follows.

$$\text{Metabolizable energy} = \text{Respiration} + \text{Net production}$$

In laboratory, respiration can be measured by measuring oxygen consumption, and CO_2 production and thus basal metabolic rate can be calculated. Net production can be measured by the growth of individuals in the population and the reproduction of new animals. It is usually

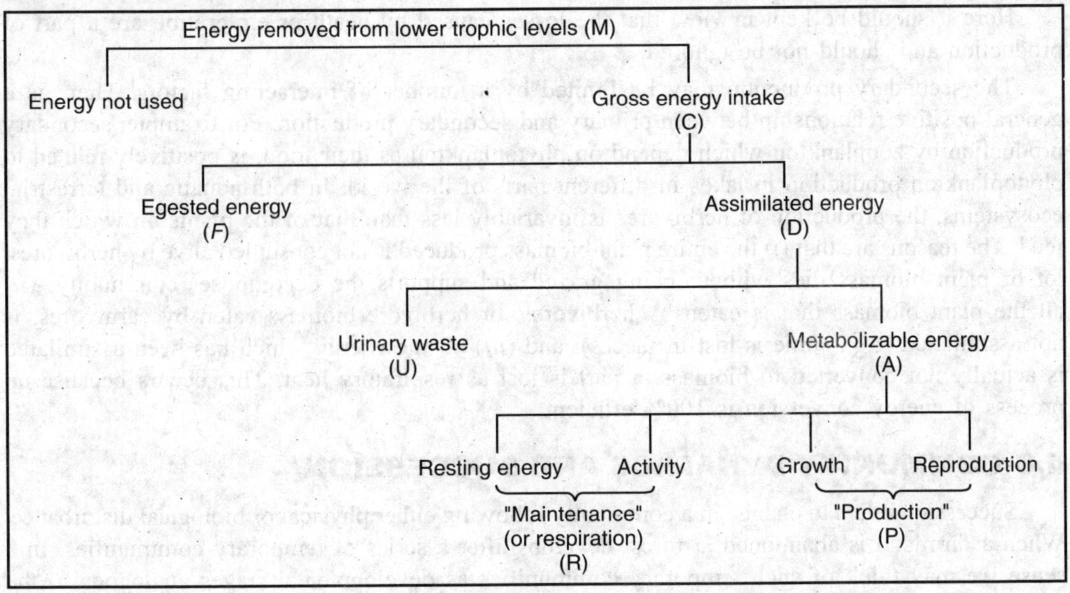

Fig. 4.9 Flow of food energy in an animal (After Krebs, 1994)

measured as units of biomass and converted to energy by determination of the caloric value of a unit weight of the species. Fig. 4.10 shows that the production is the sum of growth and natality.

$$\text{Production} = \text{Growth} + \text{natality}$$
$$= 20 + 10 + 10 + 10 + 10 + 30 - 10 - 10 = 70 \text{ units of biomass}$$

The production can also be calculated as follows:

$$\text{Production} = \text{Net change in biomass} + \text{losses in mortality}$$
$$= 30 + 40 = 70 \text{ units of biomass}$$

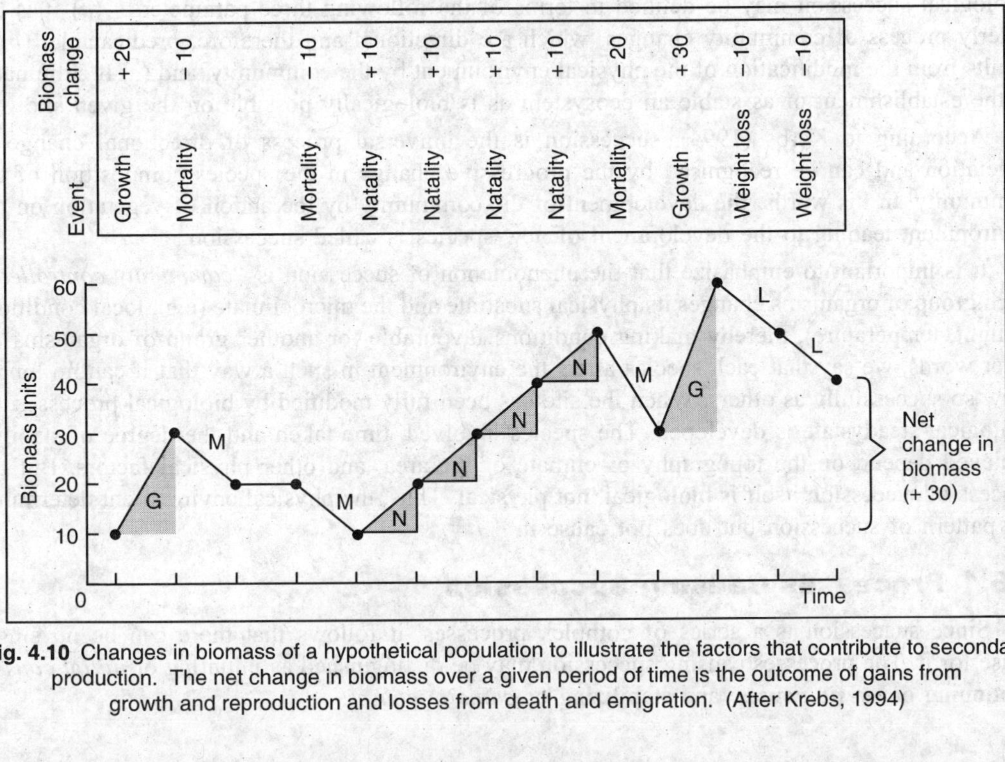

Fig. 4.10 Changes in biomass of a hypothetical population to illustrate the factors that contribute to secondary production. The net change in biomass over a given period of time is the outcome of gains from growth and reproduction and losses from death and emigration. (After Krebs, 1994)

Here it should be kept in view that the losses caused by death or emigration are a part of production and should not be ignored.

The secondary production may be limited by a number of interacting factors. There is a general positive relationship between primary and secondary production. For example, secondary production by zooplankton which depend on phytoplankton as their food, is positively related to phytoplankton production in lakes in different parts of the world. In both aquatic and terrestrial ecosystems, the production of herbivores is invariably less than that of the plants on which they feed. The reasons are that (*i*) the entire plant biomass produced is not consumed alive by herbivores, lot of plant biomass dies without being grazed and supports the decomposer community; (*ii*) all the plant biomass that is eaten by herbivores, or herbivore biomass eaten by carnivores, is not assimilated (*e.g.*, some is lost in faeces); and (*iii*) all the energy which has been assimilated is actually not converted to biomass, a part is lost as respiratory heat. This occurs because no process of energy conversion is 100% efficient.

4.6 COMMUNITY DYNAMICS AND SUCCESSION

Succession refers to change in a community following either physical or biological disturbance. When a farmland is abandoned, a forest develops after a series of temporary communities. In a sense we may think of such temporary communities as developmental stages analogous to the life history stages through which many organisms pass before reaching adulthood.

The concept of succession was largely developed by the botanists Warming (1909) and Cowles (1899), who studied the stages of sand dune development. It has been further elaborated by Clements (1916, 1936) who proposed a theory of plant succession and community development called the *monoclimax hypothesis.* Later on Tansely (1939), and Daubenmire (1966) proposed the *polyclimax hypothesis.*

Succession is a unidirectional change in the community structure. In simple words, succession my be defined as natural change in the structure and species composition of a community. Ecological succession may be defined in terms of the following three parameters : (*a*) It is the orderly process of community changes, which are directional and therefore, predicatable, (*b*) It results from the modification of the physical environment by the community, and (*c*) It culminates in the establishment of as stable an ecosystem as is biologically possible on the given site.

According to Kerbs (1994), succession is the universal process of directional change in vegetation and can be recognised by the progressive change in the species composition of the community. In his words, the development of the community by the action of vegetation on the environment leading to the development of new species is called succession.

It is important to emphasize that the phenomenon of succession is *"community controlled"*. Each group of organisms changes its physical substrate and the microclimate (e.g., local conditions of light, temperature), thereby making conditions favourable for another group of organisms. In other words, we say that each species alters the environment in such a way that it can no longer grow so successfully as others. When the site has been fully modified by biological processes, an ecological steadystate is developed. The species involved, time taken and the degree of stability achieved depend on the topography or climate of the area, and other physical factors. But the process of succession itself is biological, not physical. Thus, the physical environment determines the pattern of succession but does not cause it.

4.6.1 Processes Causing Succession

Since succession is a series of complex processes, it follows that there can be no single cause for it. The processes causing succession may be distinguished as initiating or *initial causes,* continuing or *ecesic causes,* and stabilizing or *climatic causes.*

The *initial causes* produce the bare area or destroy the original population in areas already vegetated. The deposition of sediment as alluvial fans at the mouth of a river illustrates the former; and the wandering sand dune covering a forest, the latter. The *ecesic causes* produce the essential character of vegetational development, *i.e.*, the successive plant populations. They have to do with the interaction of vegetation and habitat and are directive to the highest degree. The *climatic causes* determine the nature of the climatic climax, *i.e.*, the end point of succession. They have a profound effect in determining the population from the beginning to end, the number and kinds of stages as well as the reactions of the successive stages.

While the process of succession in the tropics is similar to that in temperate regions, the plant populations are often quite different. This is due to the climate.

4.6.2 Sere and Climax Community

In ecological terminology, the developmental stages of a community are known *seral stages,* and the final stage as the *climax community.* The initial community of the area which is replaced in time by a sequence of succeeding communities until the climax is reached, is called the *pioneer stage.* The entire series of communities that is characteristic of a given site is called *a sere* (Fig. 4.11).

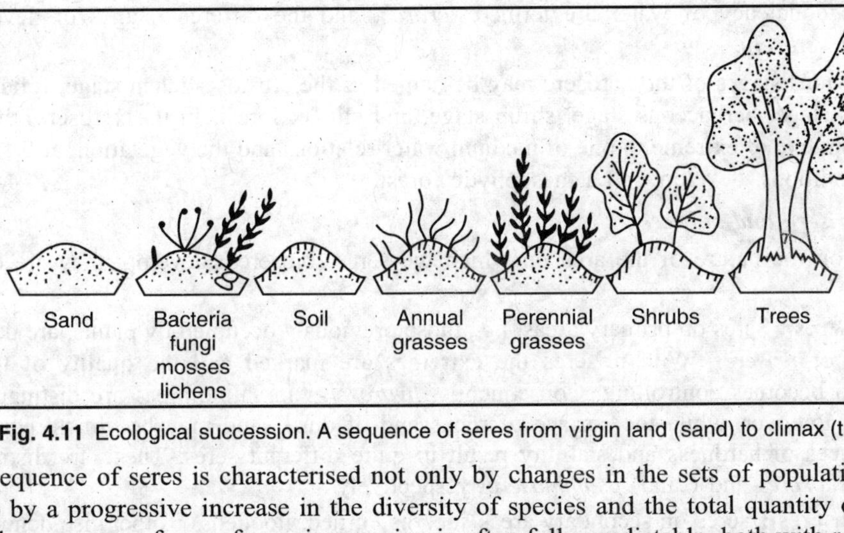

Fig. 4.11 Ecological succession. A sequence of seres from virgin land (sand) to climax (trees)

A sequence of seres is characterised not only by changes in the sets of populations present, but also by a progressive increase in the diversity of species and the total quantity of the living mass. The sequence of seres for a given region is often fully predictable, both with respect to the general types of population expected at each sere and to seral durations. On land, for example, a climax stage is often represented by a forest community. If the original physical environment is sand then the sequence of seres includes: soil-forming organisms (bacteria, lichens, mosses, fungi), annual grasses, perennial grasses, shrubs, and trees (Fig. 4.11).

Characteristic animal populations are associated with each of such seral plant populations. It can be shown that from sand to forest climax may take about 1000 years. If previously cultivated land is the starting condition, *i.e.*, if soil is already present, then a grass climax may be attained in about 50 years, a forest climax in about 200 years. We may note here that successive generations of the same plant population gradually exhaust the soil in specific ways. Moreover, the dead bodies of plants add materials to the soil which often prove to be toxic to subsequent generations of the same plants. Thus, both specific exhaustion of soil and specific additions to it promote seral succession, not only among the plants but also among the animal populations dependent on them.

4.6.2.1 Classification of Seres

Seres are classified primarily on the basis of the water content of the initial area in which they develop. The two main types are hydrosere and zerosere.

4.6.2.1.1 Hydrosere

Successions beginning in ponds, lakes, marshes, or elsewhere in water are termed *hydrarch* and the different stages of the sereis or sere constitute a *hydrosere.*

Weaver and Clements (1937) have described a number of stages through which an area covered by deep water is transformed into a forest. The various stages— submerged stage or plants, floating plants, reed-swamp plants, sedge-meadow plants, woody plants, and climax forest are nearly cross sections to a continuous development made with reference to certain points where, because of more or less pure dominance, change is most apparent. This whole developmental process in action may be found about lake margins where each stage is shown as a definite zone. The stages in the present horizontal sequence from shallow water to marginal forest become arranged in a vertical sequence as the bottom of the lake is built up, forest forming the top stratum.

4.6.2.1.2 Xerosere

Successions beginning on the bare rock, wind-blown sand, or other places where there is an extreme deficiency of water are termed *xerarch,* and the different stages of development constitute a *xerosere.*

The various stages of the xerosere may be named as the crustose-lichen stage, foliose-lichen stage, moss stage, herbaceous stage, shrub stage, and climax forest. In the xerosere, the habitat changes from one of extreme to one of medium water relations and the vegetation, at first adapted to xeric conditions, develop into a mesophytic forest.

4.6.2.1.3 Prisere and subsere

Based on the nature of the area on which the seres (bioseres) develop, they are classified as follows:

(*i*) Priseres : Seres on primary areas, *i.e.*, those previously occupied by plants, are designated as primary or *priseres.* With priseres the extremes are marked and the quality of the water content also becomes controlling. Consequently, *hydroseres* in saline areas are distinguished as *haloseres.* Moreover, while the surface of rocks and of dune sand may be almost equally dry, the differences of hardness and stability result in quite different seres. These are distinguished as (rock) *lithoseres* and (sand) *psammoseres* respectively.

(*ii*) Subseres : Seres on secondary areas such as burned, flooded, or otherwise denuded ones are termed subseres. In subseres, extreme conditions of water content are rare or they persist for a brief period only. Hence, it is sufficient to recognize only two subdivisions, hydrosere and xerosere.

4.6.3 Forces Causing Succession

Seres are sometimes classified according to the predominant force that is bringing them about. These forces are *biotic, climatic, physiographic* and *geologic* and their resultant seres are accordingly called *bioseres, cliseres, eoseres* and *geoseres.* Similarly, on the basis of these forces the succession is also called Biotic succession, Climatic succession, Physiographic succession, and Geologic succession.

4.6.3.1 Biotic succession

This succession is brought about by the forces inherent within the community and in the activities of animals and plants themselves. The most important of these activities are their reactions and coactions that produce modifications in the habitat and interrelations between species.

Important reactions involve filling in of ponds with animal and plant remains, the addition of organic nutrients to sterile soil, and the reduction in light intensity by increasing density of plants. With progressive improvement of the soil and changing light and moisture conditions, a series of new ecological dominants come into the area. When invasion of new species occurs, intense competition develops and if the invaders are strong and successful the old species disappear because a new community replaces them.

4.6.3.2 Climatic succession

With changes in climate, environmental conditions often surpass the limits of tolerance of established animals and plants. The result is the replacement of the existent community by another. A most interesting *clisere* is the one that has occurred since the northward retreat of the continental glacier of Pleistocene time. Stages in this clisere may be detected and even the relative duration of each stage measured from an analysis of the number and persistence of different kinds of pollen grains at various depths in peat bogs. To make such studies a core of peat is obtained from the deepest part of the bog by mean of a special hand auger. The lowest portion of the core is oldest, the most recently formed is as at the top. Samples of the core at various depths are suitably prepared, examined under a microscope, and the pollen grains identified are counted. The predominant kind of pollen at any level of the core represents the probable prevailing species of plants in the vicinity at the corresponding period of time.

4.6.3.3 Physiographic succession

Changes in the earth's surface bring a change in communities. Mountain-building brings the replacement of low-land communities with new ones that invade at higher elevations. As mountains erode, the eosere progresses in the opposite direction, until base-level or peneplanation is attained. The development that the eosere will undergo with continued erosion is sometimes locally apparent in the difference between upland and floodplain forests. Stages in the erosion cycle may be distinguished by examination of habitat and animal life progressively from headwaters to mouth of a stream.

4.6.3.4 Geologic succession

The first organisms to appear on earth were unicellular forms confined to the sea. At the beginning of the Cenozoic Era, ancient types of animals, including the great reptiles and many types of invertebrates, became extinct and mammals rose to predominance. It is significant that the rich development of mammals, birds and insects came after the worldwide establishment of the angiosperms with their rich nutrient seeds, fruits and grasses. Pleistocene brought a major change in the habitat of these animals, and some large mammals disappeared. The last stage of the geosere, the recent epoch, brought the dominance of man. Now only future ages will determine whether this stage is climax or new and different types of animal and plant life will some day evolve to replace man and to continue the succession into the indefinite future.

An important aspect of geosere is the dispersal of new types of animals and plants, for the importance of a new form depends on the extent of its distribution and the size of its population. Speciation in animals is initiated only when one population becomes isolated from another similar population so that interbreeding is checked. This permits variation and natural selection to proceed independently in the two groups. Isolation is usually affected by geographic barriers and generally involves occupancy of new niches and development of new coactions with other members of the community. The history of the past ecological relations of species and of the whole community is the subject matter of *palaeoecology.* The dynamic forces involved in these processes of speciation are among the forces determining the geosere.

4.6.4 Types of Succession

4.6.4.1 Primary Succession

According to Clarke (1954), ecological succession that begins on a bare area where no life has existed, or where the previous flora and fauna have been completely destroyed, is known as the primary succession. If succession proceeds from an area devoid of organisms, or from an area which has not been changed physically by organisms, it is called primary succession (Benton and Werner, 1976). The diverse places that may be available for primary succession (*i.e.*, sand bars, glacial moraines, recently formed ponds) may be classified as *xeric, mesic,* or *hydric* according to whether the initial moisture conditions are dry, intermediate or wet. Seres starting from these types of situations represent *xerarch, mesarch,* and *hydrarch* succession respectively.

A striking example of primary succession is the *hydrarch* succession in which a pond or lake and its community is converted to dry land with completely different community. Such environments may become dry by two major processes. They may fill up with materials washed into the basin and with the remains of dead organisms which once lived in the lake or the outlet may cut down its stream bed so that eventually it becomes level with the bottom of the lake.

The pond or lake may be colonized by aquatic plants and animals. Microscopic organisms will find their way to deep open water. In shallow areas, where light can penetrate near the bottom, rooted submerged aquatic vascular plants will invade. In more shallow water, rooted aquatic plants with floating leaves may become established. In a few centimetre deep water emergents (plants having most of their stems and leaves in the air) will be found. One shore, plants which can tolerate saturated soil will dominate. Each of these types of plants is said to form a community, since there are several species of animals and plants associated with each major plant type mentioned. In all the communities, organic matter and silt are constantly accumulating, decreasing the depth of the water. With this change of environment each community invades the one next to it in deeper water, and is in turn invaded by the one behind it in shallower water. Thus, as the filling process occurs, the submerged aquatic plants will be displaced by the floating rooted aquatics, and these by the emergent aquatics. As succession continues, the emergents will be replaced by plants of the saturated soil region, and these in turn by ordinary terrestrial plants. Gradually the terrestrial communities will move in where once there was open water, until finally a terrestrial climax community occupies the site.

Another situation in which primary succession may occur is in dry environment. In such case the starting point may be bare rock, which would probably be invaded first by lichens (called the *pioneers*). Thereafter, other organisms such as mosses might appear, followed by other herbaceous plants, then by shrubs, and finally by trees. Sometimes trees or shrubs can be pioneers. In some studies it has been found that conifers were pioneers and herbaceous species were not a prerequisite for invasion of an area laid bare by volcanic activity.

4.6.4.2 Secondary Succession

The term secondary succession refers to a community development on sites previously occupied by well developed communities, or it is the succession on sites where nutrients and conditions of existence are already favourable, such as abandoned croplands, plowed grasslands, and cut-over forests.

Secondary succession may start when the principal species of the community have been destroyed by fire, disease, flood or by human activities such as farming or lumbering. In some instances of secondary succession a community is established that is essentially the same as a stage in the previous primary succession. In secondary succession, the rate of change is much more rapid, and the time required for the completion of the final stage is much shorter. The end point for both types of succession will be a climax community.

In a study of secondary succession on landslides it has been found that although areas of bare rock were still mostly uncovered after 72 years; herbs, shrubs and trees became established within 9 years in areas where residual forest soil remained or collected between the rocks. After 72 years, the herb layer was virtually similar to the surrounding forest, and the tree species composition was approaching the climax.

4.6.4.3 Autotrophic succession

Succession that begins in a predominantly inorganic environment and is characterized by early and continued dominance by autotrophic organisms, is called the autotrophic succession. The primary and secondary successions both come under the autotrophic succession and are widespread in nature. A model for autotrophic type of ecological succession is given in Table 4.2.

Table 4.2 Showing a model for autotrophic type of ecological succession (after Odum)

Ecosystem characteristic	Trend in Succession: early stage to climax or young to old nature
	Species Structure
Species composition	Changes rapidly at first, then more gradually.
Number species of autotrophs	Increases in primary and often early in secondary succession, may decline in older stages.
Number species of heterotrophs	Increases until relatively late in sere.
Species diversity	Increases initially, then becomes stabilized or declines in older stages.
	Organic Structure
Total biomass	Increases
Nonliving organic matter	Increases
Chlorophylls	Increase during early phase of primary succession, little or no increase in secondary succession.
Pigment diversity	Increase at least early in the sere.
	Energy Flow (Community metabolism)
Food chain relationships (foodwebs)	Become more complex
Gross production	Increases during early phase of primary succession, little or no increase during secondary succession.
Net community production	Decreases
Community respiration	Increases
P/R ratio	$P > R$ to $P = R$

4.6.4.4 Heterotrophic Succession

This succession is characterized by early dominance of heterotrophs. It occurs in the special and rare case where the environment is predominantly organic as, for example, in a stream heavily polluted with sewage. Other examples of heterotrophic succession include the successions involved in rotting of dead trees, reduction of animal dung, invasion and destruction of plant galls. In these cases green plants are usually not involved. Instead, the organisms in heterotrophic successions use up the energy of the dead organic matter.

In this type of succession, energy is maximum at the beginning and gradually declines as succession proceeds until additional organic matter is added from outside. In contrast, energy flow does not necessarily decline in the autotrophic type of succession but is usually maintained or increased. In heterotrophic succession, a series of different organisms (animals, fungi, bacteria and other forms) are involved and replace each other but the end point is utilization of all the energy and dispersion of the community.

4.6.4.5 Autogenic Succession

Succession that results from changes brought about by the organisms themselves is called *autogenic succession.* The change from an abandoned agricultural field to a mature forest is an example of autogenic succession. In autogenic succession, the principal force of change comes from within the community. Therefore, autogenic succession occurs on time scales commensurate with the life span of the organisms in the community (Ricklefs and Miller, 1999).

4.6.4.6 Allogenic Succession

Succession that results from factors external to the community is called allogenic succession. In allogenic succession, the principal force of change comes primarily from outside the community. Such external forces may include climate change, changes in temperature and other environmental factors, or other types of massive disturbances. Allogenic succession occurs on a time scale which is in accordance or proportionate with the time scale of the disturbance. Allogenic succession resulting from climate change may occur over thousands of years. Seasonal changes in temperature and light intensity in freshwater lakes are responsible for succession of their phytoplankton communities.

4.6.5 Attributes of Succession

There are four major structural and functional attributes of ecological succession, which are as follows:

1. There is an increase in structural complexity, or the kinds of animals and plants change continuously with succession, no matter what is their taxonomic group.
2. The diversity of species tends to increase with succession. Increase in diversity of heterotrophs is especially striking. The variety of microbes and heterotrophic plants and animals is likely to be much greater in the later stages of succession than in the early stages. Maximum diversity of autotrophs in many ecosystems seems to be reached earlier in succession.
3. Biomass and standing crop of organic matter increase with succession. According to Odum, the increase in amount of, and the change in, organic structure are two of the main factors bringing about the succession of species. The enlargement of organic structure is, of course, related in a cause-and-effect manner to the increase in species diversity. As biomass increases, stratification and zonation create many new habitat niches such as litter, humus, bark and dead wood in a forest.
4. Species, biomass, and the *P/R* ratio (the ratio of gross production of the community to total respiration of the community, *i.e.*, the respiration of both animals and plants) continue to change long after the maximum gross primary production possible for the site has been achieved. Thus, a decrease in net community production and a corresponding increase in community respiration are two of the most striking and important trends in succession. As indicated in Table 4.2, gross production may be expected to increase in the early stages of primary succession, with little or no change in later stages or in secondary succession.

4.6.6 Examples of Animal Succession

According to Elton (1965), the animal succession may be best studied in very simple communities. It can be studied experimentally in the laboratory by making a hay infusion in water and leaving it open to the air for several weeks. Bacteria are the first organisms to become abundant, since they live upon the decaying vegetable matter. The various protozoans appear, and it is possible to see a whole animal community being gradually built up, as each new species arrives and multiplies and fits into its proper niche. In the hay infusion, the bacteria are followed by small ciliate protozoa of the *Paramecium* type, which subsist upon bacteria and also by absorbing substances in water. Then there are larger hypotrichous ciliates, which prey upon bacteria and smaller ciliates. Eventually the whole culture may degenerate due to a decrease in food material for the bacteria and therefore for the animals feeding on them. On the other hand green autotrophs, in the form of small algae may arrive and colonise the culture. These will be able to subsist for a long time, and may change the character of the whole community by providing a different type of food. This is heterotrophic type of succession where *R* exceeds *P* (or where *P* may be ultimately zero), but on the appearance of algae an equilibrium is approached and *P*/*R* ratio tends to approach 1.

Succession may also be observed in other cases. The kinds of insects that infest granaries change radically from month to month in regular fashion. Here no climax is reached, however, because the stored grain does not renew itself and does not become available again for further generations of beetles, as the soil nutrients do in the pond, land or forest.

Animals also play an important role in ecological succession in the sea. According to Wilson (1925), who studied succession on bare areas on the seashores, the pioneers were colonial diatoms, which formed the first community. These were followed by a group of colonial hydroids (mostly *Obelia*), and the latter were then replaced by a sea weed (*Ectocarpus*), which became dominant species for about four moths. Further stages were foreshadowed, and it appeared that the whole would ultimately develop into a climax species of other sea weeds (chiefly kelp). The interesting thing about this sere is the fact that the second stage in succession is formed by sessile animals, and is sandwitched between two plants stages. The zonation of animals and plants in the intertidal zone in other regions is often an alteration of animal and plant dominance, *e.g.*, *Balanus* or *Mytilus* and seaweeds, in the temperate region. On coral island, succession may consist almost entirely of a series of animal zones with only an occasional plant zone formed of calcareous algae.

4.6.7 Successional Processes

Succession is strongly influenced by three processes (Connell an Slatyer, 1977). These are :

(*i*) ***Facilitation:*** Changes in the abiotic environment that are imposed by the developing community and allow other species to invade. Facilitation is prominently observed in early stages of primary succession.

(*ii*) ***Inhibition:*** Species of one seral stage resist invasion by later successional species such that invasion is only possible following disturbance or by replacing individuals as they die, *e.g.*, green algae (*Ulva*) on intertidal boulders. Inhibition is a more common feature of secondary succession.

(*iii*) ***Tolerance:*** Late successional species invade because they are able to tolerate lower resource levels and can outcompete early successional species for limited resources like light and space, *e.g.* old field succession. Thus, highly competitive species which are tolerant of low resource levels will replace opportunistic good colonizers and dominate the climax community.

The tolerance model suggests a predictable sequence of species replacement based on their strategies for resource exploitation. Tilman (1988) in his *resource ratio hypothesis of succession* proposed that the relative availability of two resources—light and nutrients, determines the species dominance. During succession, light levels decrease as a result of shading, while nutrient levels increase with soil formation and litter accumulation.

4.6.8 Rate of Succession and Climax

One of the important questions pertaining to succession is—why a climax is reached? Probably the answer is simple—there are no more successful species to replace the climax ones. However, species that are climax in one region may be replaced by even more tolerant species in other regions, showing that there is nothing unique about the climax species. Sometimes different initial conditions may lead to the same climax. This is a mathematical property of replacement processes, but the purely biological explanation is that replacement (in tree species) imparts a premium on dense canopy and tall trees (to compete successfully in the shade). As long as both the initial conditions are such that the succession replacement is directed towards more and more shade, it follows that both are likely to end up with the same shade-tolerant species.

It has been suggested by the ecologists that the *degree of succession expected is proportional to the degree of modification of the habitat by the climax species.* The following example illustrates this statement. In harsh environments, such as extreme deserts where the permanent vegetation consists of widely spaced bushes, the first colonizers are often these same species. Thus the presence of low, widely spaced individuals does little to modify the physical environment. In contrast, where conditions are favourable, as in forests, the first species to colonize a bare area are hardly ever those, which constitute the climax community. In this way, the trees of a dense forest modify the environment, making it darker, damper and less windy.

It is thus current theory that the *stability of the climax community is a function of its species diversity* and that the aging phenomenon or succession in an ecosystem is well described as an evolution towards high diversity—a large number of ecological niches and its counterpart of a large number of species.

4.6.8.1 Climax Theories

The final stage of succession is called the climax or climax community (Clements, 1936; Shimwell 1971). It is the final or stable community in a successional series. It is self-perpetuating and in equilibrium with the physical and biotic environment. Climax communities undergo changes in structure as a result of birth, death and growth processes in the community. There are following theories of the climax.

4.6.8.1.1 Monoclimax Theory

According to the monoclimax theory of succession (Clements,1936), every region has one climax community toward which all communities are developing. He believed that *climate* was the determining factor for vegetation and the climax of any area was solely a function of its climate. Various terms such as *subclimax, disclimax, postclimax,* and *preclimax* are used to describe the deviations from the climatically stabilized climax. These communities, controlled by topographic, edaphic (soil), or biotic factors are regarded as exceptions by the supporters of the monoclimax view.

4.6.8.1.2 Polyclimax Theory

This theory was proposed by Tansley (1939) and later supported by Daubenmire (1966). The polyclimax theory of succession holds that many different types of vegetation as climax communities may be recognised in a given area. These will be climaxes, controlled by soil moisture, soil nutrients, activity of animals and other factors. According to this theory, climate is only one of the several factors, any of which may have a controlling influence on the structure and stability of the climax. This allows many climaxes in a climate region and is, therefore, called the polyclimax theory.

The difference between this theory and the monoclimax theory is largely a matter of emphasis on which factor is responsible for the stability of a climax. According to Krebs (1994), the real difference between two theories lies in the time factor of measuring relative stability. The climate varies on an ecological time scale as well as on a geological time scale. Succession in a sense, then, is continuous because we have a variable vegetation approaching a variable climate.

4.6.8.1.3 Climax-pattern Theory

Whittaker (1953) emphasized that a natural community is adapted to the whole pattern of environmental factors in which it exists; the major factors are: genetic structure of each species, climate, site, soil, biotic factors (activity of animals), fire, wind, availability of plant and animal species, and chances of dispersal. According to this theory, climax communities are patterns of populations varying according to the total environment. There is thus no discrete number of climax communities and no one factor determines the structure and stability of a climax community.

Whereas the monoclimax theory allows for only one *climatic climax* in a region and the polyclimax theory allows *several climaxes,* the climax-pattern hypothesis allows a *continuity of climax types* varying gradually along environmental gradients and not clearly separable into discrete climax types.

4.6.8.1.4 Climax as Vegetation

According to Egler (1954) one can say that "climaxes" in a broad sense are nothing more than *totality of vegetation,* itself. He, thus, favours the study, of vegetation; as it is, with careful observations to explain and interpret past, present, and future conditions of particular communities.

We may conclude from these theories that the end point of succession is climax which is in itself not completely stable. The climate of an area has overall control on the vegetation; but within each of the broad climatic zones there are many modifications caused by soil, topography, and animals which lead to many climax situations. Climax communities do not necessarily represent a halt to successional change.

4.6.9 Significance of Succession

The achievement of a measure of stability or homeostasis, rather than a mere increase in productivity, in a fluctuating physical environment may well be the primary purpose of ecological succession when viewed from the evolutionary viewpoint. A mature community with its greater diversity, larger organic structure, and balanced energy flows is often able to buffer the physical environment to a greater extent than the young community.

The principles of ecological succession are of the greatest importance to mankind. Man must have early successional stages as a source of food, since he must have a large net primary production to harvest. In the climax community, because production is mostly consumed by animal and plant respiration, net community production in an annual cycle may be zero. On the other hand, the stability of the climax species and its ability to buffer and control physical forces such as water and temperature are desirable characteristics from the viewpoint of the human population. The only way man can have both a productive and a stable environment is to ensure that a good mixture of early and mature successional stages are maintained, with interchanges of energy and materials. Excess food produced in young communities helps feed older stages that in return supply regenerated nutrients and help buffer the extremes of weather, such as storms and floods.

In the most stable and productive natural situation there is usually such a combination of successional stages. A favourable situation exists in many terrestrial landscapes where productive croplands on the plains are intermingled with diverse forests and orchards on the hills and mountains. Ecologically speaking, the crop fields are young communities because they are maintained as such by the constant labour of the farmer, while forests represent older, more diverse, and self-sustaining communities that have lower rates of net production but do not require the constant attention of man. It is important that both types of ecosystems be considered together in proper relation. If the forests are destroyed merely for the temporary gain of wood, water and soil may wash down from the slopes and reduce the productivity of plains.

However, succession is not always predictable and unidirectional. For example, cyclic succession occurs in most communities on a small scale. In nature, even an old, seemingly stable community is in a state of flux and is a shifting mosaic rather than a steady state. Although succession is undoubtedly an important and widespread process, other influences and processes may dominate the physical and biological structure of many communities. According to Elton, succession of communities plays an important part in the slow dispersal of animals. The intricate problems of competition between different species of animals can be studied in series of changing communities. Thus, it becomes clear that succession has a very significant role in shaping the destiny of animal and plant communities.

4.7 FRESHWATER COMMUNITY

4.7.1 Introduction

Freshwater community consists of an array of organisms depending on the physico-chemical and biological characteristics of the freshwater environment. Freshwater habitats are divided into two major categories, lotic (lotus = washed, or running water), and lentic (lenis = calm, or standing water) habitats. Lotic habitats are those existing in relatively fast running streams, springs, rivers and brooks. Lentic habitats are represented by the lakes, ponds, and swamps.

The above classification of the freshwater environments is based on two conditions: *currents* and *the ratio of the depth to surface area.* Since lakes and ponds often contain currents or at least wave action and since streams often harbour quiet pools or calm backwaters, the difference between lotic and lentic waters is not very precise. However, temperature, light, currents, amount of respiratory gases, and concentration of biogenic salts are important limiting factors influencing the organisms of all freshwater habitats.

4.7.2 Lentic Community

Lentic waters are generally divided into three zones or sub-habitats: littoral, limnetic, and profundal. A small pond may consist entirely of littoral zone. However, a deep lake with an abruptly sloping basin may possess an extremely reduced littoral zone.

4.7.2.1 Lake Zonation

The three major zones of a *lake* are described as follows (Fig. 4.12).

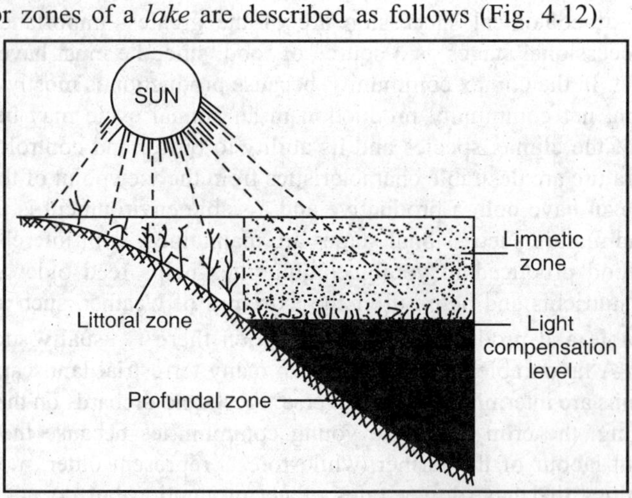

Fig. 4.12 Three major zones of a freshwater lake.

(*a*) Littoral zone : The littoral zone adjoins the shore (and is thus the home of rooted plants) and extends down to a point called the *light compensation level,* or the depth at which the rate of photosynthesis equals the rate of respiration. Within the littoral zone producers are of two main types: rooted or benthic plants, and photoplankton (plant plankton) or floating green plants, which are mostly algae.

The littoral zone is the home of greater variety of consumers than are the other zones. The zooplankton (animal plankton) of the littoral zone is rather characteristic and differs from that of the limnetic zone in preponderance of heavier, less buoyant crustacea which often cling to plants or rest on the bottom when not actively moving their appendages. Important groups of littoral zooplankton are large, weak-swimming species of *Daphia* and *Simocephalus,* some species of copepods, many families of ostracods and some rotifers.

The nekton of littoral zone is often rich in species and numbers. Adult and larval diving beetles and various adult Hemipetra are conspicuous. Various Diptera larvae and pupae remain suspended in the water, often near the surface. Pond fish, frogs, turtles, and water snakes are almost exclusively the members of the littoral zone community. Tadpoles of the frogs are important primary consumers, feeding on algae and other plant material. *Periphyton* of the littoral zone exhibit a zonation paralleing that of the rooted plants, but many species occur almost throughout the littoral zone. Among the periphyton forms, for example, pond snails, damselfly nymphs and climbing dragonfly nymphs, rotifers, flatworms,bryozoa, hydra, and midge larvae rest on, or are attached to stems and leaves of the plants.

Another group containing both primary and secondary consumers may be found resting or moving on the bottom or beneath silt or plant debris—for example, sprawling odonata nymphs (which have flattened rather than cylindrical bodies), crayfish, isopods, and certain mayfly nymphs. Descending more deeply into the bottom mud are burrowing odonata and ephemeroptera, clams, true worms, snails, chironomids (midges), and other diptera larvae.

(*b*) Limnetic zone : The limnetic zone includes all the waters beyond the littoral zone and down to the light compensation level. The limnetic zone derives its oxygen content from the photosynthetic activity of phytoplankton and from the atmosphere immediately over the lake's surface. The atmospheric source of oxygen becomes significant primarily when there is some surface disturbance of water caused by wind action or human activity. The community of the limnetic zone is composed only of *plankton, nekton,* and sometimes *neuston* (organisms resting or swimming on the surface).

Phytoplankton producers consist of diatoms, green algae, blue-green algae, and algae- like green flagellates, chiefly the dinoflagellates. The limnetic zooplankton consist of few species but the number of individuals may be large. Copepods, cladocerans, and rotifers are generally of first importance; but their species are largely different from those found in the littoral zone. The limnetic nekton consists almost entirely of fish. In ponds, the fish of the limnetic zone are the same as those of the littoral zone, but in large bodies of water a few species may be restricted to the limnetic zone.

(*c*) Profundal zone : The bottom and deep water area of a lake, which is beyond the depth of effective light penetration is called the profundal zone. In north-temperate latitudes, where winters are long and severe, this zone has the warmest water (4°C) in the lake in winter and coldest water in summer. The major community consists of bacteria and fungi and three groups of animal consumers: (*a*) blood worms, or haemoglobin containing chironomid larvae and annelids, (*b*) small clams, and (*c*) phantom larvae, or *Chaoborus* (corethra). The first two groups are benthic forms, the last are plankton that regularly move up into the limnetic zone at night and down to the bottom during the day.

All the animals of the profundal zone are adapted to withstand periods of low oxygen concentration, whereas many bacteria are anaerobic. Large numbers of bacteria in the bottom ooze constantly bring about decomposition of the organic matter (plant debris, animal remains, and excreta) that accumulates on the bottom. Eventually, the organic sediments are mineralized and nitrogen and phosphorus are put back into circulation in the form of soluble salts. In this way, the profundal zone provides rejuvenated nutrients, which are carried by currents and swimming animals to other zones.

4.7.2.2 Thermal Stratification of a Lake

Lakes in temperate latitudes exhibit marked seasonal temperature changes which may be described as follows:

Winter**:** During winter the coldest water forms ice at 0°C (32°F) and floats at the surface. The water at increasing depth below the ice is progressively warmer and denser. The heaviest water, at the bottom of the lake, has a winter temperature of 4°C and throughout winter the water remains relatively stable.

Spring**:** Following the ice melt, the surface water gradually warms to 4°C. At this point, the water column is nearly isothemarl, *i.e.*, all the water is of uniform temperature and density. Hence, the strong spring winds cause considerable stirring, which results in a complete mixing of water, dissolved oxygen, and nutrients from the lake surface to the lake bottom, a phenomenon known as the *spring overturn or spring turnover.* As the spring progresses, however, the surface waters naturally become warmer and lighter than the water at lower levels, as a result, the lake becomes thermally stratified into the following three zones (Fig. 4.13).

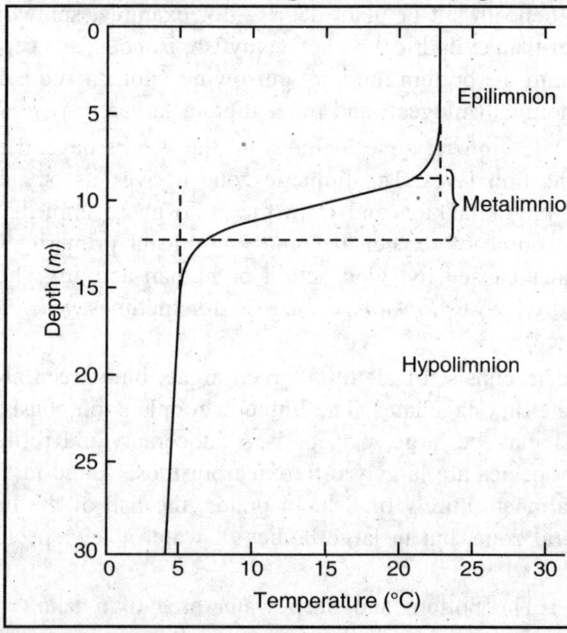

Fig. 4.13 Typical thermal stratification of a lake into three strata.

(*a*) *Epilimnion*: The upper stratum, which usually has the highest dissolved oxygen concentration and is characterized by a temperature gradient of less that 1°C per metre of depth, is the *epilimnion* (literally the "upon-lake" or the "upper lake"). This stratum contains more or less uniformly warm, circulating, and fairly turbulent water.

(*b*) *Hypolimnion*: The lower stratum of water characterized by a temperature gradient of less than 1°C per metre of depth is the *hypolimnion* (literally the "lake below" or "bottom lake"). This part contains more dense, cooler, and relatively quiet water.

(*c*) *Metalimnion*: It is the transitional stratum of marked thermal change between the epilimnion and hypolimnion. The middle layer of the lake, characterized by a temperature gradient of more than 1°C per metre of depth is the *thermocline.* The term thermocline refers to the plane or surface of maximum rate of decrease of temperature in the metalimnion (Wetzel, 1983).

The depth at which the thermocline forms is not fixed, being determined by the degree of solar heating, the transparency of the water, and the morphometry of the lake, but wherever it forms it effectively divides the lake into two layers, the upper epilimnion, and the lower hypolimnion. This division is not merely an interesting physical phenomenon but it has consequential effects on the ecology of the lake. The epilimnion is well lit and oxygenated with sufficiently high temperatures to promote algal productivity and hence to support zooplankton and fish. When nutrients are in ample supply, algal growth is accelerated and blooms may occur. By contrast the hypolimnion is cold, dark and becomes progressively deoxygenated as the decaying remains of organisms sink down from the epilimnion. Conditions in the hypolimnion may become so extreme that anoxia follows after which the biological productivity becomes least. Under less extreme conditions the epilimnetic material provides an energy source for benthic invertebrates. The sinking of dead algae and zooplankton from the epilimnion not only contributes to the potential deoxygenation of the hypolimnion but also prevents immediate recycling of nutrients. Nutrient depletion may become so high that algal growth is limited.

Summer: Unless the lake is very clear and permits photosynthesis, the hypolimnion frequently becomes depleted of oxygen in summer because of the biological oxygen demand of bacterial decomposers, the reduced photosynthetic activity, and the minimal mixing with upper waters as a result of density differences (Fig. 4.14).

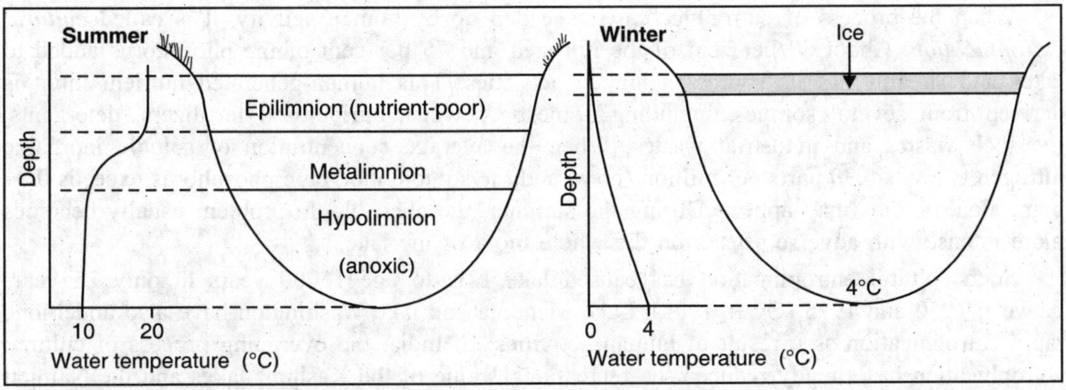

Fig. 4.14 The temperature gradients produced in a lake in summer, and in winter. In summer, lakes often stratify into a layer of warm water (the epilimnion) which floats on denser cooler water (the hypolimnion). In winter the lake is warmest at the bottom, usually 4°C, and gradually becomes cooler towards the ice layer at the surface.

Autumn: At the end of thermal stratification the surface waters of the epilimnion gradually cool as a result of conduction, evaporation and convection. Now the lake attains temperature uniformity from top to bottom. Because the water is now also of uniform density, it becomes well mixed by wind and wave action, a phenomenon known as the *fall overturn* or *fall turnover.* As a result, the nutrients, dissolved oxygen, and plankton become uniformly distributed.

Winter: As the winter approaches, the lake gets colder until the water attains a uniform temperature of 4°C at which it has maximal density. As the surface cools below it becomes lighter. Eventually the surface water may freeze at 0°C. During the winter season, the ice cover forms on the surface and in such ice-bound lakes there exists an *inverse stratification* of water temperature, with the coldest water (ice) at the surface and the warmest water (4°C) on the bottom.

Lakes undergoing complete circulation in spring and autumn separated by thermal summer stratification and winter inverse stratification are called *dimictic lakes* (Wetzel, 1983). Such lakes are quite common among temperate lakes of moderate size.

4.7.2.3 Lake Eutrophication

Eutrophication is the process by which lakes are enriched naturally or, more usually, by human activity. As a result, an *oligotrophic lake* ("underfed" or "nutrient poor") may become *eutrophic* ("well-fed" or "nutrient rich") in due course of time. The key nutrients responsible for eutrophication are *nitrogen* and *phosphorus.* Excess nutrients come from a variety of sources of which sewage effluents and agricultural runoff are more important.

4.7.2.3.1 Natural Eutrophication

The process of lake aging characterized by nutrient enrichment is known as *natural eutrophication.* This process gradually converts the original oligotrophic lake into a eutrophic lake. Natural fertility of lakes is increased by addition of nutrient rich loads of sediment and organic matter from the catchment area. This permits the production of larger quantities of phytoplankton and algal blooms, and other aquatic vegetation, including aquatic weeds, water hyacinth, water fern, and water lettuce, which in turn provide more ample food for herbivorous zooplankton and fish. When these plants and animals die, their bodies settle at the lake bottom and accumulate alongwith the sediment brought by tributary streams. The plant and animal remains then gradually decompose, thus releasing nutrients that could be channeled into the living bodies of future generations of organisms. As the natural process of lake ageing continues over the years, the original oligotrophic lake is gradually converted into an eutrophic lake.

4.7.2.3.2 Cultural Eutrophication

When the process of eutrophication is speeded up by human activity, it is called *cultural eutrophication.* About 80 per cent of the nitrogen and 75 per cent of the phosphorus added to lakes and streams has its source in human activities. This human-generated nutrient input is derived from several sources, including domestic sewage, agricultural fertilizers, detergents, livestock wastes and industrial wastes. When the average concentration of soluble inorganic nitrogen exceeds 0.30 parts per million (ppm) and the soluble inorganic phosphorus exceeds 0.01 ppm, algal blooms may appear. During the summer, the algal bloom problem usually becomes more intense with adverse effects on the whole biota of the lake.

Such cultural entrophication has caused lake Erie to age 15000 years in only 25 years between 1950 and 1975 (Owen, 1985). Lake Mendota and lake Washington have also undergone rapid eutrophication as a result of human activities. In India, the evergoing process of cultural eutrophication has already reduced the recreational value of the Kashmir lakes and the Nainital lake is also undergoing rapid eutrophication, especially as a result of sewage, domestic waste and detergent discharges.

4.7.2.3.3 Effects of Eutrophication

The adverse effects of nutrient pollution on a lake are many and may be summarised as follows :

(*i*) Eutrophication destroys the aesthetics of the lake, rendering it repulsive to swimmers and sport enthusiasts. In other words, the recreational values of the lake are reduced.

(*ii*) The algal blooms impair water quality by giving it a bad taste and odour. If the lake is a source of drinking water, considerable expense may be involved in improving its quality.

(*iii*) Toxic gas hydrogen sulphide that emanates from rotting algae has foul smell. Some of the blue-green algae release chemicals that are poisonous to other organisms, including fish and humans.

(*iv*) Dense algal blooms at the lake surface reduce penetration of sunlight to the lake bottom. As a result, the deeper waters contain less amount of dissolved oxygen, which is further reduced by decomposing algae and organic matter accumulated at the lake bottom. The oxygen depletion often causes *winterkill* of fish in northern lakes.

(*v*) The rooted weeds gradually reduce the lake size and also interfere with fishing, navigation, irrigation, and the production of hydro-electric power.

(*vi*) As the clean water becomes turbid and dirty, the game fish such as the Trout and the Mahseer are replaced with trash fish.

(*vii*) The lake basin is gradually filled in by sediments (sediment pollution) rendering the lake shallow, and in due course of time the lake becomes "extinct" as a result of eutrophication.

4.7.3 Community of Running Waters

The biotic community in the streams and rivers is quite different from that in the ponds and lakes. The differences in the community are largely due to differences in the physical and chemical conditions of their environments. Important basic characteristics of a stream are as follows:

(a) Current: The presence of definite and continuous current is the main characteristic of a stream. The current velocity, which may vary from 0.5 to 2 metre per second or more, depends on the stream gradient. The greater the *discharge* (volume/time), the greater the current velocity (distance/time) and also the amount of suspended material that is transported.

The rate of flow, in turn, influences a number of physical and chemical factors such as the temperature and dissolved oxygen concentration, which act directly on the biota. The current is a major limiting factor in streams. It might be assumed that plankton would be absent from fast-flowing streams, since such organisms are largely at the mercy of the current. Plankton in small streams, if present, originates in lakes, ponds, or backwaters connected with streams and is soon destroyed as it passes through rapids of streams. Only in slow moving parts of streams and in large rivers, plankton is able to multiply and become an integral part of the community.

(b) Land-water interchange: Streams are usually relatively shallow and, therefore, have a large surface compared to their depth. Land-water interchange is relatively more extensive in streams, resulting in a more open ecosystem. This means that the streams are more intimately connected with the surrounding land than are most lakes and ponds. For this reason, a significant portion of the stream's nutrients falls into them from their banks in the form of terrestrial leaves, grasses and other debris. In fact, many of the primary consumers living in a stream feed on detritus and *allochthonous material* (*i.e.*, outside the system) rather than on green aquatic vegetation (*autochthonous production*). The producers of a stream can supply only a fraction of the energy required by its animal consumers.

(c) *Oxygen*: The third important characteristic of a stream, which is also a difference between a stream and a lake, is the high amount of dissolved oxygen. The reasons for this are many: flowing water, the relative shallowness of the ecosystem, and the large surface area exposed to the atmosphere. For these reasons the waters of a shallow, fast-moving stream will have higher oxygen levels than deep, sluggish rivers. In a stream, the photosynthetic production of oxygen is not as important as it may be in a lake or pond. Because of trubulence, the stream water is usually well aerated and problem of oxygen depletion, as seen in the hypolimnion of a deep lake, is almost absent. But stream animals are very sensitive to decreases in the oxygen content of the water. If a stream becomes polluted with excessive load of oxygen-demanding organic matter, such as the domestic sewage or the industrial waste, the resulting oxygen-depletion may cause serious problems such as a massive fish kill.

4.7.3.1 Stream Zonation and Communities

Streams and rivers have a *longitudinal zonation* or profile (Fig. 4.15) instead of *horizontal* in lakes and ponds. Illies and Botosaneanu (1963) have recognised two major subdivisions of a river course: the steep and torrential upper course (*"rhithron"*) and the flat, slow-flowing lower course (*"potamon"*). However, a true rhithron having water temperature below 20°C, may be often absent in the tropical rivers. By contrast, temperate rivers are characterised by relatively lengthy rhithron zones.

4.7.3.2 Rhithron

A rhithron is characterised by the presence of (*i*) steep, narrow and shallow *riffles* or rapids, and (*ii*) flatter, wider and deeper reaches, termed *pools*. Intermediate areas of moderate current often found in larger streams and rivers are termed *runs*. Riffles have high, turbulent flow, coarse bottom of boulders, rocks or pebbles and limited attached vegetation. Pools have lower flow, bottom of somewhat finer material and some rooted vegetation. Low temperatures and high values of dissolved oxygen are important characteristics of the rhithron zone.

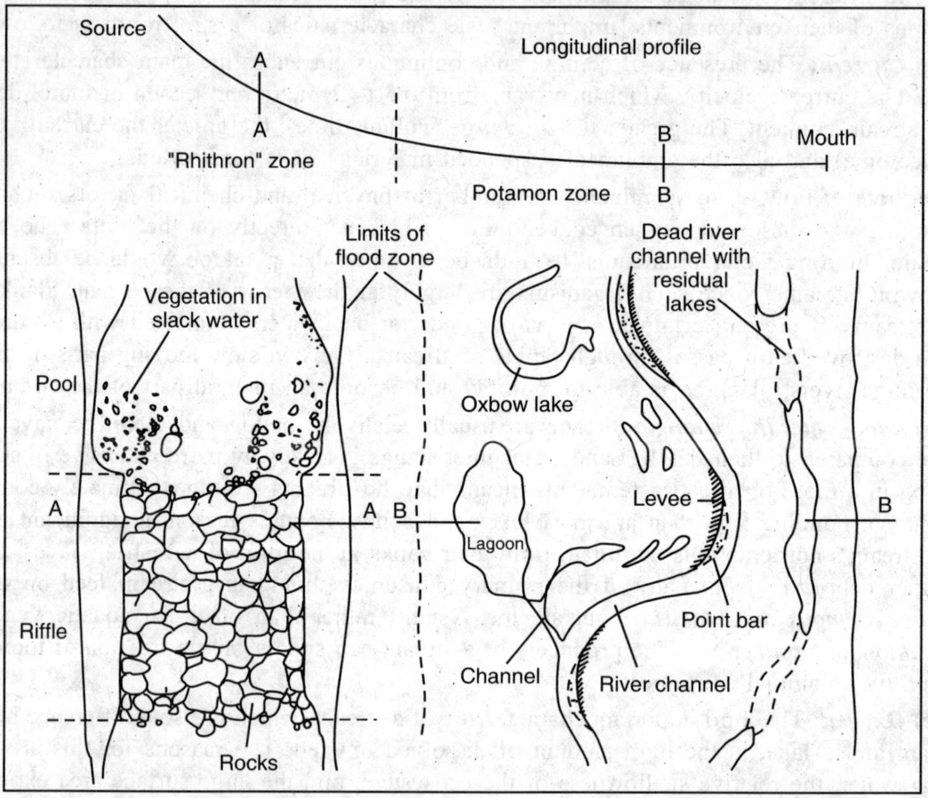

Fig. 4.15 Diagrammatic representation of a river course showing Rhithron (A–A) and Potamon (B–B) zone.

4.7.3.2.1 Communities of Rhithron

Biotic communities of the rhithron zone consist of plankton, periphyton, "aufwuchs", nekton and a variety of benthos or bottom-dwelling forms. During periods of spate there is no phyto - or zooplankton, although during low water transient blooms may occur as flow drops and pools become isolated.

In rhithron zone of relatively undisturbed stream, the main primary producers are the attached algae and diatoms (*e.g.* filamentous green algae like *Ulothrix* and *Cladophora,* and diatoms like *Nizschia* and *Gomphonema*). Those organisms that attach firmly to a substrate (submerged rocks, plants, debris) but do not penetrate it (in contrast to plants rooted in the bottom or certain parasites) are collectively called *"aufwuchs"* (German word proposed by Ruttner, 1953). They comprise all attached organisms except the macrophytes in contrast to the English equivalent *"periphyton"* (meaning "on plants") which includes the plants and animals adhering to parts of rooted aquatic plants. The diatoms and small algae are sometimes referred to as *microphytes.* Besides, in sunny areas productive streams may have luxuriant growth of algae like *Cladophora* which is a *mesophyte* (intermediate between microphytes and higher plants or *macrophytes*).

The zooplankton of the rhithron zone mainly consists of ciliate protozoans (*e.g. Vorticella*), rotifers (*e.g. Asplancha, Brachionus, Keratella, Philodina*), and crustaceans (*e.g., Daphnia, Bosmina, Cyclops, Diaptomus, Gammarus pulex*).

There are also many insect species adapted to life on the stream bottom where they shelter among or cling to the rocks. Commonly encountered insects in streams are:

(*a*) The mayflies *Ephemeroptera* (*e.g., Baetis rhodani, Rhithrogena, Ecdyonurus, Ephemerella, Caenis, Ameletus, Epeorus*).

(*b*) The stonefiles *Plecoptera* (*e.g., Nemoura, Protonemura, Perlodes, Diura, Arcynopteryx, Perla, Isoperla, Leuctra, Brachyptera*).

(*c*) The caddisflies *Trichoptera* (*e.g., Hydropsyche, Rhyacophila, Philopotamus, Drusus, Limnephilus, Agraylea*).

(*d*) The two-winged flies *Diptera (e.g., Simulium* (blackfly) *Chironomus riparius,* (bloodworms), *Antocha, Liponeura*).

(*e*) The dragonflies *Odonata* (*e.g. Libellula*).

Some worms (*e.g.*, flatworm *Dugesia,* roundworm *Dolichodorus,* annelid worm *Tubifex*), leeches (*Haemopis, Glossiphonia, Erpobdella*), molluscus (*e.g. Bithynia, Dreissena*) and crustacea (*e.g. Gammarus*) may also occur among the benthic forms.

The nekton of the rhithron zone consists of water bugs *Hemiptera* (*e.g., Notonecta, Micronecta, Gerris, Corixa*), beetles *Coleoptera* (*e.g. Dytiscus*) and fish. The resident fish species in rhithron zones are entirely *rheophilic* and belong to two main groups. Firstly, there are those species, which live among rocks of the bottom and are found mainly in the riffles. These are of small size and are adapted to grip or cling to the substrate (*e.g., Chiloglanis, Astroblephus, Glyptothorax, Pseudecheneis*). Other species such as *Mastacembelus* have pointed snout and eel-like shape that enable them to get into the holes in the rocky bottom. The second group of fish is those species, which are adapted to swim fast as to resist the current, and some of them even move against it (*e.g., Barbus, Salmo, Barilius, Schizothorax, Tor*). Because of the severity of the habitat, the diversity of resident species tends to be low.

4.7.3.2.2 Adaptations of Rhithron Community

Most animals living in the rhithron zones of streams are easily recognizable from their adaptations for living in swift current (Fig. 4.16). Some important adaptations are as follows :

(*i*) *Streamlined bodies*: Almost all stream animals possess more or less egg-shaped body which offers minimum resistance to water flowing over it. Best examples of hillstream fish with streamlined body are trout (*Salmo*), *Barilius* and *Schizothorax* species.

(*ii*) *Flattened bodies*: Many stream animals possess flattened bodies, which enable them to find shelter under stones. Some Ephemeroptera like *Epeorus, Rhithrogena,* and fishes such as *Noemacheilus* (Cobitidae) and *Pseudecheneis* (Sisoridae) are the best examples.

(*iii*) *Hooks and suckers*: These enable riffle animals to grip smooth surfaces. For example, *Bibicocephala* and *Liponeura* have a row of ventral suckers.

(*iv*) *Adhesive apparatus*: Some hillstream fishes, *e.g., Glyptothorax, Garra,* and *Pseudecheneis* possess a well developed adhesive apparatus on the ventral surface.

(*v*) *Sticky undersurfaces*: Some snails and flatworms are able to adhere to the substrate by their sticky undersurfaces.

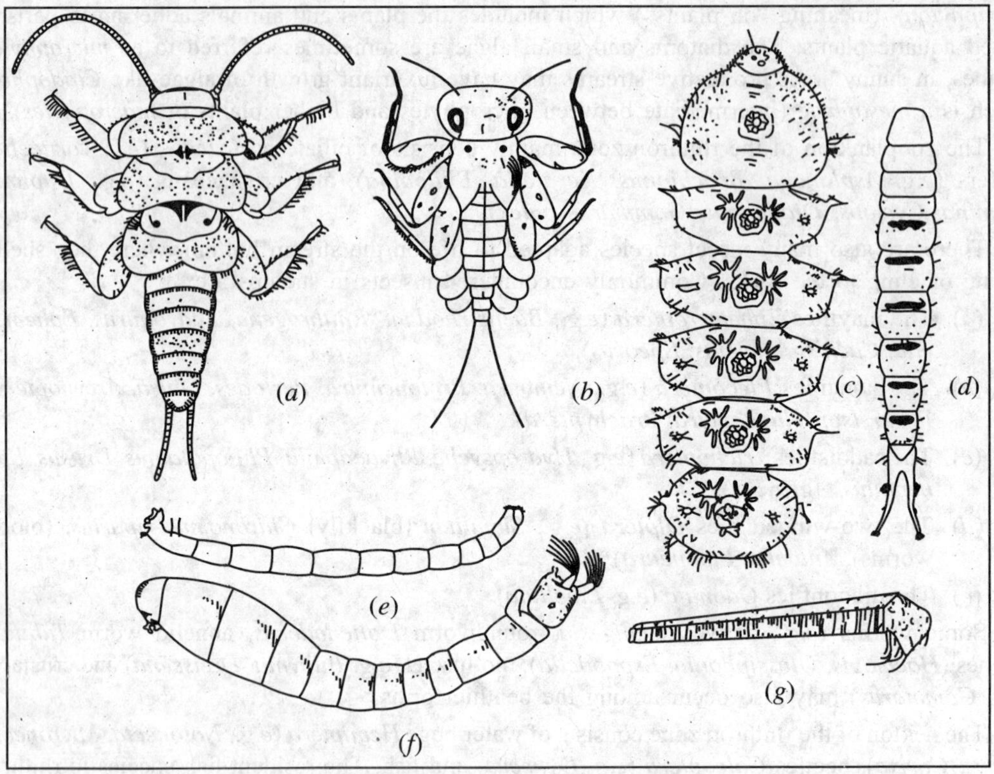

Fig. 4.16 Some lotic animals showing their adaptations. (*a*) stonefly nymph, (*b*) mayfly nymph, (*c*) *Blepharocera,* (*d*) *Antocha,* (*e*) *Chironomus,* (*f*) *Simulium,* (*g*) *Caddisfly* larva in a case of plant material (after Needham and Needham, 1972, A Guide to the study of fresh water biology)

(*vi*) ***Positive rheotaxis***: Stream animals orient themselves upstream and, if capable of swimming movements, move against the current (*rheo,* current; *taxis,* arrangement). This is an inherent behaviour pattern and is just as important an adaptation as the morphological features given above.

(*vii*) ***Positive thigmotaxis***: Many stream animals have an inherent behaviour pattern to cling close to a surface or to keep the body in close contact with the substratum (*thigmo*, touch or contact). Almost all the larvae of mayflies and stoneflies and many fishes exhibit this type of behaviour.

(*viii*) ***Permanent attachment***: Some organisms of the lotic environment remain permanently attached to a firm substrate such as a stone, log, or leaf mass. Green algae *Cladophora,* encrusting diatoms such as *Nitzschia, Gomphonema, Melosira* and some aquatic mosses (*Fontinalis*) are good examples. Such organisms are known as "aufwuchs".

4.7.3.3 Potamon

The potamon zone of a river is the flat, slow-flowing lower course. Zonation within the potamon is both longitudinal and lateral (Fig. 4.15). Longitudinally, there is a repetition of differing habitats associated with the meanders of the channel. Laterally, there is the distinction between the main channel and its floodplain. The floodplain is normally an area of relatively flat land flanking the main channel. The channel at each band is deeper by the outer, concave bank where the current is fastest, whereas the inner, convex bank consists of a sandy or muddy point bar. The plain itself contains many types of water body, some of which retain water throughout the inter-flood period. Because of deposition of silt, such features show a succession from open lagoon, through vegetation-lined pools and heavily vegetated swamps to dry land.

In short, the potamon zone is environmentally more complex and differs from the rhithron zone in the following features:

(*i*) Geomorphology of the channel
(*ii*) Physico- chemical qualities of the water
(*iii*) Nature of substratum
(*iv*) Biota; including floating, emergent and submerged vegetation
(*v*) High bacterial density

4.7.3.3.1 Communities of Potamon

The above mentioned physico-chemical, geologic and biotic features have a great bearing on the communities of the potamon zone. The community consists of the populations of the following groups :

(*i*) Plankton: The occurrence of plankton is closely related to the flow conditions. During the floods planktonic organisms may be present but are rare, whereas during the dry seasons algal blooms may develop within the lentic waters of the plain and also in the main channel. In short rivers these are generally confined to backwaters. In longer rivers the time taken for individual masses of water to travel downstream is sufficient to allow the development of plankton. In rivers whose flow has been slowed by other hydraulic work, plankton also develops to a greater degree. However, the contribution of the free plankton to the primary production is slight.

(*a*) *Algae and diatoms*: Algal species may be categorised into three groups on the basis of their sensitivity (Bilgrami and Datta Munshi, 1985). The first group comprises such algae which exhibit maximum pollution tolerance and can be used as *pollution indicators.* They include *Oscillatoria limosa, Microcystis aeruginosa, chlorella vulgaris, Stigeoclonium tenue, Ankistrodesmus falcatus, Scenedesmus quadricauda, Synedra ulna, Navicula viridula, Euglena viridis, Phacus caudatus* and *P. viridis* and are usually confined to the sites of urban and industrial discharge. The second group includes forms recorded from clean as well as moderately polluted zones. Important of them were *Oscillatoria princeps, Phormidium uncinatum, Anabaena circularis, Closterium acerosum, Pediastrum duplex, Scenedesmus dimorphus, Gomphonema parvulum, Cymbella turgida, Navicula cuspidata* and *Synedra acus.* The third group includes highly sensitive clean water forms such as *Merismopedia glauca, Oscillatoria subbrevis, Phormidium calcicola, Pediastrum simplex, Melosira ambiguans,* and *Gomphonmea* species.

(*b*) *Zooplankton*: Zooplankton of the potamon zone consists of members of protozoa, rotifera, cladocera and copepoda. Forms like *Filinia longiseta, Brachionus angularis, B.calyciflorus, B. forficula, B. quadridentata, Keratella tropica* and *K. cochlearis* may be present in clean as well as moderately polluted waters. But forms like *Brachionus rubens, Platiyas polyacanthus and Rotatoria rotatoria* are characteristic of polluted parts of the potamon zone.

(*ii*) Nekton: Some insects, beetles, and fish dominate the nekton of potamon zone, although forms which are not true nekton such as *Kachuga kachuga, Trionyx gangeticus, Lissemys punctata, Gavialis gangeticus, Platanista gangetica* have been also reported from the Ganga and other rivers.

Two main groups represent fish communities of the potamon zone. Firstly, there are fish species adapted to resisting low dissolved oxygen concentration. Their adaptations may be in the form of auxiliary respiratory organs for using atmospheric oxygen as in *Channa, Clarias, Erythrinus* and *Notopterous,* or may be physiological as with *Carassius,* or even behavioural as with many cyprinodonts. The same species often have a capacity to tolerate high temperatures.

The other group of fish includes the species which usually live in the main channel and are fast swimmers such as *Mystus* (*Aorichthys*) species, *Labeo* species, *Catla catla* and *Cirrhinus mrigala.*

(*iii*) Benthos: The benthos of the potamon is relatively poor as unstable mud bottoms, heavy siltation and seasonal desiccation do not favour well settled bottom living organisms. The benthic forms may include some molluscs (*e.g., Lymnaea, Bithynia, Corbiula*), chironomids, ephemerptera, plecoptera, trichoptera and worms and leeches.

(*iv*) Macrophytes: The presence of macrophytes is a characterestic feature of the potamon zone. Most aquatic macrophytes are angiosperms (flowering plants), but aquatic ferns, mosses, liverworts, and even the large algae of the Charophyceae may be present.

The freefloating macrophytes such as *Salvinia* (water fern), *Pistia* (water lettuce), *Eichhornia crassipes* (water cabbage) may be quite common. In temperate and subtropical parts much smaller free-floating plants such as the duckweed *Lemna* may be present in backwaters. Rooted macrophytes may have all or part of their vegetative and sexually reproductive parts above the water or may be completely submerged. Some examples of rooted macrophytes of streams are *Nymphaea* (waterlily), *Potamogeton* (pond weed), *Ceratophyllum* (coontail), *Myriophyllum* (milfoil) and the water crowfoot (*Ranunculus*). The eel grass (*Zostera*) and widgeon grass (*Ruppia*) are the common rooted macrophytes in estuaries.

Special Features of the Community Structure of Rivers

The community structure of a river is largely determined by the physical environment. However, biotic interactions also play their role. Traditionally, most of the rivers have three community types. These are: disturbance-controlled, competition-controlled and predation-controlled. Which of the three will dominate in a given area is determined by a number of interacting forces, but the intensity of disturbance is considered as the overriding factor. If the intensity of disturbance or environmental stress such as discharge and frequency of floods (or flash floods) is high, then this disturbance is the chief determinant of community structure. As environmental stress is reduced, competition will start to operate. However, predation is a major determinant of the community structure at low levels of environmental stress. It must be emphasized that the communities are not mutually exclusive (extreme disturbance to epilithic algae may be tolerable perturbation to an otter), neither are they fixed.

Thus, disturbance affects ecological diversity of the riverine community. Fluctuations in water level, discharge, construction of dams and barrages and other anthrophogenic pollutants, industrial effluent, pesticide discharges from adjoining fields and changes in riparian vegetation, also affect the community of rivers.

4.8 MARINE COMMUNITY

4.8.1 Introduction

Covering about 70 per cent of the earth's surface ocean contains many communities. The life extends to all depths of the ocean, although life is much denser around the margins of continents and islands. However, all the oceans are interconnected. But temperature, salinity, and depths are the chief barrier to free movements of marine organisms. The sea is in continuous circulation because air temperature differences between equator and poles set up strong winds which, together with rotation of the earth, create definite currents. The sea is dominated by waves of many kinds and by tides. The sea is salty, and salinity and temperature represent the two of the important limiting factors in the sea. The character of the radiation, as well as its intensity varies with depths. Besides the above important features of the sea which are of major ecological interest and greatly influence the marine community, the sea is the most fiercely competitive environment on the earth. For this reason marine organisms lead "eat-or-be-eaten" mode of existence.

4.8.2 Zonation of Ocean

Marine habitats are separated into several divisions, the two largest of which are the *neritic* and the *oceanic zones.* (Fig. 4.17)

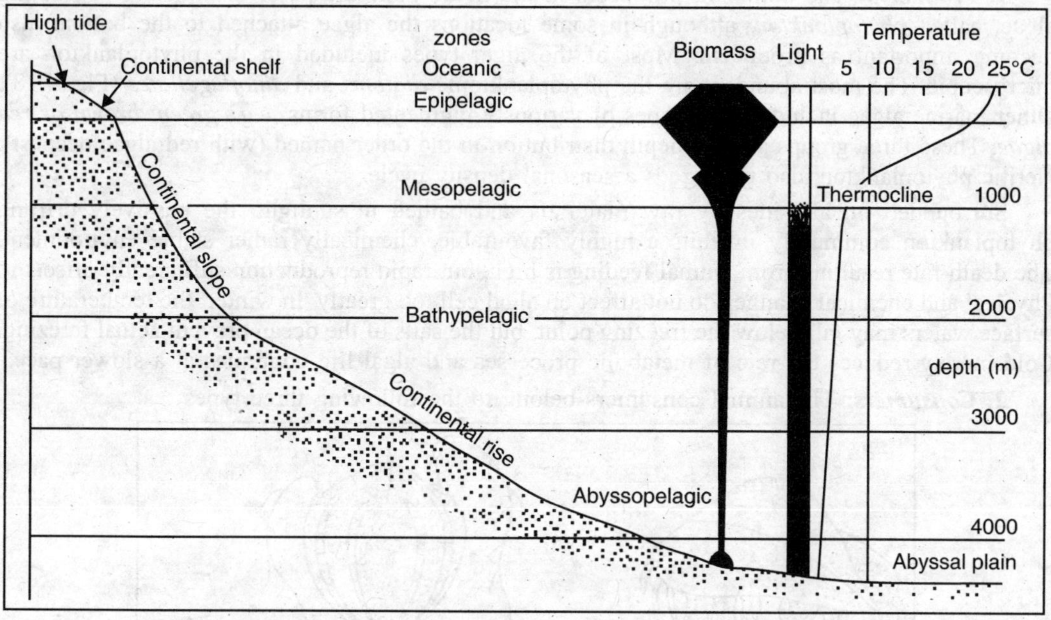

Fig. 4.17 Zones of the ocean. Modified from Marshall (1971).

4.8.2.1 Neritic Zone

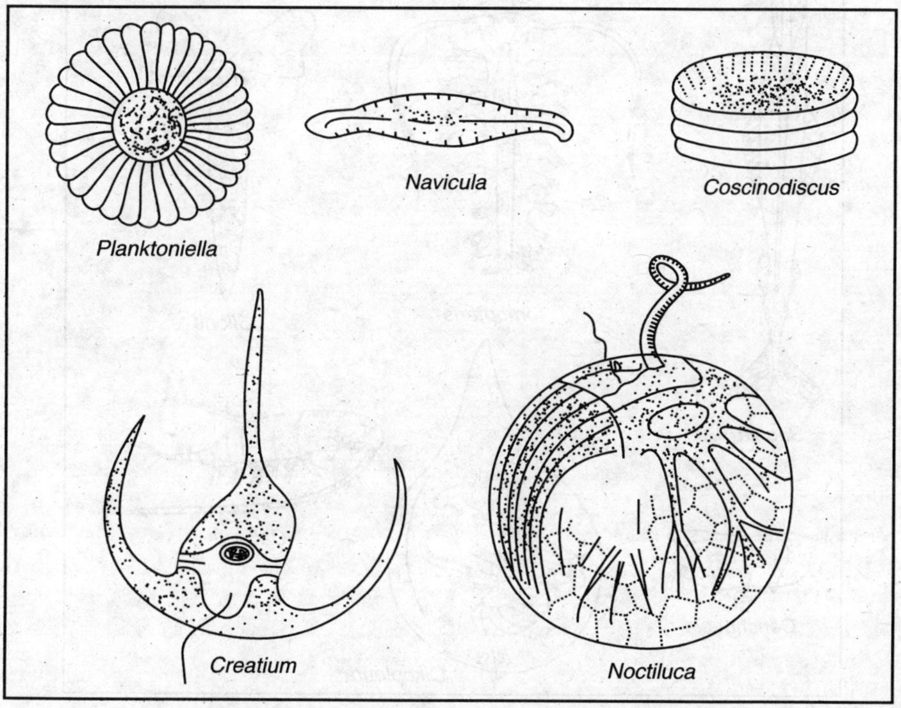

Fig. 4.18 Some representative phytoplankton of the sea.

The neritic zone ("near shore") extends from the high tide mark to the edge of the continental shelf. Thus, it is the shallow water zone on the continental shelf. The biotic communities of the nertic zone consist of the following groups.

1. *Producers*: The dominant producers in the nertic zone are passively drifting or floating algae, called *phytoplankton,* although in some locations the algae attached to the bottom also become important as producers. Most of the algal types included in the phytoplankton are microscopic. The most abundant are the phytoplanktonic *diatoms* and *dinoflagellates* (Fig. 4.18). Other marine algae include many types of variously pigmented forms, *e.g., green, brown or red algae.* These three groups show a depth distribution in the order named (with red algae deepest). Neritic phytoplankton also undergoes a seasonal density cycle.

Surrounded on all sides by raw materials and bathed in sunlight, the passively drifting phytoplankton community inhabits a highly favourable, chemically rather stable environment. The death-rate resulting from animal feeding is high, but rapid reproduction sufficiently offsets it. Physical and chemical changes do not affect on algal cell too greatly. In winter, the temperature of surface waters may fall below the freezing point, but the salts of the ocean prevent actual freezing. Cold merely reduces the rate of metabolic processes and algal life continues at a slower pace.

2. *Consumers*: The animal consumers belong to the following three types :

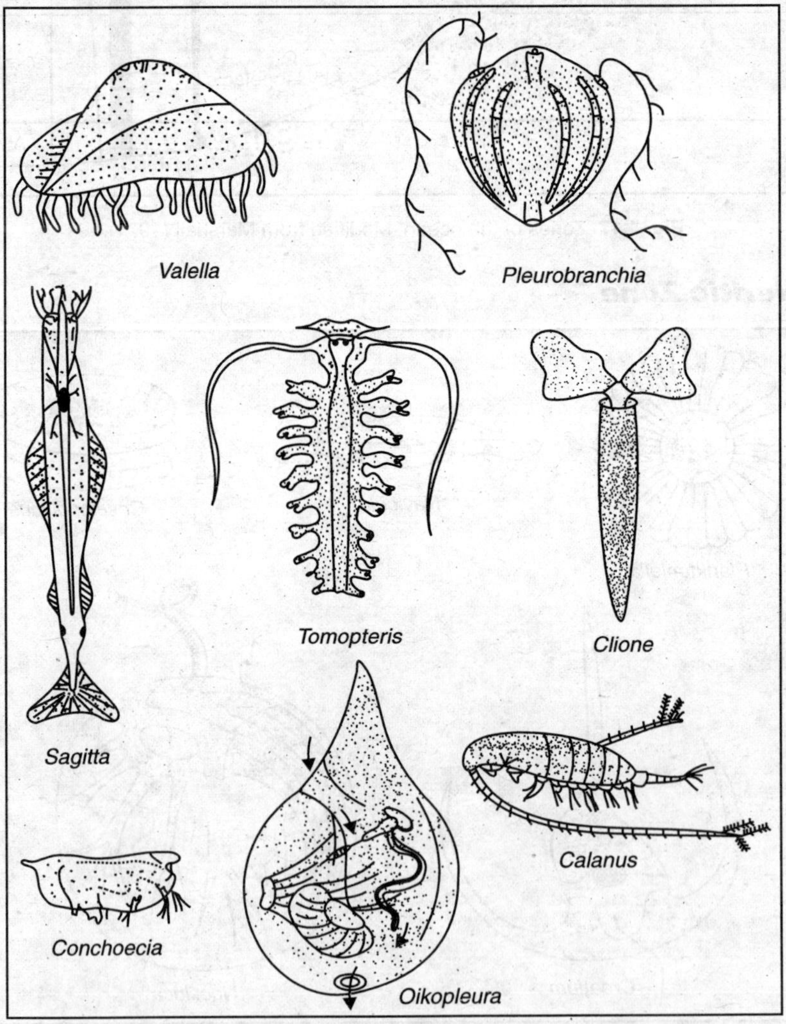

Fig. 4.19 Different animals which make up the holoplankton of the sea.

***(i) Zooplankton*:** Living side by side with the photosynthetic phytoplankton in the open waters are the small nonphotosynthetic, free swimming, or floating forms. Several different phyla are represented in the zooplankton. Organisms which remain for their entire life cycle in the plankton are called *holoplankton,* or permanent plankton (Fig. 4.19). The mature forms of protozoans (green flagellaters, rhizopod forminifera, actinopod radiolaria, and ciliate tintinnids), coelenterates (Pleurobranchia and many siphonophores), certain flat worms (*Sagitta*), segmented worms (*Tomopteris*), molluscs (*pteropods*), arthropods (copepods, *calanus*), and tunicates (*Oikopleura*) are most important permanent components of the zooplankton. Other forms are in larvel stages and form a considerable portion of the zooplankton. They are called *meroplankton,* or temporary plankton (Fig. 4.20) and grow into large organisms that will later become part of either the nekton or the benthos.

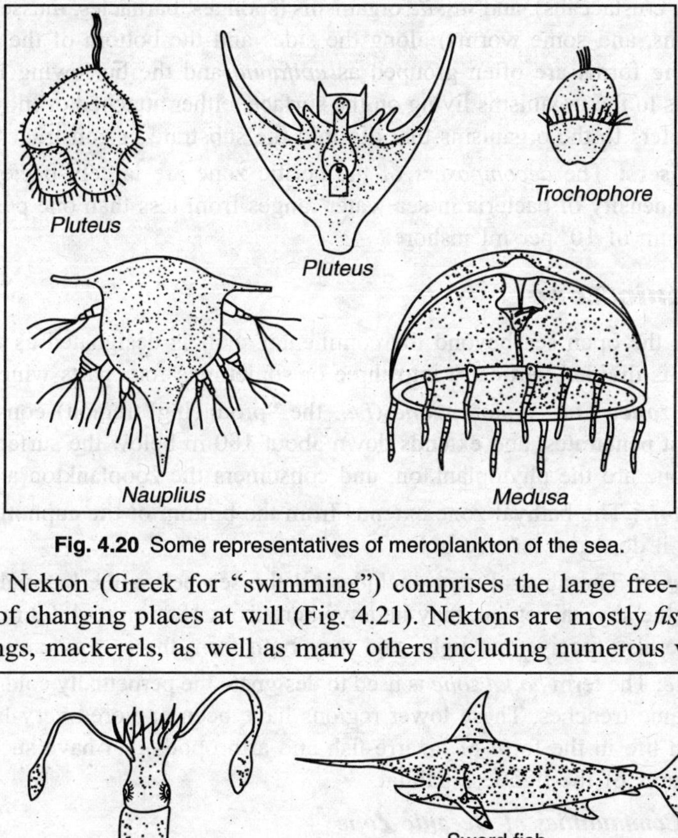

Fig. 4.20 Some representatives of meroplankton of the sea.

***(ii)* Nekton:** Nekton (Greek for "swimming") comprises the large free-swimming forms that are capable of changing places at will (Fig. 4.21). Nektons are mostly *fish* such as sharks, flying fish, herrings, mackerels, as well as many others including numerous varieties of small

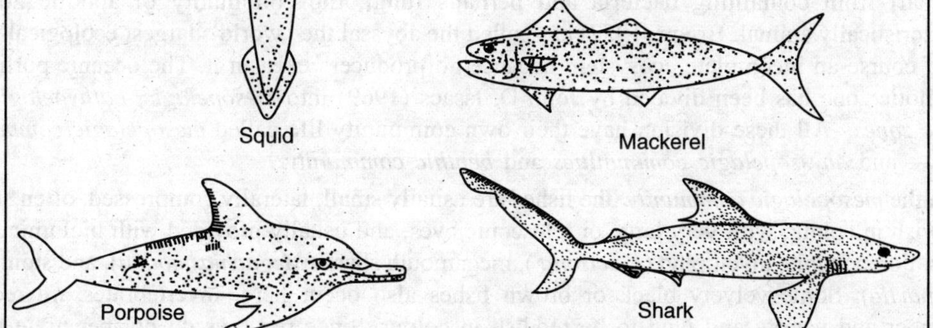

Fig. 4.21 Some representatives of nekton of the sea.

species. The warm temperate or tropical waters of shallow seas and coastal waters have numerous varieties of coral reef fishes such as butterfly fish (*Chaetodon*) damsel fish (*Abudefduf*), parrot fish (*Leptoscarus*). However, the coastal and shallow sea fauna of temperate regions contains less number of species than that of warm waters. The sea also contains estuarine euryhaline fishes such as mullets, sciaenids and Bombay duck. Intertidal fishes such as pipe fish (*Siphostoma*), cling fish (*Heteroclinus*), butter fish (*Pholis*) and mudskipper (*Periophthalmus*) also from an interesting group. Nekton also include *mammals* (seals, porpoises, dolphins, and whales), certain *arthropods* (larger crustacea), *molluscs* (squids), and marine *birds* (penguins, pelicans).

***(iii)* Benthos:** Benthos (Greek for "depth") consists of *crawling, creeping* (crabs, lobsters, certain copepods, amphipods, other crustaceans, many protozoans, snails, echinoderms, some bivalves, and some crustaceans), and *sessile* organisms (sponges, barnacles, mussels, oysters, corals, hydroids, bryozoans, and some worms) along the sides and the bottom of the ocean basin. The sessile and creeping forms are often grouped as *epifauna* and the burrowing forms as *infauna.* The epifauna refers to the organisms living on the surface, either attached or moving freely on the surface, infauna refers to the organisms that dig into the substrate or construct tubes or burrows.

***(iv)* Decomposers:** The *decomposers* of the neritic zone are largely bacteria. According to Zobell (1963), the density of bacteria in sea water ranges from less than one per litre in the open ocean to a maximum of 10^8 per ml inshore.

4.8.2.2 Oceanic Zone

The region of the open sea beyond the continental shelf is designated as the *oceanic zone.* The oceanic zone is divided vertically into three or sometimes four parts which are as follows:

***(a)* Euphotic zone:** The *euphotic zone* (*i.e.,* the "producing" region) consist of the waters into which sunlight penetrates, and extends down about 180 m below the surface. The producers of the euphotic zone are the phytoplankton; and consumers the zooplankton and nekton.

***(b)* Bathyal zone:** The bathyal zone extends from the bottom of the euphotic zone to a depth of about 1800 m, it does not receive light.

***(c)* Abyssal zone:** The abyssal zone includes all the sea below the bathyal. The area of the ocean "deeps", or the abyssal region, may lie anywhere from 1800 m to 4500 m. Like the bathyal zone, abyssal zone also receives no light, and they both form the *aphotic zone.*

***(d)* Hadal zone:** The term *hadal zone* is used to designate the perpetually cold and dark supreme depths of the oceanic trenches. These lower regions have been explored very little, although we know they contain life in the form of bizarre fish and arthropods that have successfully adapted to the great pressures of the lightless habitat.

4.8.2.2.1 Biotic Communities of Oceanic Zone

Apart from containing bacteria and perhaps fungi, the community of aphotic zone is characteristically animal. Bruun (1957) has called the abyssal the "world's largest ecological unit". It is of course an incomplete ecosystem because no producers exist in it. The oceanic portion of the aphotic zone has been divided by John D. Issacs (1969) into *mesopelagic, bathypelagic* and *benthic zones.* All these division have their own community life called *mesopelagic community, bathy* — and *abyssopelagic communities* and *benthic community.*

In the *mesopelagic community* the fishes are usually small, laterally compressed, often silvery or greyish in colour, with very large or telescopic eyes, and usually provided with bioluminescent organs. For example, myctophid (*Electrona*), megamouth shark (*Megachiasmodon*), and stomiatoid (*Bonapartia*). Some velvety black or brown fishes also occur here. Invertebrates are reduced in number and variety and tend to be reddish in colour. Since red rays do not penetrate to the depths where these animals live, they are essentially invisible.

The bathy–and *abyssopelagic communities* are considered as one. The fishes are slender and dark coloured. For example, gulper (*Eurypharynx*), gulper eel (*Saccopharynx*) and angler (*Linophryne*). Pelagic invertebrates include a few endemic species of radiolarians, jelly fish, ctenophores, nemertinians, ostracods, copepods, amphipods, shrimps, and squids. Red colour is more common than at intermediate depths.

Benthic fishes lack swim bladders. Some of them show interesting specialisation of the eyes and some have luminons organs. Some examples are tripod fishes (*Bathypterois* spp) and lizard fish (*Bathysaurus*). Other benthic animals of the ocean commonly have flat bodies, very long legs, or other means of distributing their weight over the loose, flocculent ooze. Many species rise above the ooze on stalks. The fragile glass sponges, long stemmed crinoids, and long legged crabs are possible only in very quiet waters that occur at great depths. Skeletons of all animals are fragile because of the difficulty of forming lime at low temperatures. Animal abundance decreases with depth, but even at 8300 m in the hadal zone, some twenty species have been found, chiefly holothurians, polychaetes and sea anemones.

Some of the deep sea animals are capable of traversing the whole ocean from bottom to surface. They may, therefore, feed directly on the rich food supplies in surface waters. But the bulk of the deep sea nekton is adapted to particular water pressures only, and some animals are rigidly confined to limited pressure zones at specific depths. Such animals must, therefore, obtain food either from the dead matter drifting down from the surface, or from within the nekton itself. The latter condition makes the deep sea the most fiercely competitive environment on earth. The very structure of the animals underlines their violently carnivorous, "eat-or-be-eaten" mode of existence. For example, most of the fishes have enormous mouths equipped with long, sharp teeth, and many can swallow fish larger than themselves (Fig. 4.22).

Since the environment of the deep sea is pitch-black, one of the critical problems for these animals is to find food. A highly developed pressure sense provides one solution. Turbulence in the water created by nearby animals can be recognized and depending on the nature of the turbulence, may be acted upon either, by flight or by approach. Another important adaptation of deep sea animals is bioluminescence. Many of the deep sea animal posses light producing organs. The light patterns emitted may include a variety of colours and probably serve partly in species recognition (identification of mates). Another function of the light undoubtedly is to warn or to lure. Some of the bioluminescent lures, as in lantern fishes, help to catch the prey.

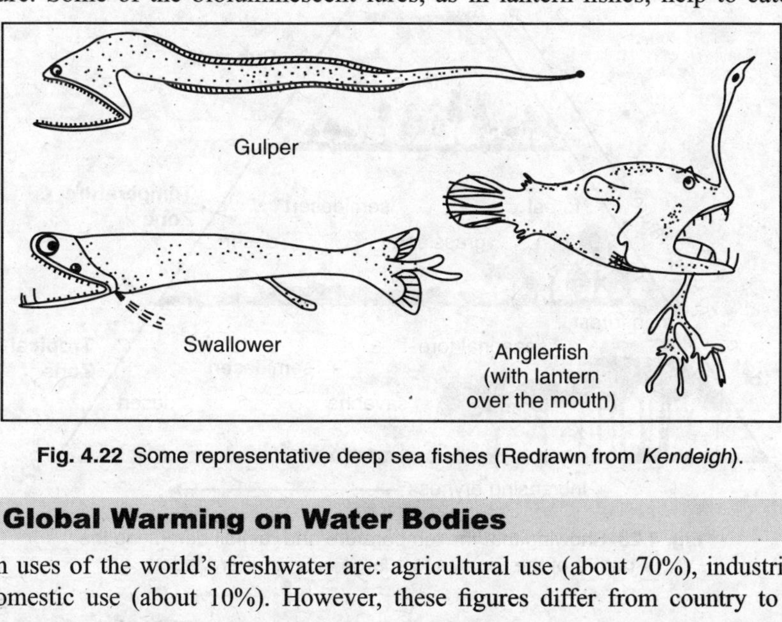

Fig. 4.22 Some representative deep sea fishes (Redrawn from *Kendeigh*).

Effect of Global Warming on Water Bodies

The main uses of the world's freshwater are: agricultural use (about 70%), industrial use (about 20%), and domestic use (about 10%). However, these figures differ from country to country. For example, in US, the agricultural use is 41% only. But it is certain that global warming is going to affect the uses of water as well as the total quantity of water in lotic and lentic water bodies and oceans.

Global warming can increase global rates of evaporation; it can disrupt water supplies and thus food supplies, some areas will get more precipitation and some less; river flows will change; and coastal areas may be submerged creating problems for humans and other species.

The UN Report, *Global Environment Outlook*-4, released in October, 2007 has warned that ecological overshoot is higher in developed countries. As a result, the biodiversity register of the planet is becoming thinner and one in ten of the world's large rivers is running dry every year before reaching its natural end, the sea.

4.9 TERRESTRIAL COMMUNITY

4.9.1 Introduction

Terrestrial community is naturally quite different from the aquatic community. The amount of water determines the habitability of a particular landmass. A comparison of a rain forest with a desert makes this point clear to us. Besides water, the sustaining foundations of all terrestrial life are air, and, directly or indirectly soil. Like air, soil is itself a home for a vast majority of terrestrial organisms. By creating the conditions necessary for the survival of all other terrestrial organisms, soil becomes a major agency, which transforms terrestrial environments into life sustaining habitats. The physical conditions, *e.g.*, temperature, wind, humidity, sunlight of the terrestrial environments are not as uniform as in the aquatic habitats. Hence, terrestrial habitats often show greater extremes, as of temperature, and sharper discontinuities, as between opposite sides of a mountain range, than are found in the ocean, for example.

4.9.2 Terrestrial Life Zones

Since rainfall and temperature vary with geographic latitude and altitude, they divide the soil covered land surface into a number of distinct habitat zones, or *biomes* (Fig. 4.23). A *biome is,*

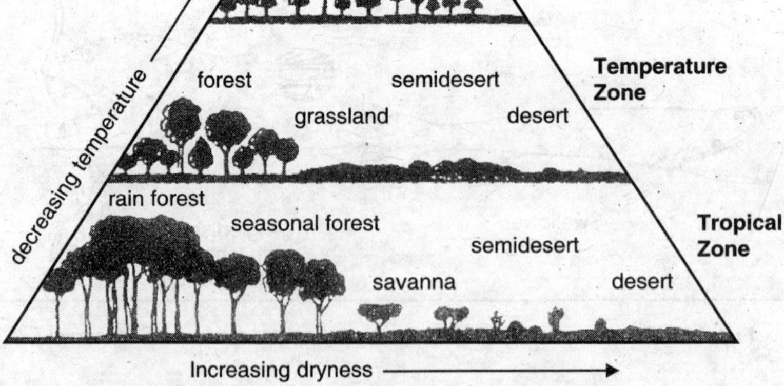

Fig. 4.23 Showing that the temperature and rainfall determine the nature of the Biomes to a large extent (after Mader)

therefore, the largest terrestrial ecological unit, which is characterized by interactions of flora, fauna, and abiotic substances. The climate is considered to be a part of the abiotic component. A biome gets its name from its climax or dominant plant form. Therefore, the term 'biome' may also be defined as one of the major categories of the world's distinctive plant assemblages. For example, a *grass biome* is an area in which grasslands predominate; a *coniferous biome* is an area of cone-bearing evergreen trees.

The various zoogeographic regions contain more than one biome. The Oriental region, for instance, contains a tropical rain forest, a tropical deciduous forest, a tropical scrub forest, temperate deciduous and rain forest, temperate grassland, mountains, and desert. Many of the similar biomes are also found in Australian, Ethiopian, and Neotropical regions.

4.9.3 Latitudinal and Altitudinal Life Zones

Progressive changes in temperature, solar radiation, and precipitation from the equator toward the poles and similarly from the low-lands to the mountain peaks control the distribution of certain major vegetation types and these, in turn, are accompanied by characteristic sets of plants and associated animal species. As a result, we may recognise a series of *life zones* or *terrestrial habitats* extending across the land from the tropics to the polar regions, and also, though on a much smaller scale, extending up mountain slopes from the warm lowlands to the alpine conditions of high elevations or altitudes.

Thus, horizontal sequence of biomes between equator and pole is repeated more or less exactly in a vertical direction, along the slopes of mountains (Fig. 4.24). For example, on a high mountain in the tropics, the sequence of biomes from mountain base to snowline is the tropical

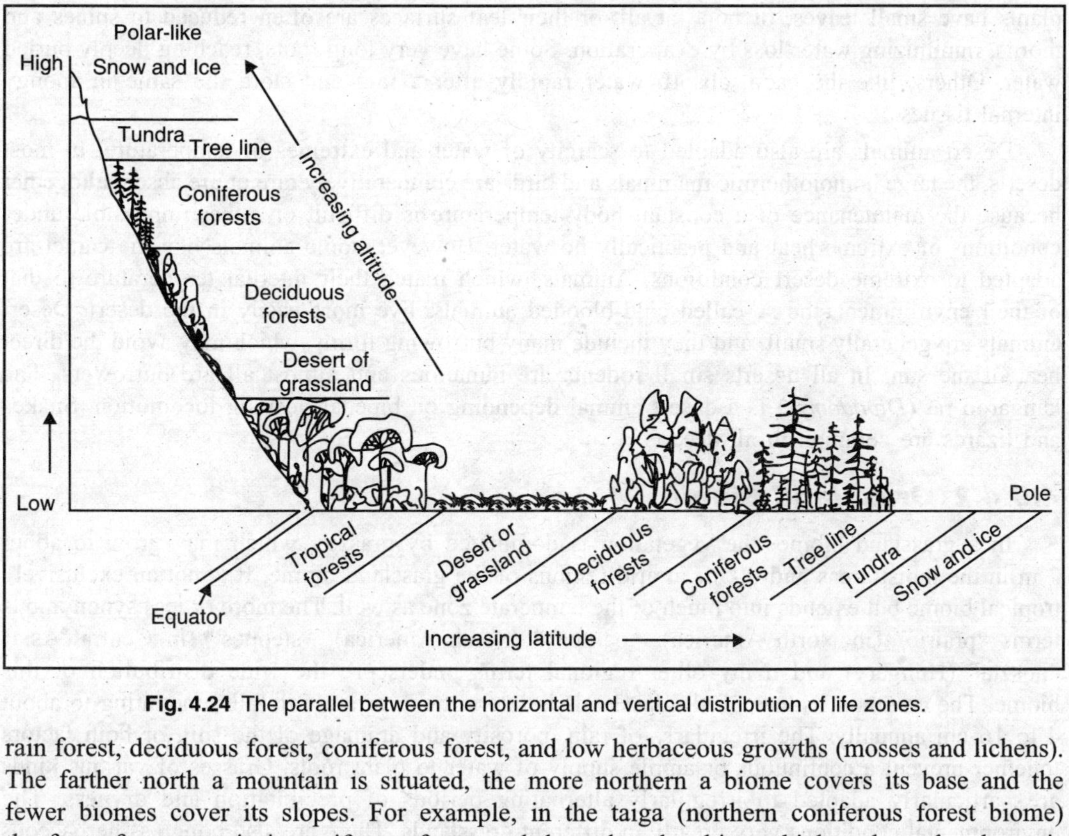

Fig. 4.24 The parallel between the horizontal and vertical distribution of life zones.

rain forest, deciduous forest, coniferous forest, and low herbaceous growths (mosses and lichens). The farther north a mountain is situated, the more northern a biome covers its base and the fewer biomes cover its slopes. For example, in the taiga (northern coniferous forest biome) the foot of a mountain is coniferous forest and the only other biome higher along the slopes is the zone of low shrubby plants. Conditions above timber line on high mountains resemble those of the tundra with some differences in lengths of day and night, temperatures, and atmospheric pressures. The vegetation is similar in general appearance and in mountains of the temperate zone both vegetation and much of the fauna may consist of species related to those in the arctic. Ptarmigan and varying hares, related to arctic hares and also turning white during winter, extend far south of the tundra in alpine environments. Alpine insects are often closely related to arctic species. On upper parts of very high tropical mountains, even the general aspect of fauna and flora is like that of the arctic tundra but the species are more distinct. In this way, the terrestrial habitat zones which are spread over thousands of miles latitudinally are telescoped altitudinally into a few thousand feet.

4.9.4 Major Biomes

The major biomes of the world as follows:

4.9.4.1 Desert Biome

A desert usually has less than 25 cm of rain per year. Deserts are also characterized by intense sunshine and very hot days (40ºC and upward) at least during summer; and the evaporation rate is very high. Nights are generally cold, even in summers, and daily variations in temperature reach extremes found in no other environment.

Desert life is usually well adapted to the dry weather. Most annual plants in the desert are small. They grow rapidly, bloom and produce seeds all within a few days after a rain. Since the growing season is greatly restricted, such plants live relatively small. Many perennial desert plants have small leaves, or none at all, or their leaf surfaces are often reduced to spines and thorns, minimizing water loss by evaporation. Some have very long roots, reaching deeply buried water. Others, like the cacti, absorb water rapidly after a rain and store the same in spongy internal tissues.

Desert animals are also adapted to scarcity of water and extremes of temperature. In most deserts, the large homoiothermic mammals and birds are comparatively rare or are absent altogether because the maintenance of a constant body temperature is difficult or rather impossible under conditions of extreme heat and practically no water. However, some animals like the camel are adapted to extreme desert conditions. Animals, which match their internal temperature to that of their environment, the so-called cold-blooded animals, live more easily in the desert. Desert animals are generally small, and they include many burrowing forms, which may avoid the direct heat of the sun. In all deserts small rodents are numerous and almost all are burrowers. The kangaroo rat (*Dipodomys*) is a desert animal depending on bipedal, leaping locomotion. Snakes and lizards are common in all deserts.

4.9.4.2 Grassland Biome

In a grassland biome, the vegetation is dominated by grasses, which may grow to about 2 m in the moist areas and 0.2 m in arid regions of the grassland biome. It is not an exclusively tropical biome but extends into much of the temperate zone as well. The more or less synonymous terms "prairie" (in North America), "pampas" (South America), "steppes" (in Central Asia) "puszta" (Hungary) and many other regional terms underscore the wide distribution of this biome. The common feature of all glasslands is intermittent, erratic rainfall, amounting to about 4 to 16 cm annually. The irregularly of rain, porosity and drainage of the soil, or both factors together prevent a continuous or ample supply of water to plant roots. Grasses of various kinds are particularly adapted to irregularly alternating periods of precipitation and dryness. The environmental conditions vary greatly in different grasslands. There are also nongrass herbaceous species, which are called *forbs*.

Grassland biome probably supports more species of animals than any other terrestrial habitat. In all grasslands, the primary consumers are the large grazing mammals like the bisons, pronghorns (*Antilocapra americana*) and zebra (*Equus zebra*). African glasslands support large herds of zebras and several species of grazing antelopes. The grassland ungulates are *cursorial.* Hares and rodent are also common primary consumers in the grasslands. Many rodents, like the prairie dogs and other ground squirrels or the pocket gophers, are burrowing or *fossorial* animals. Asutrialian grasslands have herbivores very different in appearance and relationships but ecologically similar. These are large grazing cursorial kangaroos and small, burrowing, rodent-like pouched "mice". Predators are adapted to the herbivore prey: wild dogs, lions, and the like preying on the ungulates; weasels, snakes, and others on the smaller herbivores. Herbivorous insects such as locusts and grasshoppers are also numerous. Grasslands also support some herbivorous predacious birds.

4.9.4.3 Rain Forest Biome

They occur in those tropical and subtropical parts where torrential rains fall practically every day and where a well-defined rainy season characterizes the winter. Rain forests exhibit a communal coexistence of up to several hundred different species of trees. They cover much of Central Africa, South and Southeast Asia, Central America, and South America. However, in contrast to rain forests in the tropics, the species diversity of temperate rain forests is quite low. A tropical rain forest generally has a hundred or more species of trees, and as many as 500 have been observed in one such forest. Two trees of the same species seldom stand near each other. However, the actual species present may be totally different in rain forests found in widely separated regions of the earth.

Tropical rain forest biome receives about 200 cm of precipitation during a year. The productivity of this biome is more than that of any other terrestrial biome. Because of high temperature and abundant moisture, plant litter decomposes quickly and the vegetation immediately takes up the nutrients released. A striking feature of a tropical rain forest is the *vertical stratification* of plant communities. Trees in such forests are normally so crowded together that they form a continuous overhead canopy of branches and foliage, which cuts off practically all the sunlight, much of the rain water and wind. As a result, the forest floor is very humid and quite dark and, therefore, plants that require only a minimum of light populate it. Apart from the forest trees themselves climbing *Lianas* and *Epiphytes* are quite characteristic of the tropical rain forests. Rooted in the dark forest floor, lianas are climbing vines, which use the standing trees as supports upon which they climb toward the canopy where they spread their leaves in the light. Epiphytes grow on other plants. Orchids, ferns, and many other epiphytes form veritable aerial gardens among the high branches of the trees of rain forests.

In rain forests animal communities too are stratified vertically into different habitats found between canopy and ground. A much larger proportion of animals lives in the upper layers of the vegetation than in temperate forests where most of the life is near the ground level. For example, 31 to 59 species of mammals in British Guiana are arboreal and 5 are amphibious, leaving only 23, which live on the ground. In addition to the arboreal mammals (monkeys, rodents, squirrels), there is an abundance of chameleons, iguanas, geckos, arboreal snakes, frogs, and birds. Ants, termites, beetles and the Orthopetra, as well as butterflies and moths are ecologically important. Frogs may be present in large numbers. In the Old World rain forests, ground-dwelling herbivores include musk deer, small forest antelopes, and forest pigs. In both hemispheres partly arboreal carnivores, especially cats such as leopard and jaguar, kill the herbivores. Symbiosis between animals and epiphytes is widespread. Many animals of the rain forest are nocturnal.

4.9.4.4 Deciduous Forest Biome

In the temperate zones such as Europe, eastern Asia, southern Canada and eastern part of United States, the most characteristic biome is the deciduous forest. Tropical deciduous forests also occur in many tropical parts of the world. The fundamental climatic conditions of a deciduous forest biome are cold winters, warm summers, and well spaced rains bringing about 75 to 100 cm of precipitation per year. The biome is also characterized by seasonal temperature variations, which are greater than daily variations. Winter makes the growing season discontinuous, and the flora is adapted to this. The temperate deciduous forests cover many parts of the United States, the British Isles, Central Europe, China and south-eastern Siberia. Similar forests also occur in the Temperate Zone of South America, but they are not so widespread there.

The term "deciduous" implies the most obvious characteristic of this biome and the most obvious adaptation to it (*i.e.*, trees shed their leaves and hibernate). Half the year or somewhat more is the growing season, when perennial plants put on their leaves and are active, while annual plants go through the whole cycle from seed to seed. The rest of the year the trees are bare. Common trees of the deciduous forest are beech, tulip, sycamore, maple, oak, hickory, elm, poplar, and birch. Chestnuts were also formerly common. A deciduous forest differs from a rain forest in that trees are spaced at considerable distance from one another and there are only few species of trees. Compared with the hundreds of tree species in a rain forest, there may be only about 10 or 20 in a deciduous forest.

The most striking herbivorous mammals of a deciduous forest are the browsing deer, mainly the whitetailed deer in North America and other species in Eurasia and South America. In Eurasia, wild pigs (or boars) are also found in this biome, but they do not occur native in America. The principal carnivores are the large cats including puma, mountain lion, cougar, or panther (all one species, *Felis concolor*) ranging into most of the environments of North and South America. Foxes are also common in them. The arboreal martens are locally as common here as in the taiga, and the raccoon (absent in Eurasia) is especially abundant in deciduous forests of North America. Throughout the world these forests are also rich in tree squirrels. Among mammals of the North American deciduous forests, over a third of the species are mainly arboreal. Tree-nesting birds are also abundant and woodpeckers are characteristic of this biome. The leaf and mold-covered forest floor supports many species of invertebrates and fungi.

4.9.4.5 Taiga Biome

North of the deciduous forests and the grasslands across northern Europe, Siberia, and Canada, stretches the *taiga* (northern coniferous forest biome). It is also called the boreal forest biome. This is a biome of long, severe winters and of growing seasons limited largely to the few months of summer. The vegetation is extremely frost-tolerant, as temperatures may fall to — 60ºC during the winter. The precipitation is in the range of 40-100 cm. Hardy conifers, spruce in particular, are most representative of the flora; and moose, wolves, and bears of the fauna.

The taiga is largely a zone of forests, which differ from other types of forests in that they usually consist of single species of coniferous tree. Thus, over a large area, spruce, for example, may be the only kind of tree present. Among other coniferous trees, alder, birch, and juniper thickets are common. They might be found in an adjacent equally large area. Many of the larger herbivorous vertebrates, such as the moose (elk), snowshoe hare, and grouse depend on broad leaved developmental communities of spruce forest. The seeds of conifers provide important food for many animals such as squirrels, siskins, and crossbills. In taiga, seasonal periodicity is pronounced and populations tend to oscillate. The snowshoe hare-lynx cycles are classic examples. Smaller mammals are much more varied than in the tundra. Black bears, wolves, and martens are more common in this biome that elsewhere. Fishers, wolverines, lynxes and some rodents such as the northern vole are practically confined to it. Squirrels and birds also thrive in coniferous forests. Most of the birds here, however, are summer breeders and migrate southward in the autumn. The many species of insects remain dormant during the severe winters.

4.9.4.6 Tundra Biome

In Asia, Europe, and North America a vast northern zone encircling the Arctic Ocean is known as the *tundra.* This biome lies north of the taiga. The tundra has the arctic climate, which is cold, and there may be continuous night during the winter season and continuous daylight, of comparatively low intensity, during the summer. Some distance below the surface, the ground is permanently frozen. This is called *permafrost.* Above the ground, frost can form even during the summer; plants often freeze solid and remain dormant up to the growing season. The latter is very brief, as in the deserts; but in the tundra the chief limiting factor is temperature, and not water supply. However, alpine tundra does not contain permafrost.

Precipitation is generally less than 50-60 cm but in low lying areas soils may remain saturated with water during most of the growing season. Plants are low, ground hugging forms, and trees are absent. Lichens (especially reindeer moss), mosses (especially *Sphagnum*) coniferous and other shrubby growths, and herbs with brilliantly coloured flowers, are characteristic of the habitat.

The warm-blooded animals of this biome are caribou, reindeer, musk ox, arctic hare, arctic fox, lemming, and polar bear. They are well protected by fur. Some of the resident birds, like the ptarmigan (an arctic grouse-like bird), and mammals, like the snowshoe hare, turn white in winter. White is protective colouration in a snowy environment and also minimizes heat loss by radiation. Musk oxen and caribou (wild reindeer) are large herbivores, which depend mainly on the abundant moss and lichens. Arctic hares and lemmings (small, rat-like rodents) are numerous and are preyed on by arctic foxes. Polar bears are amphibious, frequenting coasts and ice flows but also wandering inland on the tundra. Insects, especially flies, are so numerous as to be one of the major drawbacks of the tundra from the human point of view. Their eggs and larvae are particularly cold resistant and the adults appear by the billions on warmer summer days. Migratory birds, especially waterfowl, are conspicuous during the short summer. Well marked oscillations, or cycles, in population density of some animals are characteristic of tundra communities. For example, when lemmings abound, predatory birds, such as owls are abundant and breed. Whereas few predators breed at all during the years of prey scarcity.

However, the life does not end at the northern margin of the tundra but extends farther into the ice and bleak rock of the soilless polar region. Polar life is almost exclusively animal, and it is not really terrestrial in any way but is based on the sea (*e.g.* walrus, seals, penguins).

However, some authors also describe the so-called *alpine tundra biome.* This occurs above the tree line as in the Rocky Mountains of North America and on the Tibetan Plateau of Central Asia. The alpine and arctic tundra have some similarities as well as differences. Alpine tundra does not contain permafrost; it has warmer and longer growing seasons resulting in higher productivity, less severe winters, and higher species diversity than arctic tundra.

MODEL QUESTIONS

1. Define 'biotic community'. Give an account of the characteristics of an ecological community.
2. Write an essay on interdependence in a biotic community.
3. Describe energy flow in a community in relation to the laws of thermodynamics.
4. Define net primary production. Which factors affect primary production in all natural communities?
5. Define secondary production. How it is measured?
6. Define succession. Comment upon the statement that the phenomenon of succession is 'community controlled'.
7. What is climax community? Give a review of the theories of climax.

8. Write the difference between:
   (*a*) Autotrophic and Heterotrophic succession
   (*b*) Autogenic and allogenic succession
9. Describe the main zones of a lake. How a lake ecosystem is different from a river ecosystem?
10. How thermal stratification of a lake helps in proper functioning of a lake ecosystem?
11. Define 'eutrophication'. How it affects lakes and other water bodies?
12. How the biotic community of running water is different from stagnant waters? Give an account of the adaptations of rhithron community.
13. Describe the role of benthos and macrophytes in a potamon community.
14. Give an account of the zonation of ocean. How the deep sea organisms are adapted to their environmental conditions?
15. Define the term 'biome'. Give a brief account of the major biomes of the world.
16. Write notes on the following:
   (*a*) Ecotone (*b*) Food web (*c*) Energy flow in a community
   (*d*) Climax community (*e*) Eutrophication (*f*) Deepsea community
   (*g*) Desert biome

MULTIPLE CHOICE QUESTIONS

17. Which of the following community shows highest NPP?
   (A) Cultivated land (B) Open ocean
   (C) Desert (D) Tropical rain forest
18. Which of the following is an example of hydrarch succession?
   (A) Open field being converted into forest
   (B) Grassland being converted into forest
   (C) Lake being converted into dry land
   (D) All of them
19. Who proposed the monoclimax theory of succession?
   (A) A.G. Tansley (B) F.E. Clements
   (C) C.W. Clark (D) R.H. Whittaker
20. The key nutrients responsible for eutrophication are :
   (A) Nitrogen and Phosphorus (B) Nitrogen and Calcium
   (C) Phosphorus and Sulphur (D) Magnesium and Iron
21. Algal blooms are caused by :
   (A) Green algae (B) Blue-green algae
   (C) Diatoms (D) All algae
22. Which of the following is an example of stonefly that occur in fast flowing streams?
   (A) *Baetis* (B) *Rhyacophila*
   (C) *Simulium* (D) *Perla*
23. Which of the following is a macrophyte?
   (A) *Aorichthys* (B) *Anabaena*
   (C) *Ceratophyllum* (D) *Brachionus*
24. Which of the following are biological indicator of good quality water?
   (A) Water cabbage (B) Stoneflies
   (C) Blue-green algae (D) Blackflies

25. Vertical stratification of plant communities is a characteristic of :
(A) Grassland biome (B) Tundra biome
(C) Tropical rain forest biome (D) All the biomes

26. Biomes are not homogeneous because of the following that influence their biotic communities :
(A) Local topography (B) Geology and nature of soil
(C) Temperature and water (D) All the above

27. Which of the following may be called "marine desert?
(A) Sea coast (B) Open sea
(C) Estuary (D) Dry river

28. Why community composition changes as one moves up a mountain ?
(A) Due to increasing altitude
(B) Due to climate change
(C) Due to changes in soil and moisture content
(D) Due to all the above factors

29. Why community composition changes as one moves down into the abyssal depths of the ocean?
(A) Due to changes in temperature and pressure
(B) Due to absence of primary producers
(C) Due to more competition among consumers
(D) Due to all the above factors

30. Who gave the concept of 'open community'?
(A) F.E. Clements (B) H.A. Gleason
(C) J.R. Runkle (D) R.E. Ricklefs

31. The part of the earth in which life exists is known as :
(A) Lithosphere (B) Biosphere
(C) Atmosphere (D) Hydrosphere

32. The biggest realm is :
(A) Nearctic (B) Palaearctic
(C) Australian (D) Ethiopian

33. Which of the following is a social insect?
(A) Termite (B) Locust
(C) Louse (D) Mosquito

34. Which of the following is a deep sea fish?
(A) *Tor putitora* (B) *Carassius auratus*
(C) Angler fish (D) Mackerel

35. The communities of rhithron include the following except :
(A) Stone fly (B) Caddis flies
(C) Trout (D) *Lymnaea*

5

CHAPTER

Ecosystem

5.1 INTRODUCTION

The ecosystem is the basic structural and functional unit of ecology. The term ecosystem was first proposed by Tansley in 1935. Traditionally, an ecosystem is defined as any unit that includes all the living organisms (the community) in a given area interacting with the physical environment so that a flow of energy leads to clearly defined trophic structure, biotic diversity and material cycling between the abiotic and biotic components within the ecosystem. (Ramakrishnan, 2001). The ecological study of ecosystems or "*Ecosystem ecology*" is considered as the most important aspect of ecology.

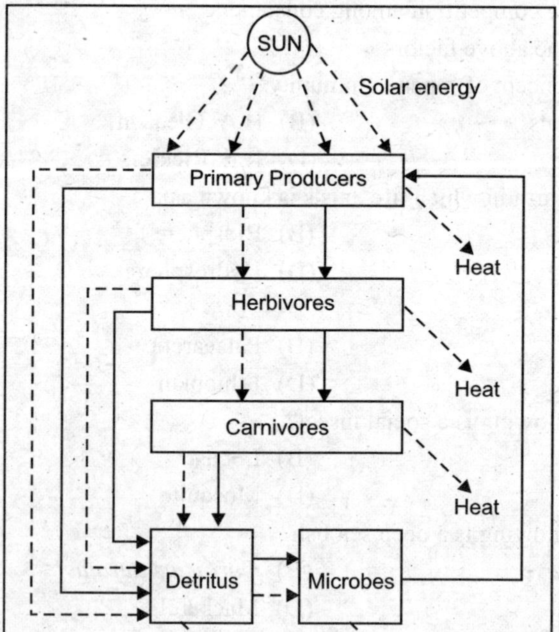

Fig. 5.1: Diagrammatic representation of energy flows (dashed lines) and material flows (solid lines) in an ecosystem. The energy flows include solar radiation, chemical energy transfers and radiation of heat into space (from DeAngelis, 1992)

The central theme of ecosystem concept is that the living organisms of a community not only interact among themselves but also have functional relationship with their nonliving environment.

This structural and functional system of communities and their environment is called an ecosystem. It comprises the interactions of the organisms and their environmental factors, involving the flow of energy and cycling of materials (Fig. 5.1). Thus all natural ecosystems are self-sustaining since there is continuous flow of energy and nutrient cycling, while all human managed systems are more or less dependent upon energy subsidy that is external to the system.

5.2 ECOSYSTEM FUNCTIONING

Each ecosystem has spatial and temporal dimensions and perturbation is now considered as an integral component of ecosystem functioning. All ecosystems are subjected to a variety of perturbations, which contribute towards their resilience and sustainability. Ecosystem's response to perturbation results in ecosystem fragility. Some ecosystems are more fragile then others. According to Ramakrishnan (2001), the ecosystem of Cherrapunji forest is extremely fragile.

In a balanced condition, ecosystem functioning is self- regulating and self-sustaining. This dynamic nature of ecosystem is dependent upon a number of factors including flow of energy, cycling of materials and perturbations, both intrinsic and extrinsic. Ecosystem is now recognised as a dynamic concept with structural heterogeneity based on at least four functional compartments or phases (Holling, 1995). A rapid 'release phase' (Fig. 5.2) consisting of tightly bound resources is replaced by a 'reorganisation phase' followed by an 'exploitative phase', which is gradually transformed into 'conservation phase' or the so-called climax phase. In conservation phase, ecosystem consolidates resources largely through slow growing species that conserve resources. During the slow change (as indicated in the figure by more distance between arrows) from exploitative phase to conservation phase, stability and connectedness increase with accumulation of biomass and nutrient pool. With tightly bound resources in climax phase, a small perturbation

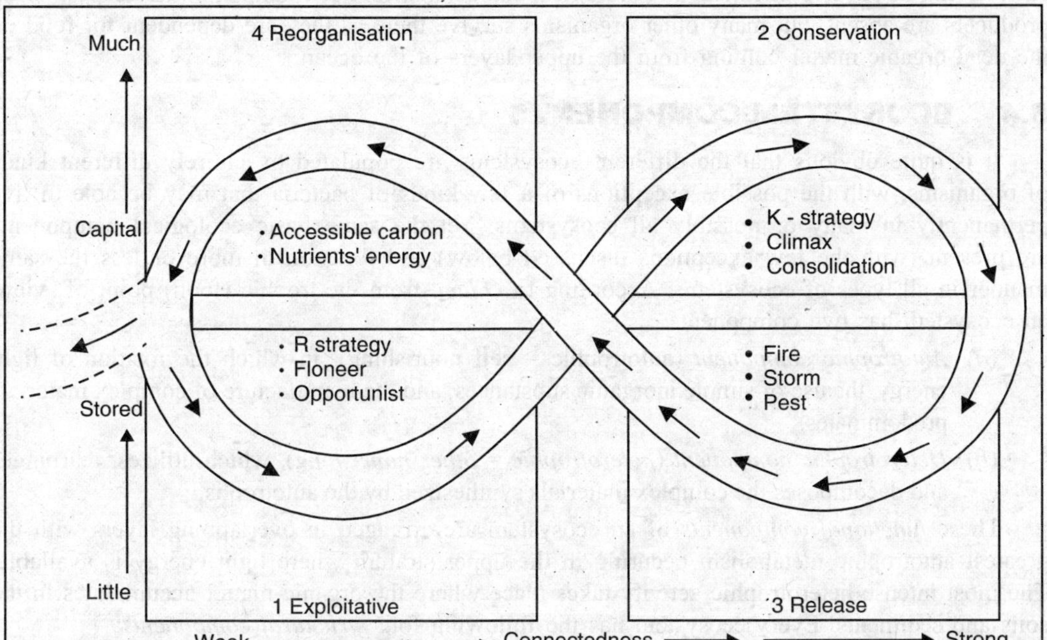

Fig. 5.2% Inter-connectedness of the four major phases and the flow of events of ecosystem organisation and function. The arrows indicate the speed at which recovery of the system occurs after a perturbation, starting with the 'release' phase, of flow; closeness of the arrows indicate rapid change. Y axis shows the amount accumulated capital of carbon and nutrients; X axis shows the degree of connectedness between variables within the system. The exit cycle is indicated on the left, where a flip is likely into a less or more productive and organised system (after Holling, 1995)

may bring about rapid changes, which may lead to release of resources in the biomass, and the disturbed part of the ecosystem then gradually returns to the conservation phase. But larger perturbations between the exploitative and reorganisation phase lead to breakdown of the ecosystem, as indicated by an arrow on the left side of (Fig. 5.2). The control of ecosystem function by nutrient flux and the conditions of the physical environment is called *Bottom-up control.* The regulation of ecosystem function via trophic interactions is called *Top-down* control.

The science of control (cybernetics) of ecosystems is quite complex. Both positive and negative feedbacks control ecosystems. De Angelis *et, al.* (1986) gave an account of positive feedback in natural ecosystems. For example, species populations grow exponentially until some control is imposed on them from outside. Negative feedback operates at sublevels in ecosystems. For example, when nutrient release exceeds a certain level, negative feedback inhibits further release or when a population attains enormous size, various events are started that control reproduction. Thus, ecosystems are self- regulating entities.

5.3 ECOSYSTEM TYPES

Depending upon the species diversity and the manner in which they are organized, we can recognize the ecosystem types such as aquatic, marine, forest, grassland and agro-ecosystem. These typologies are determined not only by the species composition but also by the physiognomic characteristics and soil and climatic conditions. Freshwater ecosystems are usually named after the size and nature of the freshwater body such as pond, lake, river. A wetland ecosystem, looking outwardly, presents an intermediate condition between an aquatic and a terrestrial ecosystem. The largest ecosystem on this earth is 'marine ecosystem', which consists of several subdivisions, each having its physico-chemical and biological characteristics. For example, in the deepest ocean producers are absent, but many other organisms survive there as they are dependent for food on the dead organic matter coming from the upper layers of the ocean.

5.4 ECOSYSTEM COMPONENTS

It is quite obvious that the different ecosystems are populated by entirely different kinds of organisms, with the possible exceptions of a few kinds of bacteria that may be able to live permanently in many or possibly all ecosystems. Yet the same basic ecological components are present (with the few exceptions discussed below) and function in more or less the same manner in all types of ecosystems. According to *Odum,* from the trophic (food) point of view, an ecosystem has two components:

(*a*) *Autotrophic component* (autotrophic = self nourishing), in which the fixation of light energy, the use of simple inorganic substances, and the manufacture of complex materials predominates.

(*b*) *Heterotrophic component* (*heterotrophic = other nourishing*), which utilizes, rearranges and decomposes the complex materials synthesized by the autotrophs.

These *functional components* of an ecosystem are arranged in overlapping layers with the greatest autotrophic metabolism occuring in the upper stratum where light energy is available. The most intense heterotrophic activity takes place where the organic matter accumulates in the soils and sediments. Every ecosystem has the following four *structural components*:

5.4.1 Producers

Producers or the autotrophic organisms are those living members of the ecosystem that utilize sunlight as their energy-source and simple inorganic materials like water, carbon dioxide and salts to produce their own food. Producers are largely photosynthetic plants and their kind varies with the kind of ecosystem. For example, in a dense forest the trees are the most important producers.

In lakes and ponds, the producers are rooted or large floating plants and microscopic plants (phytoplankton), usually the algae. In the ocean, the rooted and floating algae are the producers. However, whether large or small, rooted in the soil or floating in the water, all producers support an array of consumers and decomposers.

5.4.2 Consumers

Consumers are those living organisms that ingest other organisms and are, therefore, called heterotrophs. They derive their food directly or indirectly from the producers or the green plants. The food is then digested, *i.e.*, broken down to simple substances which are metabolized in the consumer's body and the waste products of their metabolism are released to the environment. The vast majority of consumers are the animals. The *primary consumers* in an ecosystem are *herbivores,* which feed directly on the producers (green plants). However, primary consumers also vary with the kind of the ecosystem. For example, a deer or a giraffe is a primary consumer in a forest ecosystem, while a cow or a goat is a primary consumer in a grassland or crop ecosystem. Microscopic animals, such as the protozoans and certain crustaceans, which feed on the floating algae of ponds and seas, are the primary consumers in their habitats. The *secondary consumers* (carnivores), such as the predaceous insects and game fish in a pond, feed on the primary consumers. There are, in most ecosystems, some organisms that eat other carnivores like snake eats a frog, or a bird eats all types of fishes including carnivores. These are called the *tertiary consumers.* Some ecosystems may have the *top carnivores* like the lion and vulture, which are not killed or rarely killed and eaten by other animals. Another important type of consumers are the bottom-living *detritivores,* which subsist on the rain of organic detritus from autotrophic layers above.

The consumer animals ingest plant products or other organic materials as food, break them in their digestion process and produce the new types of organic materials like the proteins and animals tissues. Therefore, animals are also busy in the process of organic production. Based on this concept, the green plants are called *primary producers* and the heterotrophic organisms as the *secondary producers.* However, producers and consumers differ in the natural ecosystems (Table 5.1).

Table 5.1: The biotic and abiotic components of some ecosystems

S. No.	Example of ecosystem	Producers *(Autotrophs)*	Consumers *(Heterotrophs)*	Decomposers *(Saprophytes or Saprotrophs)*	Abiotic substances *(Materials and energy)*
1.	Pond or Lake	Rooted & floating plants and algae (phytoplankton)	Zooplankton (*Daphnia, rotifers*) insects, water bugs, water snails, tadpoles, beetles, fish, benthic organisms, worms and birds	Bacteria, fungi and flagellates	Water, minerals/ salts, organic matter, energy (light/heat/ chemical) dissolved gases.
2.	Graessland or crop	Grasses & crop plants	Earthworm, beetle, rabbits, grasshopper, frog, snake, vulture, hawk, cow, goat and other mammals.	–do–	–do– plus humus and atmospheric gases

3.	Forest	Forest trees, herbs and shrubs	Millipede, centipede, frog, snake, bird, deer, giraffe, lion, tiger, vulture, eagle and mammals.	–do– molds, mushrooms and earthworms.	–do– plus humus and atmospheric gases
4.	Sea	Phytoplankton (diatoms, dino-flagellates and algae) & other vegetation (producers are absent in deep sea)	Zooplankton, nekton, worms, benthos, crustaceans, fish, molluscs, birds and mammals.	Various micro-organisms	As in pond or lake ecosystem

5.4.3 Decomposers

Decomposers or the microconsumers are also called the *saprobes* or *saprophytes.* They are also the living components, chiefly the bacteria and fungi, that break down the complex compounds of dead protoplasm of producers and consumers, absorb some of the decomposition products, and release simple substances usable by the producers. Molds and mushrooms (*Agaricus*) of the forest floor are the largest of the decomposers that are visible to a naked eye. In a pond, the aquatic bacteria, flagellates and fungi are distributed throughout, but they are especially abundant in the bottom, where dead bodies of plants and animals accumulate.

The role of decomposers is very important in an ecosystem, for without their activity the entire cycle of minerals will get blocked. Decomposers are sometimes also called detritivores, particularly when they break down animal and plant material called detritus.

5.4.4 Abiotic Substances or Components

The abiotic component consists of two things: the *materials* and the *energy.* The materials are like water, minerals atmospheric gases, and salts. They also include some organic matter, such as the amino acids and other products of decay of living things. The quantity of abiotic materials like the minerals present at any given time in an ecosystem is termed as the *standing state* similar to the *standing crop,* which refers to the quantity of biotic components. A small portion of these nutrient materials circulates through other living components of the ecosystem and through their death and decay returning to the soil and the environment. The whole process is called *mineral circulation* or the *biogeochemical cycle.* A much larger portion of the materials is held in reserve in the environment as well as in the organisms themselves.

The *energy* is in the form of light, heat, and stored energy in chemical bonds. While the materials circulate, the solar energy is continuously trapped by the green plants on the one hand, and lost in space through respiration and loss of heat by all types of organisms, on the other. The energy flows through an ecosystem on a one-way traffic and for the proper maintenance of any ecosystem the energy must be consistently added and transferred from organism to organism.

Thus, in an ecosystem, the first three of the four structural components are composed of living things. Green plants are the predominant producers; animals, the consumers; and fungi and bacteria, the decomposers. However, some ecosystems may lack one or another of the components. For example, in the deep oceans producers are absent. But their place is taken by the rain of dead decaying protoplasm that sinks from the surface. Other aquatic communities are sometimes devoid of consumers for short periods, and decomposers will attack the producers directly, although usually only when the plants are dead.

The four components collectively function as a whole system to maintain the stability of an ecosystem. If the balance is disturbed, the whole system breaks down. As mentioned above, for the proper functioning of an ecosystem, there must be a continuous *flow of energy* and *cycling of minerals* among the organisms of the ecosystem.

5.5 TROPHIC LEVELS

The trophic level concept was first given by Lindeman (1942). According to him, a food chain consisted of steps— primary producers, herbivores, carnivores— that he referred to as trophic levels. Trophic literally means feeding, so trophic levels are the positions or levels at which organisms of an ecosystem feed. In complex natural communities, organisms obtain their food from plants through one, two, three or four steps and are accordingly said to belong respectively to the first, second, third and fourth trophic level.

(*a*) Green plants (primary producers)- I trophic level

Green plants are autotrophs. These are the organisms that literally 'feed themselves'. They can be divided into two groups— *photoautotrophs* or photosynthetic organisms and *chemoautotrophs* which are autotrophic but non-photosynthetic (*e.g.*, the prokaryotes).

(*b*) Herbivores (primary consumers)-II Trophic level

Herbivores obtain their food directly from producers. Hence, herbivores belong to second trophic level. Herbivores can be called predators if they are larger than their prey and consume most or all of a plant. They are parasites if they live in or on their hosts for long periods of time. Thus, most mammalian herbivores are predators and most insect herbivores are parasites.

(*c*) Carnivores (secondary consumers)- III trophic level

Carnivores feed upon herbivores and thus belong to the third trophic level. However, they may be classified as first, second, third level carnivore.

(*d*) Top carnivores (tertiary consumers)- IV trophic level

Organisms belonging to fourth trophic level are top carnivores. They are rarely killed or never killed by other organisms for food, *e.g.*, vulture and lion in a forest ecosystem.

It may be noted that trophic complexity increases with the number of trophic levels and vertical connections in the food web. However, decomposers play an important role in all ecosystems. They breakdown the organic waste products and dead remains of producers and consumers into the organic substances needed by the producers. But decomposers are not assigned to any trophic level.

It should be known that classification of organisms by trophic levels is one of *function* and not of species as such. However, some organisms feed at more than one trophic level. For example, male house flies feed on nectar and plant juices while the females are blood-sucking ectoparasites. Some carnivores also feed at more than one trophic level and as such they cannot be assigned to discrete trophic levels. However, some authors have solved this problem by allocating carnivores to non-integer trophic positions, so that a carnivore might be described as a level 3.2 carnivore, for instance. Animals of successive trophic levels in a food chain tend to be larger. Of course, there are upper and lower limits to the size of food which a carnivore can eat. Generally large carnivores cannot live on very small food items because they cannot catch enough of them in a given time to satisfy their metabolic needs. The energy flow through a trophic level equals the total assimilation (A) at that level, which in turn equals the production (P) of biomass plus respiration (R).

5.5.1 Trophic Efficiency

Lindeman observed that less energy was available to each higher trophic level due to the work performed and to the inefficiency of biological energy transformations at the trophic level below. For example, of the light energy impinging on a lake ($\wedge_0$), plants use only a fraction ($\wedge_1$)—resulting in the primary production of the system. Herbivores assimilate less energy ($\wedge_2$) than plants do because plants utilize some of their own production to maintain themselves before herbivores consume them. The ratio of production on one trophic level to that on the level below it (for example, $\wedge_2 / \wedge_1$) is the biological efficiency of that link in the food chain (Ricklefs and Miller, 2000). It may be defined as the efficiency of energy transfer from a trophic level to the one above.

According to *Lindeman's law of trophic efficiency* (Lindeman,1942), the efficiency of energy transfer from one trophic level to the next is about 10%. In other words, about 10% of the net primary productivity of producers ends up as herbivores, about 10% of the net primary productivity of herbivores ends up as first-level carnivores, and so on. Thus, *Lindeman* proposed that trophic efficiencies *increase* as one ascends a food chain. However, more recent data support neither Lindeman's suggestion that trophic efficiency increases as one ascends a food chain, nor the generalisation that the efficiency of energy transfer from one trophic level to the one above is about 10%. It is now known that trophic efficiencies vary from less 0.1% to over 20%. For example, a community of herbivorous small mammals in a grassland ecosystem might have a trophic efficiency of less than 0.1%, herbivorous zooplankton feeding on phytoplankton might have a trophic efficiency of over 20%.

5.5.2 Arrangement of Trophic Levels in a Food Pyramid

In any natural community the distribution of organisms in terms of their total mass and their contribution to the food chain can be compared to the structure of a pyramid. A *food pyramid* represents several features of a natural community of organisms. The broad base of a food pyramid is always composed of photosynthetic plants. The next higher layer or trophic level in the pyramid is composed of small animals that feed on the many plants of the lowermost pyramid layer. The smaller animals in turn serve as a food source for the larger animals, which make up the next higher trophic level in the food pyramid. The successively higher trophic levels in the pyramid are typically constituted of progressively larger but fewer animals.

In general, for terrestrial food pyramids the total mass of protoplasm of any trophic level is always less than the layer below it. The fewest but largest animals usually but not always occupy the peak. By contrast in marine pyramids the microscopic phytoplankton usually consist of smaller total protoplasmic mass than the higher trophic levels because they are unicellular and rapidly consumed. However, their rate of turnover is high. In all cases, whether on land or in the sea, the amount of organic matter produced per unit of time is always greater at the basic trophic level than at the higher trophic levels of the pyramid.

5.6 ECOLOGICAL PYRAMIDS

There is some sort of relationship between the *numbers, biomass* and *energy* contents of the producers and consumers of different orders in any ecosystem. These relationships, when represented in diagrammatic ways, are referred to as *ecological pyramids,* which are of the following three types:

5.6.1 Pyramid of Numbers

The first attempt to provide a quantitative law concerning the trophic levels was given by Elton (1927) who coined the term 'pyramid of numbers'. The pyramid of numbers deals with the relationship between the numbers of primary producers and consumers of different orders. In all cases, the base of such a pyramid always represents the numbers of primary producers and the subsequent structures on this base are represented by the number of consumers of successive levels, the top representing the number of top carnivores in that ecosystem. In the ecological pyramid so formed, the higher the step in the pyramid, the lower the number of individuals and the larger their size.

The shape of the pyramid of numbers may be upright or inverted (Fig. 5.3 *a*). In a cropland ecosystem, the numbers of crop plants shown at the base of the pyramid are very large; the numbers of grasshoppers are usually less than the numbers of green plants; numbers of frogs is lesser than the number of grasshoppers and finally the number of top carnivores is the least in the series of organisms forming a food chain. But in an inverted pyramid of numbers (Fig. 5.3 *b*), the number of primary consumers (herbivorous birds feeding upon the tree fruits) is always greater than the number of primary producers (a tree), and the number of parasites living and feeding upon the bird's body is still greater.

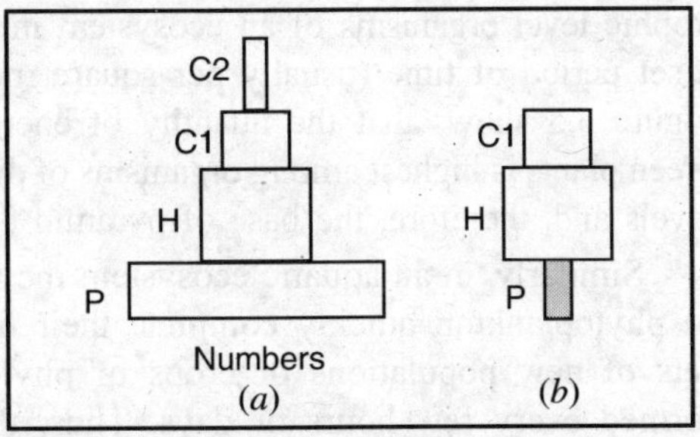

Fig. 5.3: (*a*) Upright pyramid of numbers, the number of organisms decreasing in successive trophic levels. (*b*) Inverted pyramid of numbers (for example a single tree on which many birds feed, which may be having numerous parasites)

5.6.2 Pyramid of Biomass

In order to explain the inverted nature of pyramid of numbers, ecologists proposed the idea of pyramid of biomass where the weight of primary producers forms the base. As the term suggests, a pyramid of biomass takes into account, for a given unit area, the biomass of the producers, the biomass of the herbivores, the biomass of the first-level carnivores, and so on. In (Fig. 5.4 (*a*)), two types of ecosystems are shown where the pyramid of biomass is upright. The biomass of a single tree is naturally very high than the biomass of a number of birds feeding upon the tree. Similarly, the biomass of even a very large number of bird parasites is lesser than that of the birds. Therefore, in upright pyramid the biomass decreases with successive trophic levels. The pyramid of biomass becomes upright in a case where it was inverted if the numbers were considered. However, in some instances the inverted pyramids of biomass are also met with. In a pond or lake ecosystem, (Fig. 5.4 (*b*)), the biomass of diatoms and other phytoplankton is quite negligible as compared with that of the crustaceans and small herbivorous fish that feed on these producers. The biomass of large carnivorous fish living on small fishes is still greater. In fact, this is the condition in most aquatic ecosystems, including the sea.

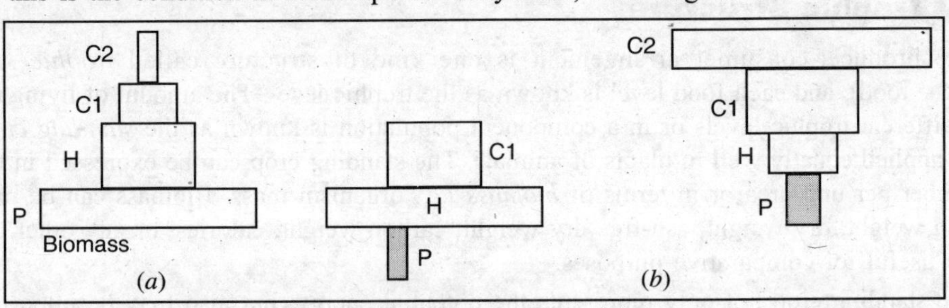

Fig. 5.4: (*a*) An upright pyramid of biomass, the biomass decreasing with successive trophic levels. (*b*) The inverted pyramid of biomass as in the sea is due to low biomass of primary producers. P. producer; H. herbivore (primary consmer); C1 and C2, small carnivore and large carnivore respectively

5.6.3 Pyramid of Energy

A pyramid of energy shows the flow of energy from one trophic level of a community to the next. The shape of pyramid of energy is always upright or triangular because in this the *time factor* is always taken into account. The pyramid of energy represents the total amount of energy utilized by different trophic level organisms of an ecosystem in unit area over a set period of time (usually per square metre per year). Figure 5.5 shows that the quantity of energy trapped by green plants is highest among organisms of different trophic levels and, therefore, the base of pyramid is broad.

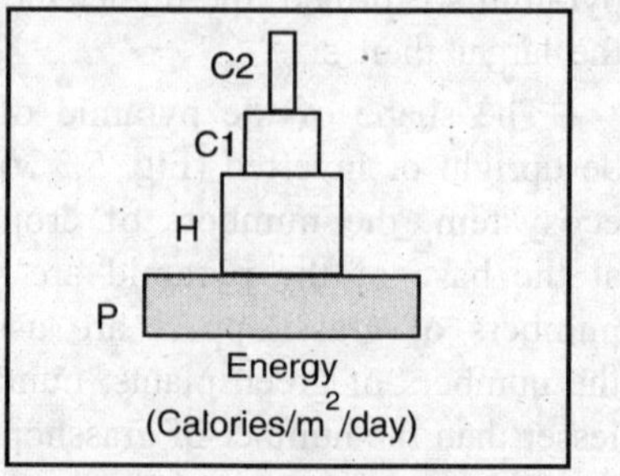

Fig. 5.5: The pyramid of energy. There is always a drop in the amount of energy at each level of a food chain.

Similarly, in an aquatic ecosystem the large numbers of phytoplankton quickly complete their life cycles and sets of new populations or crops of phytoplankton are formed every few hours or days. Thus, the cumulative energy contents that these generation after generation of phytoplankton trap in course of a year is certainly much more than trapped by only a few generations of herbivorous fishes in the corresponding time and space. The energy content trapped by the carnivores living on the herbivorous fishes is the least. Therefore, the pyramid of energy can never be inverted. This follows directly from the first law of thermodynamics, the law of conservation of energy. If primary producers, or organisms of any trophic level are destroyed, the organisms of the next higher trophic level will automatically perish for want of food or source of energy and ultimately the upright pyramidal shape is maintained.

In Florida silversprings, the energy contents of different trophic level organisms, as collected by H.T. Odum (1957), over a period of one year make an upright pyramid. The energy contents in kcal/m^2/yr from 208010 in primary producers decrease to 3368 to 383 to 21 in consumers of first, second and third orders respectively. Thus, a top carnivore for the organic production of only 21 kcal needs a very broad base of green plants equivalent to 208010 kcal of energy. However, Farlow (1976) has used the principle of pyramid of energy to answer the question of whether or not dinosaurs were poikilothermic.

The ratio of energy intake and energy of the produced biomass, *i.e.*, of input and output is called the *ecological efficiency.* This can be studied at any trophic level.

5.7 DIFFERENT ASPECTS OF ECOSYSTEM STUDY

According to Odum, the structure of an ecosystem needs to be considered from various angles in order to understand the relationship between the structure and function in an ecosystem.

5.7.1 Trophic Structure

The producer-consumer arrangement is one kind of structure called *trophic structure* (trophic = food), and each food level is known as the trophic level. The amount of living material in the different trophic levels or in a component population is known as the *standing crop.* This term is applied equally well to plants or animals. The standing crop can be expressed in terms of the number per unit area or in terms of *biomass i.e.*, organism mass. Biomass can be measured as living weight, dry weight, ash-free dry weight, carbon weight, calories, or any other unit that may be useful for comparative purposes.

The standing crop not only represents the potential energy; but also may be important as a buffer against physical oscillations and as a *habitat,* or living space for organisms. Thus, trees in a forest not only represent energy that provides food or fuel; but they modify climate and provide shelter to birds and men.

5.7.2 Biochemical Structure

The amount and distribution both of inorganic chemicals and organic materials present either in the biomass or in the environment are important factors in any ecosystem. We may consider this as the *biochemical structure* of the ecosystem. For example, the quantity of chlorophyll per unit area of land or water surface and the quantity of dissolved organic matter in water are two items of great ecological interest.

5.7.3 Species Structure

Still another aspect of great importance is the *species structure* of an ecosystem. Species structure includes not only the number and kinds of species present but also the *diversity of species*— that is, relationship between species and numbers of individuals or biomass and the *dispersion* of individuals of each species that are present in the community. As mentioned above, the amount of energy gained by the system and the mode and quantities of it transferred, affect the density and diversity of species.

5.7.4 Productivity Concept

A very important factor in the ecosystem study is the *rate of energy-trapping* by green plants, which in turn governs the rate of production of organic material from simple inorganic substances (in a given area over a given period of time). This rate of energy conversion or increase in organic biomass produced is called *primary productivity.*The total production as a result of photo- or chemosynthesis is called *gross primary productivity* (GPP). Naturally some of the primary produce is all the time lost by way of the breakdown of organic matter in respiratory metabolism of plants and animals.Thus, *net primary productivity* (NPP) is the rate of production when loss due to respiration is deducted out of the gross production rate. Primary productivity refers to total increase in weight in all parts: fruits, leaves, stems, and roots, as against the agricultural productivity which refers to useful parts as grains or fodder parts.

The efficiency of any ecosystem greatly depends upon the production rates of its primary producers. Oceans form the largest ecosystem and its productivity varies in different climatic regions. On the shores, the productivity may be 2 to 3.5 grams per square metre per day, and in deep seas only 0.5 gram. In highly productive lakes, the productivity value may be 5 to 10 grams/m^2/day and even upto 50 grams in exceptionally favourable conditions. The net productivity of crop plants (whole plant) ranges from 0.25 kg to 1 kg or a little more for rice and wheat crops per square metre per year. Sugarcane is one of the very efficient converters of solar energy and its net primary productivity value ranges from 2 to 4 kg/ m^2/year or even more. All productivity values are expressed on dry weight basis of organic matter or on the basis of energy contents in unit time and space.

5.8 SIGNIFICANCE OF ECOSYSTEM STUDY

The ecosystem study has a great bearing on the economy and welfare of the human society. The green plants utilize the inorganic materials and the energy of sun to produce numerous organic substances. A part of the latter is lost by the green plants themselves in the process of respiration and for running the machines of life all other organisms, including man, take directly or indirectly the plant produce in the form of their food. Thus, ecosystem study is of supreme importance in all types of habitats on the earth. Realizing this fact, a worldwide programme of extensive quantitative studies of diverse freshwater, terrestrial and marine ecosystems, especially with respect to their *organic productivity* and *conservation,* for the general welfare of mankind has been completed. This programme was called the *International Biological Programme* (IBP), and in most of the countries of the world ecologists have studied the various aspects of ecosystems around them.

Now ecologists and governments have become aware of the necessity of mass education about the fundamentals of interrelationship between humans and their environment. Of course, man is the most important part of the biosphere and he has ever disturbed and degraded most ecosystems of the world, making them less useful for future generations of mankind. Innumerable human practices have already rendered many ecosystems unproductive. Therefore, a basic education of natural resources, working of ecosystems, and ecological principles is very essential if we want to reduce further damage of nature and to improve our resources and use them wisely. According to Ramakrishnan (2000), traditional ecological knowledge may help us to ensure ecosystem stability and resilience.

5.9 BIOGEOCHEMICAL CYCLES

The chemical elements tend to circulate in the biosphere in characteristic paths from environment to organisms and back to the environment. These more or less circular paths of the chemical elements are known as the *Biogeochemical cycles.* These cycles of chemical elements involve *biological* organisms and their *geological* environment. Hence, they are collectively referred to as *biogeochemical.* In this term "bio" refers to living organisms and "geo" to the rocks, soil, air, and water of the earth (Odum, 1963). For each cycle two compartments or pools may be recognized. These are: (*a*) the *nutrient pool* or the *reservoir pool,* which is the large, slow-moving, generally non-biological component; and (*b*) the *exchange* or *cycling pool,* which is a small but more active portion that is exchanged (*i.e.*, moved back and forth). There are two basic types of biogeochemical cycles.

(*a*) Gaseous type

In this type of biogeochemical cycles, the atmosphere constitutes the major reservoir of the element that exists there in gaseous phase. Such cycles show little or no permanent change in the distribution and abundance of the element. The *Carbon, Oxygen* and *Nitrogen cycles* are good examples of biogeochemical cycles with prominent gaseous phase.

(*b*) Sedimentary type

In the sedimentary type of cycle major reservoir is the lithosphere, from which the elements are released by weathering. The sedimentary types are best exemplified by *Phosphorus, Sulphur,* and *Iodine cycles.* In these cycles a little portion of the supply may get lost, as in the deep ocean sediments, and thereby becomes inaccessible to organisms and to continual cycling.

The movement of those elements and inorganic compounds that are essential to life can be conveniently designated as the *nutrient cycling.* The latter may include all the three major types of cycles *i.e.*, hydrologic cycles, gaseous nutrient cycles, and sedimentary nutrient cycles.

(c) Water cycle

The *hydrologic (water) cycle* has been included in gaseous types of cycles by Odum (1963). But *Kormondy* (1969) considers it to be a separate major cycle, involving the movement of a compound, while the others involve the movement of elements.

5.9.1 Hydrological Cycle

It is a balanced continuous process of evaporation, transpiration, precipitation, surface run-off and ground water movements. Each year about 507 Tm^3 (one tetra cubic metre = one million million cubic metres) are evaporated and the same quantity of water is precipitated over the whole surface of the earth, including the land and oceans. The amount of water, which flows from the land to the sea in rivers and streams is about 44.5 Tm^3 per year and is available for the needs of man.

Solar energy evaporates water from the soil, ground surface, vegetation, water surfaces and oceans into the atmosphere (Fig.5.2). Subsequent cooling and condensation of water vapour at higher altitudes produces clouds; and precipitation as rain, hail or snow returns the water to the hydrosphere. Natural evaporation from the seas exceeds precipitation by rain into the oceans by about 9%. The latter is eventually moved as water vapour over to the land surface and so balances the hydrological cycle and provides additional water for man's needs. Thus water as rain, hail or snow is precipitated over land and water surfaces (Fig.5.2).

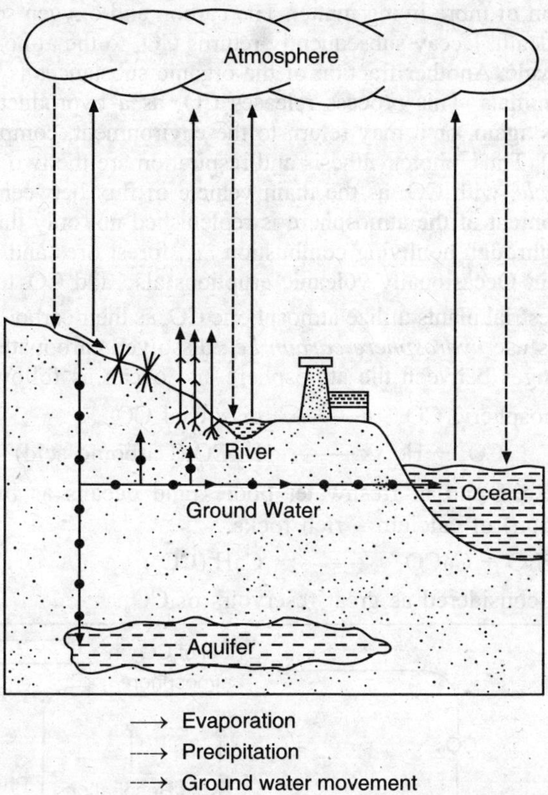

Fig 5.6: The hydrological cycle

Water on land surfaces eventually percolates into the soil as soil or *ground water*. Within the ground there is always a natural *water table*. The soil below the water table is saturated, and water is sustained by the underlying clay and rock strata. Ground water does not remain static but moves in various directions. It can move up above the water table by capillary, thus providing a continuous supply of water to the surface layers of soil, where it is absorbed by plant roots during the dry season. Some ground water moves by filtering through the interstices of the soil or substratum in any direction. Water also emerges from the ground at lower altitude levels, and flows into streams, rivers and lakes, and helps to provide man's water supplies.

In some regions of the earth water percolates through the porous rocks and forms underground reservoirs. These underground water bearing layers of porous rock, situated above impermeable rock strata, are called *aquifers*. They are important source of water, which can be extracted by sinking wells or bore holes and pumping it to the surface.

All the water precipitated on land does not percolate into the soil. Surface water or run-off flows into streams, rivers, lakes and catchment storage areas or reservoirs. Some of the ground water that is absorbed by plants passes out the leaf surfaces as water vapour by transpiration. This is an important process helping in conduction of water and dissolved minerals throughout the plant.

Thus, through the natural process of hydrological cycle, water is exchanged between the atmosphere (troposphere), land, sea all living plants and animals, and industrial plants. The water cycle is powered by energy from the sun and by gravity.

5.9.2 Carbon Cycle

The atmospheric carbon dioxide is virtually the only source of carbon which is the basic constituent of all the organic compounds. This gas is used by all plants in photosynthesis and the end products (organic substances) of this complex process are used in two ways. One fraction is used in the construction of more living matter. The carbon and oxygen so supplied by CO_2 remain in living matter until death. Decay subsequently returns CO_2 to the atmosphere, and this completes one possible carbon cycle. Another fraction of the organic substances is used as fuel in respiration by both plants and animals. This process releases CO_2 as a byproduct. Such CO_2 may now be used in photosynthesis again, or it may return to the environment, completing the second possible carbon cycle (Fig. 5.7). Thus, photosynthesis and respiration are the two major processes that drive the global *Carbon Cycle,* with CO_2 as the main vehicle of flux between atmosphere, hydrosphere and biota. The CO_2 content of the atmosphere is replenished not only through biological oxidation (respiration), but also through nonliving combustion *i.e.*, forest fires and burning of industrial fuels release CO_2 into the air. Occasionally volcanic eruptions also add CO_2 to the atmosphere.

It is clear that terrestrial plants utilize atmospheric CO_2 as their carbon source for photosynthesis, whereas aquatic plants use *hydrosphere carbon i.e.*, dissolved carbonates. These two subcycle are linked by CO_2 exchanges between the atmosphere and ocean as follows:

$$\text{Atmospheric } CO_2 \longleftrightarrow \text{dissolved } CO_2$$

$$CO_2 + H_2 \longleftrightarrow H_2CO_3 (\text{Carbonic acid})$$

The carbon is also found in freshwater bodies and oceans as bicarbonate resulting from weathering (Carbonation) of calcium – rich rocks.

$$CO_2 + H_2O + CaCO_3 \longleftrightarrow CaH_2(CO_3)_2$$

Thus, oceans are considered as great reservoirs of CO_2.

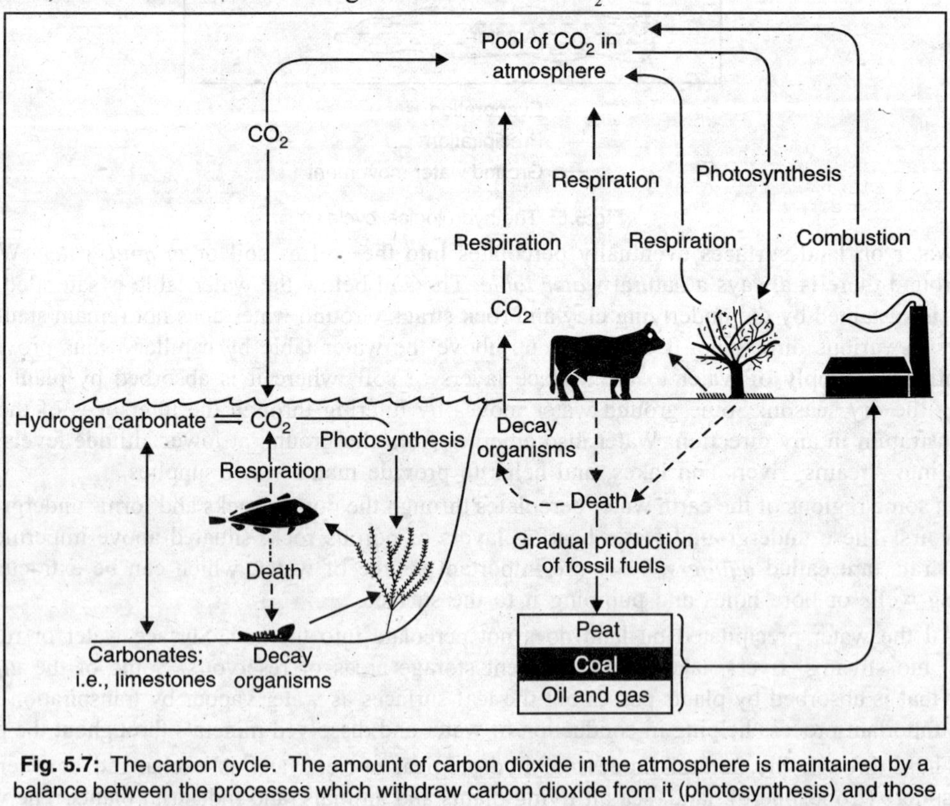

Fig. 5.7: The carbon cycle. The amount of carbon dioxide in the atmosphere is maintained by a balance between the processes which withdraw carbon dioxide from it (photosynthesis) and those which add carbon dioxide to it (respiration and combustion). Solid arrows represent the flow of carbon dioxide. (Based on P. R. and A. H. Ehrlich, Population, Resources, Environment, Freeman)

Effect of anthropogenic activities on the carbon cycle

In order to meet energy requirements and increasing economic growth, the burning of fossil fuels and clearing of vegetation has become a much faster activity than it is naturally done by nature or was earlier observed. This in turn has added to increase in the earth's average temperature by adding excess carbon to the atmosphere.

Computer simulation models of the earth's climate suggest that increased concentrations of atmospheric CO_2 and other gases could promote the planet's natural green house effect warming the lower atmosphere (troposphere) and earth's surface. The *global warming* could disrupt global food production, wildlife habitats, change temperature and precipitation patterns and increase the average sea levels in various parts of the world, besides melting polar-ice caps and flooding of various areas.

5.9.3 Oxygen Cycle

The atmosphere contains about 21% oxygen. The atmospheric oxygen enters the living organisms, as a gas required in respiration. During this process carbon dioxide and water are formed.

$$C_6H_{12}O_6 + 6\ O_2 \longrightarrow 6\ CO_2 + 6\ H_2O + \text{energy}$$

The metabolic water thus formed is added to all other water present in living organisms and as such it may undergo three possible fates. Some of it may be excreted and so added to the water content of the environment. Another part may be used as a building material for the formation of more living matter. A third possible fate of the water within organism is its consumption as a fundamental raw material along with the carbon dioxide in photosynthesis. In this process the oxygen is liberated as shown by the following equation:

$$6\ CO_2 + 6\ H_2O + \text{light energy} \longrightarrow C_6H_{12}O_6 + 6\ O_2$$

Such free oxygen may now again be used in respiration or it may be returned to the environment as molecular atmospheric oxygen, completing the cycle (Fig. 5.8). Thus oxygen enters organisms only through respiration and leaves through photosynthesis. In intervening steps the oxygen is incorported in water, and in this form it can interlink with the water cycle or indirectly with carbon cycle.

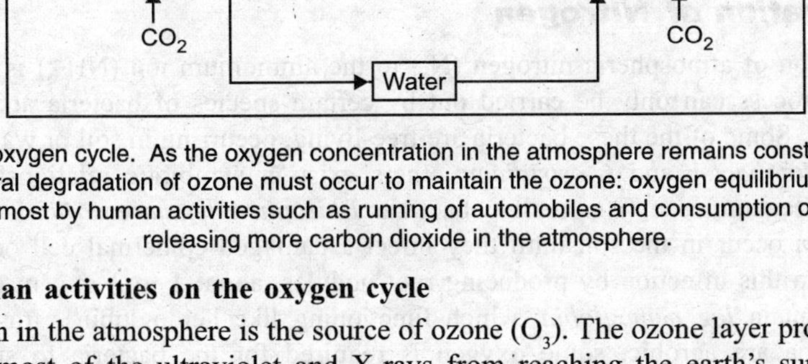

Fig. 5.8: The oxygen cycle. As the oxygen concentration in the atmosphere remains constant at 21%, it is likely that natural degradation of ozone must occur to maintain the ozone: oxygen equilibrium. The oxygen cycle is affected most by human activities such as running of automobiles and consumption of fossil fuels, thus releasing more carbon dioxide in the atmosphere.

Effect of Human activities on the oxygen cycle

The oxygen in the atmosphere is the source of ozone (O_3). The ozone layer protects organisms by preventing most of the ultraviolet and X-rays from reaching the earth's surface. The most recent factor affecting the oxygen cycle of the biosphere and the oxygen budget of the earth is the man himself. In addition to inhaling oxygen and exhaling carbon dioxide, man decreases the oxygen level and increases the carbon dioxide level by burning fossil fuels.

5.9.4 Nitrogen Cycle

Nitrogen being 79 per cent of the atmosphere, the atmospheric phase is predominant in the global nitrogen cycle. It is required by organisms in the synthesis of proteins, nucleic acids, and other nitrogenous compounds. Atmospheric nitrogen serves as the ultimate source. But aerial nitrogen, the most abundant component of air, is rather inert chemically and actually cannot be used as such by the majority of organisms. It must first be "fixed" by specialized organisms or by industrial processes. From an ecological perspective, the nitrogen cycle consists of the following stages: (*i*) Ammonification, (*ii*) Nitrification, (*iii*) Nitrogen uptake by plants, (*iv*) Fixation of Nitrogen, and (*v*) Denitrification.

5.9.4.1 Ammonification

Ammonification is a process in which the organic nitrogen of plants and animals after their death is converted to ammonium ions (NH_4) by the action of saprotrophic fungi and bacteria. The saprotrophs use the ammonia (NH_3) to synthesize their own proteins and other nitrogen-containing organic compounds.

5.9.4.2 Nitrification

Ammonium ions added to the soil by ammonification, are soon oxidised by a process known as *nitrification*. It takes place in two stages. In the first stage, ammonium (NH_4^+) is converted to nitrite (NO_2^-). This reaction involves the addition of oxygen to ammonia, giving rise to hydroxylamine (NH_2OH), which is further oxidised to nitrite. This reaction is completed by the bacteria such as *Nitrosomonas, Nitrosospira, Nitrosolobus* and *Nitrosococcus* (Hamilton, 1988).

The second stage of nitrification involves the oxidation of nitrite (NO_2^-) to nitrate (NO^-_3) by bacteria of the genera *Nitrobacter, Nitrospira* and *Nitrococcus.* The reaction proceeds by the addition of water followed by the removal of hydrogen (Hamilton,1988). The bacteria responsible for these reactions occur in the same soil. The association between *Nitrosomonus* and *Nitrobacter* has been described as one of *commensalism* (Gooday, 1988).

5.9.4.3 Nitrogen uptake by Plants

Nitrate (NO^-_3) formed in the process of nitrification is used by most plants as a mineral metabolite and may be converted by them into amino groups and other nitrogen- containing compounds. Nitrates are also added to the soil through rock dissolution and combination of atmospheric nitrogen with oxygen by lightning (nitrates so formed reach the soil by rain). However, many plants also absorb ammonuim from the soil.

5.9.4.4 Fixation of Nitrogen

The reduction of atmospheric nitrogen (N_2) to the ammonium ion (NH^+_4) is called nitrogen fixation. This process can only be carried out by certain species of bacteria and cyanobacteria (Postgate, 1988). Some of the these bacteria are free-living, occurring in soil or water (saprophytic bacteria like *Azotobacter* and *Clostridium;* others exist in symbiotic relationship with plants of the Family Leguminosae (*e.g.* nodule bacteria *Rhizobium leguminosarum).* Species of the genus *Rhizobium* occur in the soil until they infect a damaged epidermal cell or root hair. The plant responds to this infection by producing root nodules, about 1 to 3 mm in diameter. These root nodules contain *leg hemoglobin,* which functioning like hemoglobin , transports oxygen. As these bacteria are aerobic, some oxygen is required for the bacteria to survive, but too much oxygen inactivates the enzyme *nitrogenase* that catalyses nitrogen fixation. The nitrogen molecule (N_2) is very stable and 16 molecules of ATP are needed for each molecule of nitrogen that is fixed. In the soil microorganism *Klebsiella pneumoniae* a total of 17 genes, called *'nif'* genes, are known to be responsible in nitrogen fixation. Researches in biotechnology are now attempting transfer of *'nif'* genes from prokaryotes to crop plants so that yield of crops like rice and wheat may be increased. However, scientists have not yet succeeded in this attempt.

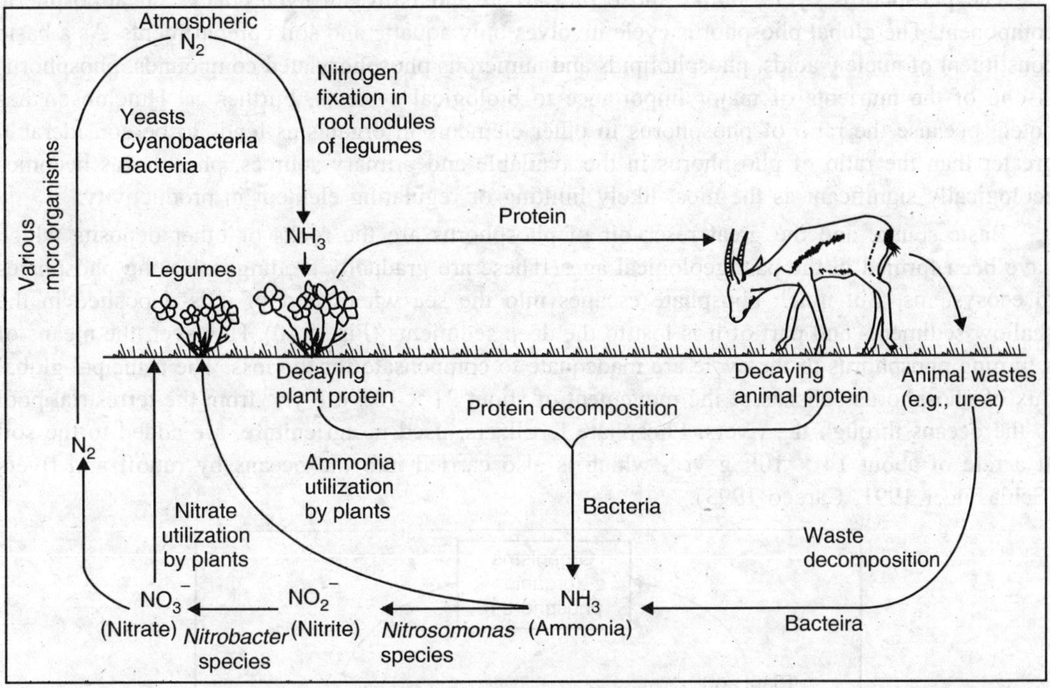

Fig. 5.9: The nitrogen cycle (after I.E. Alcamo, Fundamentals of Microbiology, 1994). The amount of nitrogen in the atmosphere is maintained by a balance between the processes that withdraw nitrogen from it (nitrogen fixation) and those which add nitrogen to it (denitrification).

5.9.4.5 Denitrification

It is a process in which the nitrate ion (NO_3^-) is reduced to nitrogen dioxide (NO_2), dinitrogen oxide (N_2O), nitrogen monoxide (NO) or nitrogen (N_2) by certain soil bacteria like *Pseudomonas denitrificans.* Thus, nitrogen is liberated into the atmosphere. Plants also lose small amounts of nitrogen to the atmosphere as gaseous ammonia, N_2O, NO_2 and NO especially when well fertilized with nitrogen (Wetselaar and Farquhar, 1980).

Thus, nitrogen cycle (Fig. 5.9) depends upon at least four different kinds of bacteria known as the decay causers, the nitrifiers, the denitrifiers, and the nitrogen-fixers and there is a regular circulation of nitrogen through the air, soil, plants and animals.

Effect of anthropogenic activities on the nitrogen cycle

1. The humans add large amount of NO (nitric oxide) by burning fossil fuel at high temperature. The NO in nature can convert to NO_2 (nitrogen dioxide gas) and HNO_3 (nitric acid) that can return to earth's surface as *acid rain.*
2. Addition of N_2O (nitrous oxide) by the action of *anaerobic bacteria* on livestock wastes and inorganic fertilizers used in soil, can warm the atmosphere and harm the ozone in the stratosphere. The NO_3 (nitrate) in inorganic fertilizers can leach into the soil and enrich the groundwater. This contaminated groundwater may prove to be harmful for infants and kids.
3. Addition of excess nitrogen in agricultural uses, resulting in surface runoff and also in discharges from sewage system, are also affecting aquatic ecosystems.
4. The harvest of nitrogen-rich crops, irrigation of crops and burning of forest and grasslands also removes nitrogen from top soil.

5.9.5 Phosphorus cycle

The phosphorus cycle, unlike those of Carbon and Nitrogen cycles lacks an atmospheric component. The global phosphorus cycle involves only aquatic and soil compartments. As a basic constituent of nucleic acids, phospholipids and numerous phosphorylated compounds, phosphorus is one of the nutrients of major importance to biological systems. Further, as Hutchinson has noted, because the ratio of phosphorus to other elements in organisms tends to be considerably greater than the ratio of phosphorus in the available and primary sources, phosphorus becomes ecologically significant as the most likely limiting or regulating element in productivity.

Basic source and the great reservoir of phosphorus are the rocks or other deposits which have been formed in the past geological ages. These are gradually eroding, releasing phosphates to ecosystems. But much phosphate escapes into the sea where part of it is deposited in the shallow sediments and part of it is lost to the deep sediments (Fig. 5.10). However, the means of returning phosphorus to the cycle are inadequate to compensate for the loss. The principal global flux of phosphorus consists of the movement of about 21×10^{12} g P yr.$^{-1}$ from the terrestrial pool to the oceans through the rivers. Phosphate fertilizers, used in agriculture, are added to the soil at a rate of about 14×10^{12} g yr^{-1}, which is also carried into the oceans by runoff and rivers (Schlasinger 1991, Caraco 1993).

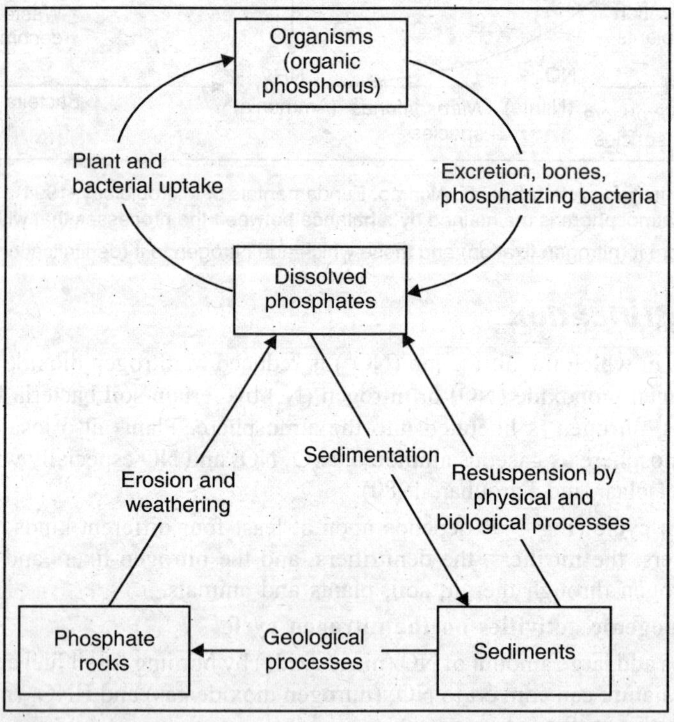

Fig. 5.10: A simplified diagram of the phosphorus cycle. (after E.P. Odum, 1993)

Much phosphate becomes lost to this central cycle by physical processes, such as sedimentation, which take it out of the reach of upwelling and major water circulation. Biological process, such as the formation of teeth and bone, and excretion also account for considerable losses from the major portion of cycle. The fish and marine birds are also important in phosphorus cycle. The latter have apparently played an important role in returning phosphorus to the cycle. In this context we can consider the tremendous deposits of *guano* (the manure of sea birds) along the western coast of South America. Although man harvests a lot of marine fish, according to Hutchsinson, only about 60,000 tons of elementary phosphorus per year is returned in this manner, compared with one to two million tons of phosphate rocks which are mined and most of which are washed away and lost.

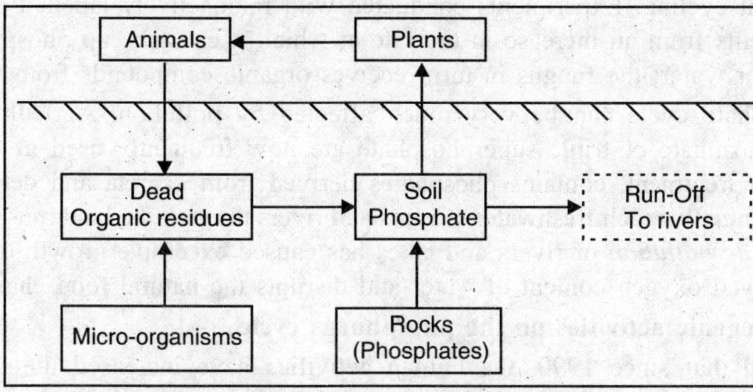

Fig. 5.11: The phosphorus cycle on land.

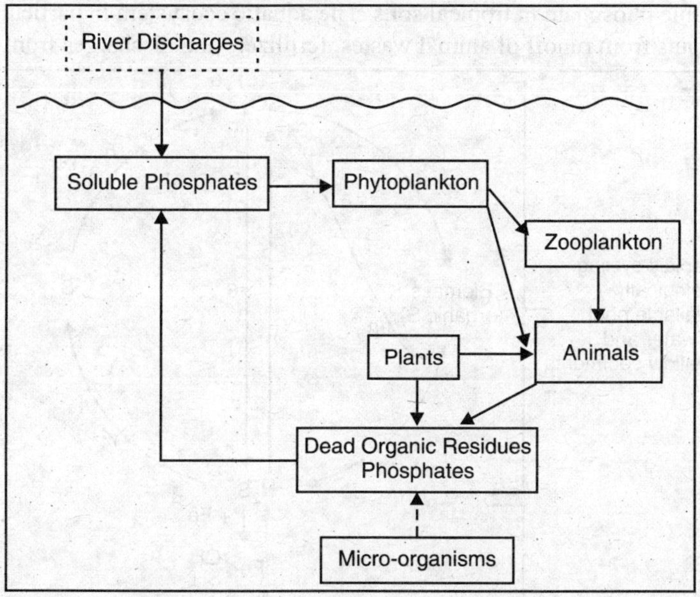

Fig. 5.12: The phosphorus cycle in water

For their nutrition plants absorb water dissolved inorganic phosphates from the soil either as dihydrogen phosphate (H_2PO_4) or as hydrogenphosphate (HPO_4) (Fig. 5.11) and convert these into organic phosphate. The latter is transferred to animal consumers and decomposers. Decomposers return phosphorus to the soil as the phosphate ion. Phosphate absorbed from the soil is returned to it in dead plant and animal organic residues, which are converted to humus by the action of soil microorganisms. Much of the phosphate in the soil is fixed or adsorbed on to soil particles but some is lost through leaching out into watercourses. In fresh water, the floating algae or phytoplankton rapidly absorb soluble inorganic phosphates and convert them into organophosphates (Fig. 5.12). Algae provide food for zooplankton which in turn are consumed by other animals. All plants and animals eventually die and in due time their organic remains or debris decay through the action of microorganisms, and phosphates are released into the water for recycling. In aquatic plants, phosphorus, limits plant growth.

Certain specialized fungi play an important role in phosphorus cycle. In many cases plant species have *mycorrhizae* (fungus roots), a mutualistic relationship between a non-pathogenic fungus and living plant root cells (Salisbury and Ross, 1985). These mycorrhizae play an important role in soil nutrient cycling. Experiments conducted with radioactively labelled elements show that the plant benefits from an increase in the rate at which it can take up phosphate and other limiting nutrients or water; the fungus in turn receives organic compounds from the plant.

Natural phosphate cycle can be very much affected by pollution. Agricultural fertilizers containing superphosphate or triple superphosphate are now frequently used in the fields; and sewage, even after treatment, contains phosphates derived from excreta and detergents. These phosphates can eventually reach freshwater streams and rivers through the land run-off and effluent discharge. *Phosphate pollution* of rivers and lakes has caused excessive growth of algae, which depletes the dissolved oxygen content of water and disrupts the natural food chains.

Effect of anthropogenic activities on the phosphorus cycle

It is estimated that since 1990, the human activities have increased the natural rate of phosphorus release into the environment by about 3.7 fold through mining large quantities of phosphate rocks to make inorganic fertilizers and detergents. The cutting of the tropical forests reduces the available phosphate in tropical soils. The aquatic ecosystem in particular get influenced by phosphorus inputs from runoff of animal wastes, fertilizers and discharges from sewage systems.

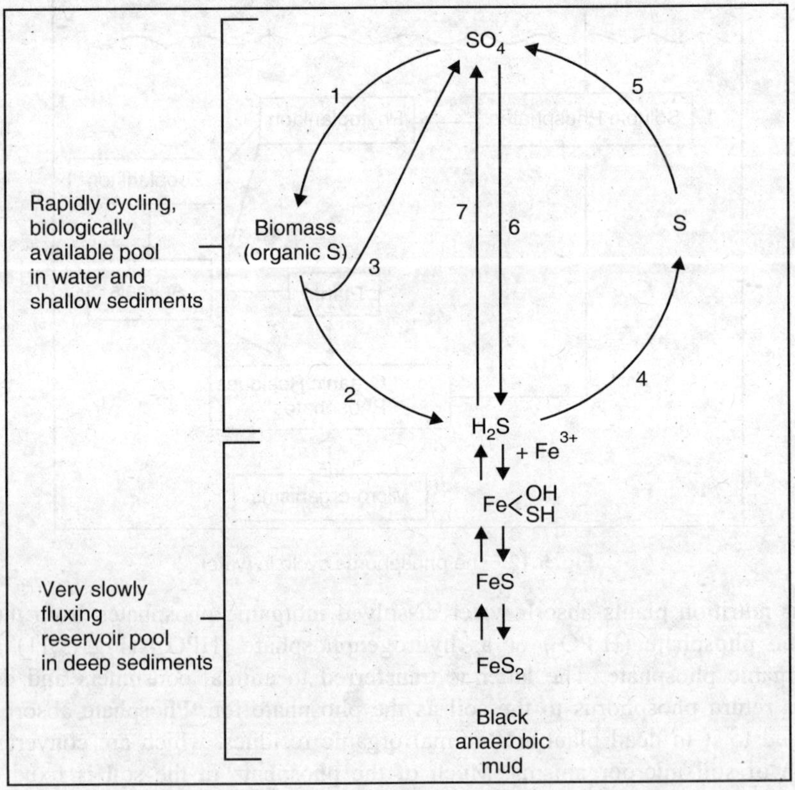

Fig. 5.13: The sulfur cycle. Step 1 is primary production by plants, specialized microorganisms, carry out steps 2–7 : 2, decomposition by heterotrophic microorganisms; 3, animal excretion; 4, 5, colorless, green, and purple sulfur bacteria; 6, anaerobic sulfur-reducing bacteria, *Desulfovibrio;* 7, aerobic sulfide-oxidizing bacteria, *Thiobacilli.* Step 8 represents the conversion of phosphorus from an unavailable to an available form when iron sulfides are formed, illustrating how cycling of one vital element can affect another. (after E.P. Odum 1993)

5.9.6 Sulphur Cycle

The global sulphur cycle is a good example to illustrate linkage between the air, water, and soil. Like the nitrogen cycle, it also illustrates the key role played by microorganisms. There are at least four major inputs of sulphur into the atmosphere from land— volcanic activity, soil dust, industrial activity and activity of sulphur bacteria which releases H_2S into the atmosphere. Sulphur also enters the atmosphre from the oceans in three ways— deep sea hydrothermal vents, biogenic gas (dimethyl sulfide) and sea water whipped by winds forms aerosols containing SO_4.

Plants and animals require a continuous supply of sulphur and its compounds in order to synthesize some amino acids such as cysteine and methionine, and proteins (Fig. 5.13). Only a few organisms gain their sulphur requirements in such organic forms as amino acids and cystein. Inorganic sulphate (SO_4) is the major source of biologically significant sulphur. Most of the biologically incorporated sulphur is mineralized by bacteria and fungi in ordinary decomposition by species of *Aspergillus* and *Neurospora,* among others. Under anaerobic condition, however, some may be reduced directly to sulphides, including hydrogen sulphide, by bacteria such as *Escherichia* and *Proteus* as shown in the following reaction: $SO_4 \rightarrow H_2S$ (anaerobic sulphate reduction).

Some organic sulphur enters the atmosphere as sulphur dioxide (SO_2) through incomplete combustion of fossil fuels, especially coal. This is one of the major sources of air pollution today. According to Schindler (1988), anthropogenic inputs of sulphur in industrial areas may contribute 90% of the sulphur input into the air. Most of the input is in the form of SO_2 which, in the presence of water, may be converted to HSO_3 or SO_4 by the following reactions:

$$SO_2 \rightarrow SO_3 \rightarrow H_2SO_4 \rightleftharpoons H^+ + SO_4^{2-}$$

$$SO_2 \rightarrow H^+ + HSO_3$$

The deposition of SO_4^{2-}, HSO_3 and H^+ ions lowers the pH of precipatation, causing acid rain problem. Inorganic sulphur as sulphate may precipitate out, but since it is relatively soluble, it serves as a source of elemental sulphur in many ecosystems. Sulphate is also reduced under anaerobic conditions to elemental sulphur or to sulphides including hydrogen sulphide, by such heterotrophic bacteria as species of *Desulfovibrio* and *Desulfomonas* ($SO_4^{2-} \rightarrow S_2$). Colourless sulphur bacteria such as species of *Beggiatoa* (chemosynthetic bacteria) often abundant in sulphur springs, oxidize hydrogen sulphide to elemental sulphur, and species of *Thiobacillus* oxidize it to sulphate. Other species of *Thiobacillus* oxidize sulphide to sulphur, and still others oxidise sulphur to sulphate as shown in the following reactions:

$$H_2S \rightarrow S \rightarrow SO_4$$

$$H_2S \rightarrow SO_4$$

The sedimentary aspect of the sulphur cycle involves the precipitation of sulphur in the presence of iron under anaerobic conditions. Ferrous sulphide is insoluble in neutral or alkaline water, and consequently sulphur has the potential for being bound up under these conditions to the limits of the amount of iron present. Because of the thermodynamics of this ferrous sulphide (FeS) system, other nutrients important to biological systems, can also get trapped for varying periods of time. Among these are copper, cadmium, zinc and cobalt. When iron sulphides are formed in the sediments, phosphorus is converted from insoluble to soluble form (see "phosphorus release" arrow in Fig. 5.10) and thus becomes available to organisms. This is an excellent example of how one cycle regulates another.

Effects of anthropogenic activities on the sulphur cycle

The burning of coal, oil, to produce electricity and generate power as also the refining of oil, and producing some metals from ores releases SO_2 (sulphur dioxide) from natural sources and human activities is converted to SO_3 and tiny droplets of H_2SO_4. It also reacts with NH_3 to produce sulphur salts. Due to increased production of SO_2, it along with other air pollutants, can harm vegetation and aquatic life.

MODEL QUESTIONS

1. Define the term 'ecosystem'. Justify the statement that in a balanced condition ecosystem functioning is self-regulating and self-sustaining.
2. What is an ecosystem ? Describe the biotic and abiotic components of an ecosystem.
3. Define the concept of trophic levels. How the organisms of different trophic levels contribute to ecosystem functioning.
4. Define 'trophic efficiency'. How will you explain differences in trophic effeciency of producers and consumers of different levels ?
5. What is an ecological pyramid ? Describe the types of pyramids studied by you.
6. What is a biogeochemical cycle ? How water cycle is different from cycles of other elements?
7. Describe the role of microorganisms in nitrogen cycle. Why eukaryotes are unable to fix nitrogen ? Describe the effect of anthropogenic activities on the nitrogen cycle.
8. Describe the phosphorus cycle. How phosphorus cycle is different from nitrogen cycle and how it is effected by pollution ?
9. Write notes on the following:
   (*a*) Ecosystem ecology, (*b*) Role of decomposers in an ecosystem,
   (*c*) Trophic levels, (*d*) Trophic efficiency,
   (*e*) Pyramid of energy, (*f*) Carbon cycle,
   (*g*) Sulphur cycle

MULTIPLE CHOICE QUESTIONS

10. Who proposed the term ecosystem?
    (A) F.Clements (B) A.G.Tansley
    (C) C.Elton (D) E.Haeckel
11. In a forest ecosystem, which of the following are decomposers?
    (A) Bacteria and fungi (B) Soil algae and bacteria
    (C) Small green plants (D) Bacteria only
12. The pyramid of biomass is inverted in:
    (A) All ecosystems (B) Forest ecosystem
    (C) Grassland ecosystem (D) Aquatic ecosystem
13. Net primary productivity (NPP) is equal to:
    (A) GPP (Gross primary productivity)
    (B) GPP – R (Respiration)
    (C) GPP + R (D) GPP + AP (Agricultural productivity)
14. 'Nif' genes are present in:
    (A) Certain green algae (B) Eukaryotes
    (C) Prokaryotes (D) Certain grasses
15. In nitrogen cycle, which bacteria are responsible for nitrification?
    (A) *Rhizobium* (B) *Clostridium*
    (C) *Nitrosomonas* (D) *Nitrosomonas* and *Nitrobacter*

16. For their nutrition plants absorb phosphates from the soil as:
    (A) Organic phosphate (B) Dihydrogen phosphate
    (C) Dihydrogen phosphate and hydrogen phosphate
    (D) All the above
17. Bacteria convert inorganic sulphate into:
    (A) H_2S (B) H_2SO_4
    (C) SO_2 (D) None of the above
18. Which of the following limit the productivity of an ecosystem?
    (A) Solar radiation (B) Carbon dioxide
    (C) Temperature, water and soil nutrients
    (D) All of the above depending on other conditions.
19. Microbivore are minute organisms that feed on:
    (A) All organic matter (B) Bacteria and fungi
    (C) Fungi only (D) Decaying organic matter
20. Which of the following is an example of wetland ecosystem?
    (A) Mangrove swamps and marshes
    (B) any Lake
    (C) any Estuary (D) Marshes only
21. The law of the sea convention came into effect from:
    (A) 1982 (B) 1985 (C) 1980 (D) 1990
22. Largest ecosystem of the earth is:
    (A) Lake (B) Forest (C) crop (D) Sea
23. The decomposers of an ecosystem are also known as.
    (A) primary producers (B) saprobes
    (C) primary consumers (D) none of them
24. For their nutrition plants absorb phosphate as:
    (A) dissolved inorganic phosphates (B) rock phosphates
    (B) organic phosphates (D) none of them
25. Top carnivores belong to:
    (A) Second trophic level (B) First trophic level
    (C) Fourth trophic level (D) Third trophic level

6 CHAPTER

Biosphere Degradation and Environmental Problems

6.1 INTRODUCTION

Biosphere means the sphere that supports the living forms. The term "biosphere" was first used by the Austrian geologist Eduard Suess in 1857. The concept was further developed by the Russian mineralogist Vernadsky in 1926. It has been defined by Hutchinson (1970) as that part of the earth in which life exists (the envelope of life). The three subdivisions of the biosphere are: atmosphere, lithosphere and hydrosphere.

(i) Atmosphere: The atmosphere is the gaseous mantle, which envelops the hydrosphere and the lithosphere. There are no characteristic or permanent inhabitants of the atmosphere, although the air is traversed by many kinds of animals and plant propagules.

(ii) Lithosphere: The lithosphere comprises the solid components, *i.e.*, the rocky substances of the continents. The soil component of the lithosphere has been called *pedosphere* by Odum.

(iii) Hydrosphere: The hydrosphere means the sum total of all liquid components, *i.e.*, the water in oceans, lakes, rivers, and on land. Thus, hydrosphere includes two major biocycles, the freshwater and marine.

Atmosphere	Oxygen
	Nitrogen, Carbon dioxide
	Ozone & other gases
	Water vapour
Lithosphere	Soil & minerals
	Land for human use
	Water (underground) & in water bodies
	Food & medicines
Hydrosphere	Water for aquatic & terrestrial orgainsms
	Minerals & other items
	Food & medicines

Fig. 6.1: The material contributions of each of the three subdivisions of biosphere towards the maintenance of life

The inorganic metabolites required by living organisms are contributed by each of these subdivisions of the biosphere. The hydrosphere supplies water; the lithosphere supplies all necessary minerals; and the atmosphere supplies oxygen, nitrogen and carbon dioxide. Together, these inorganic materials provide all the chemical elements needed in the building and maintenance of living organic matter (Fig. 6.1). In addition to being everlasting sources of supply, the three subdivisions of the environment also influence metabolic activities of the living organisms in various other specific ways.

6.2 ATMOSPHERE

The atmosphere is the gaseous realm. It consists of a mixture of gases that completely surround the earth. Of much greater significance to metabolism, however, are the gaseous cycles of the atmosphere. The air consists mainly of oxygen (about 20 per cent), carbon dioxide (about 0.03 per cent), nitrogen (about 79 per cent), water vapour, and minute traces of inert gases. Except the inert

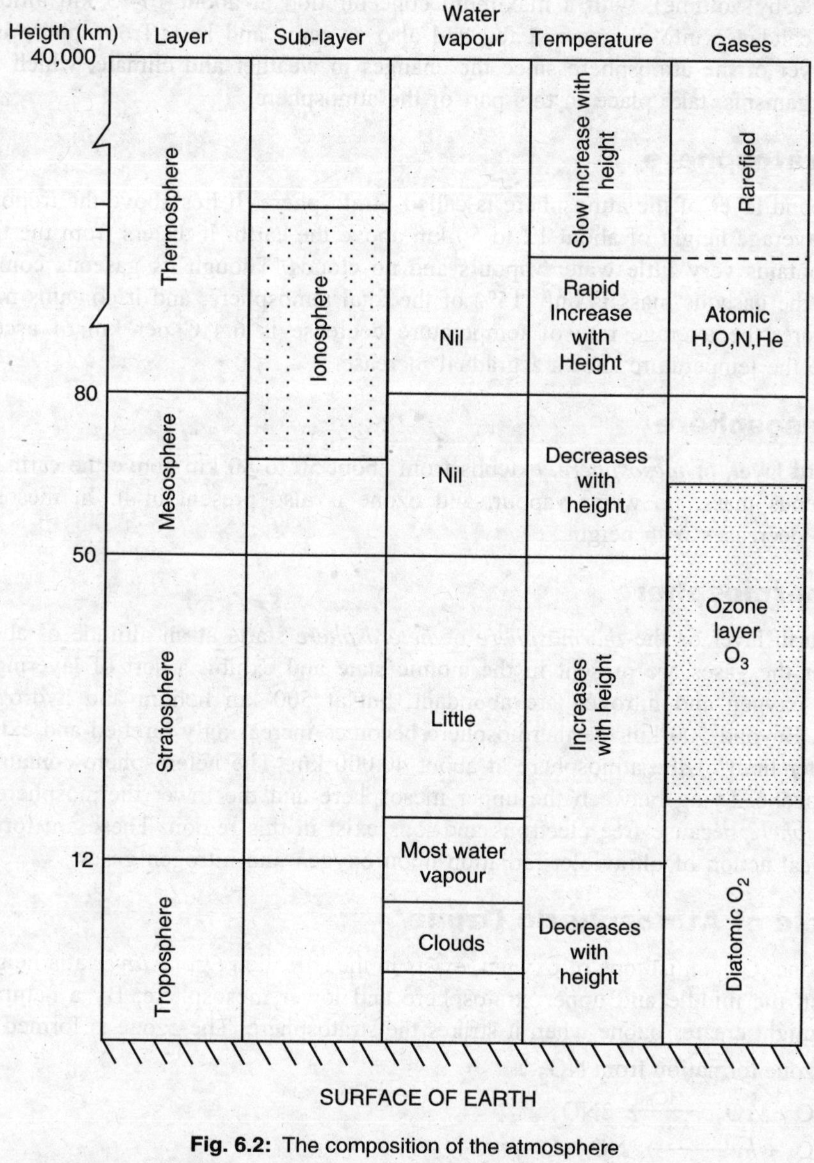

Fig. 6.2: The composition of the atmosphere

gases, all these components of air serve as metabolites and each circulates through a cycle in which the organisms play an important role. As all gases are dissolved in natural waters, the hydrosphere maintains an equilibrium with the atmosphere. The metabolic role played by important gases of the atmosphere and their cycles are discussed under biogeochemical cycles. Atmosphere is usually divided into four layers called the troposphere, stratosphere, mesosphere, and the thermosphere (Fig. 6.2).

6.2.1 Troposphere

The *troposphere* or lower atmosphere is in contact with the earth's surface, and is about 8 km high at the poles and 16 km at the equator. It consists of, in the unpolluted state, nitrogen (78.08% by volume of dry air), oxygen (20.95% by volume of dry air) and traces of other gases such as Argon (0.93%), Carbon dioxide (0.0325%), Neon (0.0018%), Helium (0.0005%), Krypton (0.0001%) and Ozone (0.000002%). The troposphere also contains varying quantities of water vapour (1-4% by volume), with a maximum concentration at about 10-15 km altitude. Water vapour can condense into clouds and fog and also as snow and hail. Troposphere is the most important layer of the atmosphere since the changes in weather and climate, which affect man and other organisms, take place in this part of the atmosphere.

6.2.2 Stratosphere

The second layer of the atmosphere is called Stratosphere. It lies above the troposphere and extends an average height of about 12 to 50 km above the earth. It differs from the troposphere in that it contains very little water vapours and no clouds. Though its gaseous composition is similar, but the gaseous mass is only 15% of the total atmosphere, and it contains more ozone. In troposphere, the average rate of temperature decrease is 6.4°C per km of ascent, but in Stratosphere the temperature shows a gradual increase.

6.2.3 Mesosphere

The third layer, or *mesosphere*, extends from about 50 to 80 km above the earth. It consists of less gaseous mass, no water vapour, and ozone is also present in it. In mesosphere, the temperature decreases with height.

6.2.4 Thermosphere

The fourth layer or the *thermosphere* or *heterosphere* starts at an altitude of about 80 km. In this layer the gases are present in the atomic state and exhibit a sort of layering. From 80 to 120 km oxygen and nitrogen are abundant, but at 500 km helium and hydrogen are the main gases. Beyond 500 km the thermosphere becomes increasingly rarefied and extends to the extreme outer limit of the atmosphere at about 40,000 km. The heterosphere contains no water vapour or ozone. Lying between the upper mesosphere and the lower thermosphere is the so-called *ionosphere* because free electrons and ions exist in this region. These are formed by the photochemical action of ultraviolet radiation upon oxygen and nitrogen gases.

6.2.5 Role of Atmospheric Ozone

The ozone (O_3), an isotope of oxygen, exists in the so-called *Ozone layer* at a height of about 15-60 km in the middle and upper stratosphere and lower mesosphere. By a natural chemical process, sunlight creates ozone when it strikes the stratosphere. The ozone is formed as follows:

1. Ozone formation from NO_2

$$NO + \tfrac{1}{2}O_2 \xrightarrow{HCs} 2NO_2$$

$$NO_2 + h\nu \longrightarrow NO + O$$

$$O + O_2 \longrightarrow O_3 \text{ (Ozone)}$$

2. Ozone formation from CO

$$CO + 2O_2 + h\nu \longrightarrow CO_2 + O_3$$

3. Ozone formation from CH_4

$$CH_4 + 4O_2 + 2\ h\nu \longrightarrow HCHO + H_2O + 2O_3$$

$$HCHO + h\nu \longrightarrow H + HCO$$

$$HCO + h\nu \longrightarrow H + CO$$

$$CO + 2O_2 + h\nu \longrightarrow CO_2 + O_3$$

4. Ozone from non-methane hydrocarbons

$$RH + 4O_2 + 2\ h\nu \longrightarrow RCHO + H_2O + 2O_3$$

The concentration of O_3 peaks during the late morning and early afternoon hours. Background O_3 concentrations have more than double in the last countury (Meehl *et.al*, 2007).

In should be noted that depletion of stratospheric ozone is a problem while excess of ozone in troposphere is also a problem. Thus, troposheric ozone is regarded as a pollutant. In fact, tropospheric ozone is the most important regional atmospheric pollutant causing risk to food production across the globe due to its phytotoxicity and prevalence over agricultural areas.

The ozone layer absorbs or blocks the *shortwave ultraviolet radiation* (Fig 6.3) that passes through it. This property gives it an important function in protecting the biosphere. It the ozone layer were not present, then the ionizing ultraviolet rays would destroy most or all the life on earth. Increased ultraviolet radiation can lead to greater incidence of skin cancer, cataracts, immune deficiencies, decreased crop yields and reduced populations of phytoplankton, zooplankton and certain fish larvae that are important constituents of aquatic food chains. Increased ultraviolet radiation can also contribute to smog, which is one of the important air pollutants. The stratospheric ozone protects oxygen at lower altitudes from being broken up by UV rays and keeps most of these harmful rays from penetrating the earth's surface.

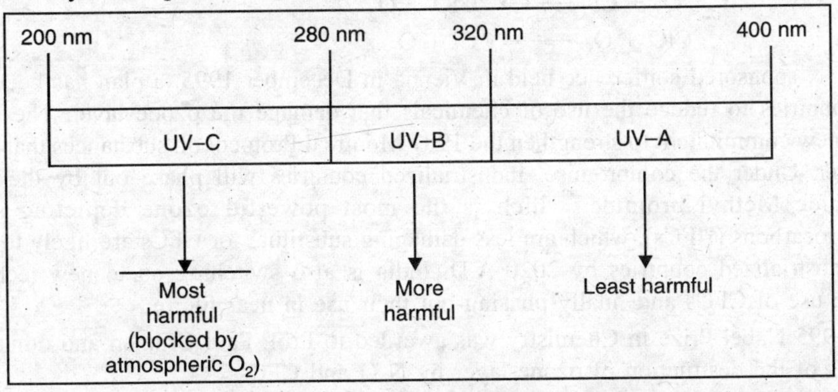

Fig. 6.3: Showing that wavelengths of ultraviolet radiation fall between 200 and 400 nanometers (nm). The UV-A (320-400 nm) is the least harmful to biological systems. The UV-B (280-320 nm) is significantly more harmful but most of its is blocked by the ozone in the stratosphere and only about 0.25% of the sun's light reaching earth's surface is UV-B. UV-B causes direct damage to DNA and proteins and is responsible for most of the ill effects of UV light on organisms. The UV-C is most harmful to organisms but it is absorbed by the oxygen in the atmosphere and thus never reaches earth's surface.

6.2.5.1 Ozone depletion

Reduction in ozone levels over the Antarctic have already been reported and the so-called Antarctic hole in ozone layer, which grew to match the area of the United States in 1986, has since expanded to cover the whole Antarctic. Recently Prof. Ernst Augstein (1987) reported that a similar hole is developing over the north pole. The Arctic hole is at present only about one-fifth the size of the growing Antarctic hole. The conclusion has been reached by comparison on the basis of many years of measurements in spring time when the ozone layer over the poles reaches its minimum. "This spring time the minimum has fallen 50 per cent in the Antarctic and 2 to 10 per cent in the Arctic, depending on the data," Prof. Augstein (1987) said. Ozone levels sink in winter mainly because of the lack of sunshine.

According to World Meteorological Organization (WMO), the hole in the ozone layer over the Antarctic grew at a record rate in October 1995, reaching 20 million km^2, or twice the size of Europe. Since the 1960s, the protective ozone layer over the Antarctic has declined by more than 65 per cent. If the present trend of depletion in ozone layer continues, the earth's inhabitants would be exposed to more ultraviolet rays, which can cause cancer, cataracts and other diseases. However, scientists still debate as to why the ozone layer is depleting, but according to many of them chlorine molecules from *chlorofluorocarbons* (CFCs) might be responsible.

CFCs (chlorofluorocarbons *e.g.*, carbon tetrachloride CCl_3, dichlorodifluoromethane CCl_2F_2) are compounds that consist of chlorine, fluorine and carbon. These gases are used as coolants for refrigerators and air conditioners, propellants for aerosol sprays and agents for producing plastic foam. As CFCs do not degrade easily in the lower atmosphere, they rise into the stratosphere where they are broken down by UV-light. In the upper atmosphere, UV-light breaks off a chlorine atom from a CFC molecule. The chlorine attacks an ozone molecule, breaking it apart. An ordinary oxygen molecule and a molecule of chlorine monoxide (ClO) are formed. A free oxygen atom breaks up the chlorine monoxide and the chlorine is free to repeat the process. Chlorine acts as a catalyst and is unchanged in the process. Consequently, each chlorine atom can destroy as many as 10,000 ozone molecules before it is returned to the troposphere. Farman (1987) has described the following reactions to illustrate how ozone is destroyed.

$$CFCl_3 + \text{UV-light} \longrightarrow CFCl_2 + Cl$$
$$O_3 + Cl \longrightarrow ClO + O_2$$
$$ClO + O \longrightarrow Cl + O_2$$

In a UN sponsored conference held in Vienna in December 1995, a plan has been approved by 110 countries to reduce the use of chemicals that damage the ozone layer. The conference agreed to new commitment to strengthen the 1987 Montreal Protocol on substances that deplete the ozone layer. Under the compromise, industrialized countries will phase out by the year 2010 the pesticide Methyl bromide, which is the most powerful ozone depleting substance. Hydrofluorocarbons (HFCs), which are less damaging substitute for CFCs are likely to be phased out by industrialized countries by 2020 A.D. India is also switching on to new technology to reduce the use of CFCs and finally phasing out their use in near future.

The 1995 Nabel Prize in Chemistry was awarded to Prof. Paul Crutzon and others for their discoveries of the destruction of ozone layer by N_2O and CFCs.

6.2.6 Greenhouse Effect

The greenhouse effect, may be defined as warming of the earth's climate owing to the increased concentrations of CO_2 and some other gases in the atmosphere (Fig. 6.4). The following greenhouse gases (GHG) are considered important: (*a*) Carbon dioxide (*b*) Nitrous oxide, (*c*) Methane, (*d*) CFCs, and (*e*) Ozone (*f*) CO and (*g*) SO_2. These gases act like the glass in a botanical greenhouse, trapping the reradiated heat near the earth's surface and warming the planet. These gases along with water vapour and clouds absorb the infrared radiation, trapping heat near the earth's surface. Important characteristics of some major greenhouse gases are given in Table 6.1.

Table 6.1: Important Characteristics of some major greenhouse (climate-changing) gases, from several sources).

S. No.	*Gas*		*Sources*		*Sinks*	*Importance for Climate*
1.	CO_2	(*i*) (*ii*)	Burning of fossil fuel Deforestation (Land-use change)	(*i*) (*ii*)	Ocean uptake Plants & photosynthesis	Absorbs infrared radiation; affects stratospheric O_3
2.	CH_4	(*i*) (*ii*) (*iii*) (*iv*)	Biomass burning Enteric fermentation Rice paddies Fertilizers	(*i*) (*ii*)	Reactions with OH Micro-organisms uptake by soils	Absorbs infrared radiations; affects tropospheric O_3 & OH; affects stratospheric O_3 and H_2O; produces CO_2
3.	N_2O	(*i*) (*ii*) (*iii*)	Biomass burning Fossil-fuel combustion Stratospheric photolysis and reaction with stratospheric O_3	(*i*)	Removal by soils, radiation effects	Absorbs infrared
4.	O_3		Photochemical reactions involving Cl and O		Catalyses chemical reactions involving NOx, ClO_x and HO_x	Absorbs ultraviolet and infrared radiation
5.	CO	(*i*) (*ii*)	Plant emissions Transport and Industries	(*i*) (*ii*)	Soil uptake Reactions with OH	Affects stratospheric O_3 and OH; produces CO_2
6.	CFCs		Industries		Broken down by UV light in stratosphere	Absorbs infrared radiation; affects stratospheric O_3
7.	SO_2	(*i*) (*ii*)	Fossil fuel and biomss burning Industrial processes and volcanic eruptions	(*i*) (*ii*)	Dry and wet deposition Reactions with OH	Forms aerosols which scatter solar radiation

6.2.6.1 Carbon dioxide(CO_2)

The CO_2 is considered as the most dominant factor responsible for the Gh effect. The troposphere or lower atmosphere contains only 0.0325% CO_2 (by volume) and its amount is controlled by a cycle, often called the carbon cycle. There are four major 'pools' or reservoirs of carbon: fossil fuels, the atmosphere, the biosphere, and the oceans.

Carbon dioxide concentration in the atmosphere has increased from 290 ppm (1860) to 350 ppm (today) and is expected to be 700 ppm in 2100. It was estimated in 1976 that the world consumption of about 5000 million tons of fossil fuels per year was contributing the equivalent of 2.3 ppm of CO_2 to the atmosphere. The net annual increase of CO_2 in the air is 0.77 ppm, which indicates that 1.6 ppm of fuel generated CO_2 is being absorbed elsewhere perhaps into the

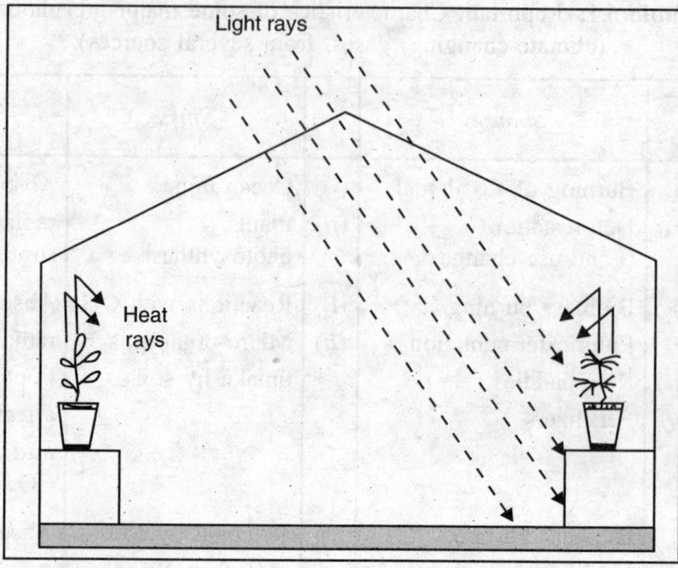

Fig. 6.4: The greenhouse effect. Light rays (short wavelength electromagnetic radiation) from the sun passes through the glass into the greenhouse. The greenhouse objects are warmed up and give out heat rays (long wavelength electromagnetic radiation or infrared radiation) which cannot get through the glass and so that heat is held inside the greenhouse. With more amount of carbon dioxide in the earth's atmosphere the same kind of thing may happen. Light rays from the sun heat up the earth's surface. However, the heat rays given out by the earth are prevented from escaping by carbon dioxide in the atmosphere. The carbon dioxide holds the heat in, just as the glass does in the greenhouse.

oceans and the biosphere. A continued increase in world fossil fuel combustion in the future can be expected to produce a further increase of CO_2 in the troposphere (about 340 ppm in 1981), and this has been estimated at 20% by the year 2000. Thus it is widely accepted that during the last couple of decades the fossil fuels may have contributed more CO_2 to the atmosphere than the terrestrial biosphere and that the oceans may not be able to absorb CO_2 that is building up in the atmosphere. Clearance of forests is another factor for the increase of CO_2.

6.2.6.2 Nitrous oxide (N_2O)

It is emitted to the atmosphere from biomass burning and artificial fertilisers, rising by 0.2% per year. Since it absorbs 259 times more heat than CO_2, it contributes to global warming by increasing the greenhouse effect. It also destroys tropospheric ozone, which protects the earth from UV radiation.

6.2.6.3 Methane (CH_4)

It is produced in a number of ways including the action of anaerobic bacteria on vegetation, decomposition of organic matter, incomplete combustion of vegetation, natural gas pipeline leaks and is rising by 2% per year. It absorbs 20-25 times more heat than CO_2.

6.2.6.4 CFCs

There are large scale industrial uses of CFCs, which include aerosol propellants (CFC-11), refrigerants (CFC-12) and a wide range of solvents. The concentration of CFCs is rising nearly 5% per year. Like nitrous oxides, CFCs trap heat 20,000 times more efficiently than CO_2 and also destroy ozone layer, thus posing a serious twofold environemental problem. The effect of CFCs is so powerful that by the year 2050 they alone could contribute more than CO_2 to global warming (Dickson and Cicerone, 1986).

6.2.6.5 Ozone

It comes mostly from hydrocarbons (HC) and nitrogen oxides (NO_x). It causes irritation to eyes and respiratory organs. It also decreases the resistance power to infections and aggravates illness.

6.2.6.6 Infra-red absorption

The CO_2 is a greenhouse gas, that is, it remains transparent to the incoming shortwave solar radiation and absorbs the outgoing longwave radiation (infra-red or IR radiation). This absorptive property of CO_2 prevents longwave IR, prticularly the earth's thermal radiation, being emitted to the outer atmospheric layers and lost into outer space (Fig. 6.4). The warming effect of this upon the atmospheric temperature has been linked to a "blanket" or 'greenhouse effect'. The increasing concentration of CO_2 in the atmosphere results in a 'heat trap' around the earth raising its temperature.

Thus, Gh effect is a phenomenon based on well established principles of infra-red (IR) absorption characteristics of gases. Besides CO_2, a number of trace gases exhibit IR-absorption properties and contribute to the Gh effect. The dominant role of CO_2 in producing the net Gh effect has been important due to its role in the evolution of the earth's climate in the past billion years, and as a major pollutant released into the atmosphere as a result of man's activities.

6.2.6.7 Impact of global warming

(*a*) *Climate change:* It is believed that increased levels of greenhouse gases that cause global warming, have affected the global climate already and these effects will increase in future. According to IPCC(1996), the world climate has warmed from 0.3 to 0.6°C during the last century. Complex computer models of global climate predict that temperatures will increase further by 1°C to 3.5°C over the next century as a result of increased concentrations of carbon dioxide and other greenhouse gases. The increase in temperature will be greatest at high latitudes and over large continents (Myneni *et al.* 1997). However, some scientists also predict an increase in extreme weather events such as flooding, regional drought and hurricanes associated with this warming (Karl *et al.* 1997). It seems likely that many species will be unable to adjust quickly to global warming and associated climate change. As a result, biological communities may suffer profoundly. More than 10% of plant species in many temperate parts will not be able to survive the new climatic conditions, they must migrate northward or die. This change has already been observed with alpine plants found growing higher on mountains and migratory birds spending longer times on their summer breeding grounds. However, the effects of global climate change on rainfall and temperature are expected to be less drastic in the tropics than in the temperate zones. But even small changes in the amount and timing of rainfall will effect species composition and plant reproduction cycles. Changes in temperature and global climate can be expected to influence the biogeochemical cycles, which have already been perturbed by anthropogenic disturbances.

(*b*) *Climate change and plant communities*: Climatic changes as a result of global warming will naturally effect the biotic communities on this earth. Some plant species may utilize the increased CO_2 concentrations and high temperature to increase their growth rates, but less adaptable species will decrease in abundance. Such unpredictable fluctuations in plant communities and associated herbivorous insect species could lead to the extinction of many rare species and great population increases in some other species. As a result, the global climate change may restructure biological communities and change the distributional ranges of many animal and plant species. Some species may be in danger of going extinct in the wild and therefore new conservation strategies including captive breeding, will have to be adopted.

(*c*) *Melting of Glaciers:* Glacial melting due to global warming leads to increased accumulation of ground water in mountains. Once the capacity of a mountain exceeds its threshold, it can burst, causing flash floods. (Ritesh Arya, 2009). Mountain flash flooding is the threat in the near future.

A study report (2012) from the NRC examined how changes to glaciers in the Hindu Kush–Himalyan region, which covers eight countries across Asia, could affect the areas river systems, water supplies and the population. The mountains in the region form the headwaters of several major river systems including the Ganga, Mekong Yantze and Yellow rivers which serve as source of drinking water and irrigation to about 1.5 billion people. The study further suggested that desert dust and black carbon, a component of soot, could contribute to rapid atmospheric warming, accelerated snowpack melting and glacier retreat.

About 90% of glaciers in Tibet are shrinking (Yao Tandong, 2013) because of black carbon pollution transferred from south Asia to the ecologically fragile Tibetan plateau. The report says that influenced by global warming the alpine glaciers of the Third Pole have seen drastic changes in recent years. The thinning and shrinkage of these glaciers pose potential hazards to people.

The JNU study on Chhota Shigri Glaciers concluded that during 2002-2010, the glacier depletion has been too such. However, glaciologists say that long term mass balance studies can give a clear picture of glacier retreat and the impact of climate change.

(*d*) *Effect on sea levels*: Rising temperatures will cause glaciers to melt and the polar ice caps to shrink. As a result of this, sea levels may rise by 0.2 to 1.5 m and flood low lying coastal areas and their biotic communities. There is evidence that this process has already started. Sea levels have already risen by 10 to 25 cm over the last 100 years, possibly due to rising global temperatures (IPCC, 1996). If the trend continues, many low lying areas may be submerged in near future. It is possible that rising sea levels may significantly change or destroy 20% -80% of the coastal wetlands. In tropical areas, mangroves will be adversely affected as seawater will be too deep in existing mangrove areas to allow the seedlings to develop. Rising sea levels are detrimental to coral reef species, which grow at a precise depth with optimum temperature and water movement. It is possible that slow growing coral reefs will be unable to keep pace with the rise in sea level and will be gradually submerged and die and only fast growing coral reef species will be able to survive. This threat to coral reefs may be further compounded by increasing seawater temperatures. Abonormally high water temperatures in the Pacific Ocean during 1982 and 1983 caused the death of symbiotic algae that live inside the coral. Subsequently, the "bleached" coral suffered a massive dieback of 70%-95% coral cover of the area to depths of 18m (Brown and Ogden, 1993).

(*e*) *Reduction of biodiversity*: As mentioned above increased temperatures, inundation of some coastal biological communities and changes in the pattern of distribution of many species over a long period of time are likely to cause reduction in *biodiversity* in aquatic and terrestrial ecosystems.

It is estimated that the Gh effect is expected to threaten 400 species of mammals, 460 species of reptiles 650 species of birds, 660 species of fresh water fish and many species of inverebrates and plants. An increase in extinction rate of some of these species and changes in their migration patterns are predicted. The global climate change is also going to affect crop fields, forests and coastal wetlands which are very rich in biodiversity. It is estimated that 11046 species of plants and animals are facing high risk of extinction today.

The project "*Climate change in our World*" is the result of cooperation between Google, Britain's Environment Ministry and the country's met office. The project shows people the reality of climate change, using estimates of both the change in the average temperature and the impact it will have on their lives all over the world.

(*f*) *Effect on Agriculture*: The global climate change may have important effects on agriculture (Rosenweig and Parry, 1994). However, the effects of this change will vary for C_3 (*e.g.*, wheat, rice, beans) and C_4 (*e.g.*, maize, millet, sugercane) plants. As temperatures increase with rising levels of CO_2, some crop plants may no longer be grown in certain regions. According to Ricklefs and Miller (2000), under the most common models of global climate change, global temperature increases will have negative effects on both C_3 and C_4 plants unless the higher levels of CO_2 in the atmosphere increase plant growth.

(*g*) *Effect on Arctic ecosystems*: Global climate change will have profound effects on arctic ecosystems. Studies on the response of arctic Tundra to elevated CO_2 indicated that the Tundra is more sensitive to global climate change than most other ecosystems on earth. According to Shaver *et. al* (1992), warmer temperatures may increase primary production, thereby increasing carbon input and soil respiration. The extent to which production may be increased is constrained by the availability of nitrogen.

(*h*) *Overall effect*: The overall effect of global warming on world climate has many dimensions, some of which are discussed above. The natural greenhouse maintains the earth's temperatures within the limits for physiological functions. But studies suggest that even a moderate increase in the average global temperature could result in significant changes in biotic communities including reduction in biodiversity both in terrestiral and aquatic ecosystems. There is a scientific consensus that anthropogenic activities affect global climate. As already indicated some major climate changing (greenhouse) gases are CO_2, CH_4, N_2O, O_3, CO, CFCs and SO_2; of them, nearly 6% rise has been recorded in the amount of CO_2. About 512 million metric tons increase has been recorded for the year 2010 making 9.1 billion metric tons from the 2009 data of 8.6 billion metric tons. Is means the *carbon footprint* is increasing. Carbon footprint is a measure of impact that human activities have on the environment and is calculated in proportion to the greenhouse gases produced and measured in units of carbon dioxide. Some of the global climate change indicators are:

(*a*) IPCC, Earthwatch, Greenpeace and IUCN have all mentioned that average global temperature is projected to increase 1.4 to 5.8°C over the period 1990-2100.

(*b*) Increase in CO_2 levels.

(*c*) Projected sea-level rise and submergence of certain coastal areas.

(*d*) Retreat of glaciers.

(*e*) Changes in hydrological cycles, rainfall patterns and flow regimes of important rivers.

(*f*) Water shortages in some parts of countries like South Africa, Brazil, China and India, etc. Upto 30% of plant and animal species could become extinct if the global temperature increase exceeds 1.5–2.5°C. Thus, among other factors, climate change has been responsible for all major extinctions reported so for. Nearly 68–77% of Indian forests are likely to undergo a sea change to another forest type by 2085. Climate change can effect some of our precious coldwater fish and may change their geographical distribution range (Singh, 2009). In respect of climate change the best thing is to believe in the words of *Andre Gide*, Nobel Prize Winning French Novelist who said "*Believe those who are seeking the truth; doubt those who find it*".

6.3 AIR POLLUTION

Air pollution is one of the major environmental problems throughout the world. According to Environment Perform Index 2014, Delhi ranked as the most polluted city and India's air quality was among 5 worst nations. The Environment (Protection) Act, 1986 of India has provisions to regulate environmental pollution (*e.g.*, the standards of quality of air, water and soil for various areas and purposes) and the maximum allowable limits of concentrations of various environmental pollutants for different areas are also laid down. These National Ambient Air Quality Standards (Tables 6.1 & 6.2) are prescribed for three distinct areas, *viz.* (*a*) industrial, (*b*) residential, rural and other areas, and (*c*) sensitive areas. The open air around us is called *ambient air*. Ambient air pollution is also referred to as *tropospheric air pollution*. Since the troposphere begins at the earth's surface, tropospheric air pollution is also called *ground level air pollution* or lower atmospheric pollution. Since humans are most easily affected by the quality of air in the lowest region of the earth's atmosphere, tropospheric air pollution has drawn attention of scientists, administrators and layman. Ambient air pollution is caused by the following three classes of pollutants:

(*a*) Criteria air pollutants
(*b*) Hazardous air pollutants, and
(*c*) Volatile organic chemicals

6.3.1 Criteria Air Pollutants

There are six pollutants, which are considered *criteria air pollutants* (CAPs) or most serious air pollutants that account for the large majority of air pollution in the lower atmosphere. These are: carbon monoxide (CO), sulphur dioxide (SO_2), Nitrogen oxides (NO_x), Particulate Matter (PM_{10}) Ozone (O_3) and lead (Pb).

Table 6.2: National ambient air quality standards (NAAQS)
(Source CPCB, 1995, all data in $\mu g/m^3$)

Pollutant	*Sulphur dioxide (SO_2)*		*Oxides of Nitrogen (NO_2)*		*Suspended Particulate Matter (SPM)*	
Time Weighted average	*Annual average*	*24 hrs average*	*Annual average*	*24 hrs average*	*Annual average*	*24 hrs average*
Industrial Area	80	120	80	120	360	500
Residential, rural and other area	60	80	60	80	140	200
Sensitive Area	15	30	15	30	70	100

Note: 24 hourly/8 hourly values should be met 98% of the time in a year. However, 2% of the time, it may exceed but not on two consecutive days.

6.3.1.1 Carbon monoxide (CO)

Carbon monoxide is an incomplete combustion product of fossil fuel and accounts about 50% of air pollution worldwide. In developed countries, 70 to 75% of CO emissions are from motor vehicles. Other sources of CO include coal, natural gas and biomass burning. Atmospheric oxidation of methane gas and other hydrocarbons also produce CO. It has 200 times greater affinity for the iron in haemoglobin than does oxygen. As a result, it reduces the normal oxygen carrying capacity of the blood. A 30 ppm concentration of CO for 4 hours is known to convert 5% of the blood's hemoglobin into carboxyhaemoglobin, thus starving the body of oxygen and causing *CO poisoning*. Other co-related illnesses are headaches, dizziness and drowsiness. In elderly people, CO problem may cause congestive heart failure.

6.3.1.2 Sulphur dioxide (SO_2)

Sulphur dioxide accounts for about 15 to 18% ambient air pollution. Sulphur dioxide is produced by burning of coal and from metal smelters. Other smaller sources are pulp and paper mills and oil refineries. Petroleum also contains sulphur, but it can be more easily removed from petroleum than from coal, so motor vehicles account for only a small percentage of sulphur dioxide emissions. SO_2 causes direct effects on humans and vegetation. It readily reacts with moisture in the lungs, other mucous membranes and eyes to form an irritating acid. Since it is so reactive, major part of SO_2 inhaled by humans is removed in the upper respiratory tract. As a result, exposure to SO_2 can aggravate already existing respiratory or heart problems. Exposure to low concentrations of SO_2 can also damage vegetation.

Table 6.3: Ambient air quality status in some cities/towns of India during 1995
Source: CPCB, 1995

Pollution level	*Annual Mean Concentration Range (μg/m³)*			
	Industrial		*Residential*	
	SO_2 & NO_2	*SPM*	*SO_2 & NO_2*	*SPM*
Low (L)	0 – 40	0 – 180	0 – 30	0 – 70
Moderate (M)	40 – 80	180 – 360	30 – 60	70 – 140
High (H)	80 – 120	360 – 540	60 – 90	140 – 210
Critical (C)	> 120	> 540	> 90	> 210

6.3.1.3 Nitrogen oxides (NO_x)

Nitrogen oxides are formed when fuel is burnt at very high temperatures, such as in industrial plants and transportation vehicles. During combustion in the engine cylinder, a reaction occurs between atmospheric oxygen and nitrogen to produce nitrogen monoxide (nitric oxide, NO). This gas passes out of the engine in the exhaust phase, cools down, and then combines with more oxygen to form nitrogen dioxide (NO_2) and nitrogen tetroxide (N_2O_4). The mixture of all these gases is generally described as nitrogen oxides (NO_x) and they account for about 6 to 10% of air pollution. Like SO_2, nitrogen oxides contribute to a number of problems. Direct exposure to nitrogen oxides irritates the eyes and lungs and aggravates asthma. Nitrogen dioxide is poisonous to plant life. Like sulphur dioxide, nitrogen oxides are precursors of acid rain. When moisture is present, chemical reactions in the atmosphere convert nitrogen oxides to nitric acid.

6.3.1.4 Particulate matter

Particulates are emitted into the atmosphere during combustion of fossil fuel. Air borne particulates are also generated from burning biomass and from windblown dirt, fertilizer, dried manure, or dried crop residues. Construction sites release large amounts of dust. Particulates in coastal areas contain high levels of chloride, which can corrode buildings and other items. Particulates account for about 10 to 13% of air pollution.

According to Indian Institute of Tropical Meteorology in Pune thick smog and haze that enveloped Delhi (2013) all day is an indicator of high concentration of pollutants. In Delhi University area the PM 10 and PM 2.5 levels were found as 450 and 270 respectively on Dec. 20, 2013 (the safe limit of PM 10 and PM 2.5 is 100 and 60 respectively). However, PM 10 levels were more than 7 times of safe limit on Dec. 18, 2013 in R.K. Puram.

The size of particulates varies greatly (0.01 to 500 μm) and they can be either dry or liquid. Very fine particles of either solid or liquid that are suspended in the air, are referred to as *aerosols* (diameter less than 1.0 μm). Sulphate particles and some soot and dust particles can be as small as 0.01 μm. Small particles, especially those in the size range of 0.01 to 50 μm. diameter are of most significance in air pollution, as they are not obviously visible. They can remain in the atmosphere for varying lengths of time and undergo chemical reactions to produce secondary air pollutants like *smog* (smoke + fog). The famous 1905 smog of London was the result of sulphur dioxide, soot and tarry materials released by the uncontrolled burning of sulphur containing coal.

The *photochemical smog*, results from the reaction of volatile organic chemicals (VOCs) and nitrogen oxides in the presence of sunlight to form ozone and other photochemical oxidants, including NO_2 and peroxyacyl nitrates (PANs)

$$\text{smog (hydrocarbons} + NO_2 + \text{sunlight)} \xrightarrow[\text{inter mediate steps}]{\text{many}} \text{PAN} + \text{HCHO} + \text{other secondary pollutants}$$

Acute effects of photochemical smog are eye irritation and headaches. It adversely effects breathing and can also damage lungs and other tissues. Photochemical smog also causes damage to plant foliage. Thus, particulate air pollution causes respiratory problems worldwide. People suffering from asthma, bronchitis and emphysema are worst affected by particulate air pollution. A recent study indicates that an increase in air particulates aggravates heart problems like ischemia and arrhythmias.

The *industrial smog* is a harmful mixture of smoke and fog. The word was first used in 1905 to describe the fog that was observed in London and totally obscured visibility for few hours. This smog resulted from SO_2, soot and tarry materials released into the atmosphere by burning of high-sulphur coal.

6.3.1.5 Ozone (O_3)

People are not easily convinced about the role of tropospheric ozone as a pollutant. Ozone is a more reactive chemical than molecular oxygen and often adversely affects both plants and animals. Ozone is not typically emitted as ozone, but it is formed from the ozone precursors, VOCs, and nitrogen oxides. Ozone is a summer pollutant as it is formed during summer when VOCs (especially formaldehyde and others) and NO_x react with atmospheric oxygen to form ozone in the summers heat and the sun's strong UV rays. Motor vehicles are a strong source of ozone precursors.

Ozone levels upto 0.06 ppm in ambient air are not considered harmful, but ozone levels from 0.12 to 0.20 ppm or more are considered unhealthful. Most ground level ozone is anthropogenic. But in the troposphere excess ozone is the problem, whereas in the stratosphere, ozone depletion is the problem. Its acute effects cause eye irritation and adversely affect lungs and respiratory membranes of nose and throat. Ozone may also increase susceptibility to infection. Ozone also adversely affects tree growth and damages forests. Sensitive crops are damaged at ozone level 0.05 ppm and estimated crop losses from ozone pollution may be 5 to 10%.

6.3.1.6 Lead (Pb)

Lead is an airborne particulate. The heavy metals lead, cadmium and mercury are quite common pollutants found at hazardous waste sites. Lead emissions are from metal mining and smelting, coal burning, electric power plants and municipal solid waste combustors. Lead in water pipes also remains in place for many years. In motor vehicles, lead-acid battery is used worldwide. The household dust may also contain lead when leaded paint flakes and crumbles into small airborne particles. Until 1970s motor vehicle exhaust was the major source of lead to the environment in many countries of the world since lead was added to gasoline as antiknock agent. As the use of lead in gasoline has been banned and many countries are using unleaded petrol, lead emissions to the air have fallen more than 90%. Blood levels of lead in 1 to 5 year old children should not be more than 4 to 5 μg per 100 ml and if the level is slightly above 10 μg per 100 ml, necessary steps to reduce lead are immediately required.

Lead poisoning of children has been described as a serious problem. Lead adversely affects the nervous system of developing foetus and small child. Boys whose bones contained high levels of lead were more likely to be juvenile delinquents. However, adults are not as sensitive to lead exposure as are children but do suffer adverse effects, including damage to the nervous system and kidney failure and chances of high blood pressure. Other potential adverse effects may be anaeamia and infertility.

6.3.2 Hazardous Air Pollutants

A large number of pollutants are grouped under the category of hazardous air pollutants (HAPs) or *toxic air pollutants.* In US, there are 189 federally regulated HAPs (Hill, 1997). Important Examples of *organic chemical HAPs* are benzene, chloroform, and formaldelyde. Other examples of organic chemical HAPs are toluene, ethylene glycol, methanol, methyl bromide, parathion, styrene and vinyl chloride. Examples of *inorganic chemical HAPs* are asbestos and arsenic and the metals (cadmium, chromium, mercury, nickel, and beryllium).

The HAP *benzene* is produced in large volumes and is also found in gasoline. It can irritate the eyes and skin, cause dizziness and headache and has also been associated with aplastic anaemia and leukemia. *Formaldehyde* (HCHO) is another important HAP, which is released from factories manufacturing furniture. It can irritate the lungs and eyes and is a carcinogen at high concentrations. The HAP *chloroform* is formed during water chlorination and other chlorine uses. It was used as an anesthetic for many years until its ability to damage the liver was discovered. Exposure to chloroform can also damage the kidneys.

The HAP *asbestos,* an inorganic chemical, is a natural fibrous mineral material widely used to insulate furnaces. Exposure to asbestos is associated with asbestosis, lung cancer and mesothelioma. When airborne fibres of asbestos are inhaled, they are permanently trapped into the lungs. The HAP *Arsenic* is a nonmetallic element used in metal alloys and glass making. Chronic arsenic ingestion is associated with skin cancer and chronic inhalation with lung cancer. The HAP *Cadmium* is released into the air by combustion of coal and mining and smelting operations. However, plants take up cadmium more readily than other metals. Because tobacoo plants concentrate cadmuim, an adult absorbs more than 90% of the cadmium inhaled through smoking. Cadmium can accumulate in kidney and liver. Because cadmium affects calcium metabolism, bone degeneration has also been observed. In laboratory studies with rodents, cadmium, depending on concentration, shows numerous toxic effects, including birth defects, and is a carcinogen.

The HAP *Mercurry* is a volatile liquid metal, sometimes called quicksilver. Mercury emitted into the atmosphere is washed out into water bodies where bacteria transform it to methylmercury. Methylmercury is more toxic than elemental mercury as most of it is absorbed in human gastrointestinal tract. The mercury level in fish may range from 0.01 to 0.5 ppm but in large ocean fish may accumulate upto 3 ppm or more. Thus, about 95% of human exposure results from eating mercury contaminated fish. Mercury, like lead, is toxic to the nervous system and foetus. High exposure to mercury in adults results in a number of toxic effects, including kidney damage, deafness and blindness and other problems of nervous system.

6.3.3 Volatile Organic Chemicals (VOCs)

These are a large number of organic chemicals that evaporate easily and some of them significantly contribute to smog formation. About 70% of the HAPs discussed earlier are among the VOCs and there are hundreds of other chemicals in this category. Motor vehicle exhausts are a major source of VOCs, emitting about 50% of the total VOCs. Other small sources of VOCs are auto maintenance shops, wood-drying and wood-painting operations and sewage treatment plants. In homes and other places, people are exposed to VOCs from solvents, paints, charcoal broiler starters, aerosol sprays and deodorants and cosmetics. Some VOCs contribute to the formation of ground level photochemical ozone. VOCs also contribute to many other problems. Some time people may develop asthma or other respiratory problems due to VOCs or ozone-containing smog.

Effects of Air Pollution

These include (*i*) acute health effects from high-level pollution as absorbed in India in 1984 in Bhopal when leakage of MIC (methyl isocyanate) killed over 2500 people and affected a large population; (*ii*) Chronic diseases like bronchitis, lung cancer and emphysema, etc. (*iii*) Conversions of haemoglobin into carboxy haemoglobin as done by CO; (*iv*) acid rain caused by SO_2 and NO_x when moisture is there in the atmosphere; and (*v*) other diseases like eye irritation, lung and skin cancer, damage to kidney, and high B.P., etc.

Toxic effects of air pollution on plants, reduction in visibility due to smoke and fog, deterioration and soiling of materials and production of haze are some other effects of air pollution. The effect of air pollution on climate has also been noticed. Solid particulates take part in cloud formation, and so urban pollution and increased water vapour emission can produce upto 10% increased cloud cover, upto 10% more wet days and increased mist, fog and smog compared to non-industrial areas. All these processes combine to increase the deposits of large sized particulates on the ground. Particulates also play a role in producing temperature changes and air movements.

6.3.4 Control of Air Pollution

During recent years efforts have been made in all the countries of the world to check and reduce air pollution. In India, ambient air quality standards have been prescribed for industrial, residential and sensitive areas (Tables 6.2 and 6.3) and the state of ambient air quality is regularly monitored in the major cities of the country. As automobile exhausts account for major part of air pollution, emission limits for diesel and petrol - driven vehicles have been fixed. The National Environment Engineering Research Institute (NEERI) has been involved in R&D for developing indigenous, lead-resistant, cost-effective *catalytic converter* for Indian conditions, under three-phased programme. Under the first phase, which has been completed, the test results indicate that the use of a catalytic converter reduces the carbon monoxide (CO) and hydrocarbons concentrations in petrol driven vehicles to about 50 to 60%.

In order to reduce the inputs of lead, now unleaded gasoline is supplied. The first phase of introducing petrol with a maximum lead content of 0.15 g/l (which was 0.56 g/l) in the four metropolitan cities of the country was implemented by the Ministry of Petroleum and Natural Gas in 1994. The recent verdict of the Supreme Court has made it mandatory to use natural gas in three-wheelers. This is an attempt to control diesel vehicular exhaust pollution. In India, efforts have also been made to control air-borne emissions from various industries. Industries have been classified under 17 categories of highly polluting industries for the purpose of monitoring and regulating pollution from them. According to the Ministry of Environment and Forests Annual Report 1996-97, 125 units were closed and others were categorised as having or not having, adequate facilities to comply with the standards. However, thermal power and fertilizer industries are defaulting in meeting air pollution standards; and sugar and pulp and paper industries are the major defaulters in complying with the norms for liquid effluents.

This brief statement of efforts made to control air pollution compels us to say that so much is yet left to be achieved. Any real control of air pollution is possible through a holistic approach because many air pollutants occur in all the three parts *viz.*— air, water, soil, of the biosphere. Reduction in biological pollutants of air and switching onto renewable resources of energy are other attempts made in this direction.

6.4 LITHOSPHERE

This subdivision of the biosphere plays two vital roles in metabolism. First, it is the only source of most mineral metabolites for all the organisms, terrestrial as well as aquatic; and second, it forms the bulk component of soil (pedosphere), required particularly by the terrestrial plants. Soil is formed from disintegration of rocks by the action of running water and weathering. Living organisms also contribute to soil formation. Their decay produces organic fractions of soil, collectively called *humus.*Humus mixed with sand and clay constitutes the soil. Soil is affected by climate (temperature and rainfall), vegetation and human activities including pollution.

6.4.1 Major Environmental Problems of Lithosphere

Major environmental problems of lithosphere may be as follows: *(i)* Soil degradation, erosion and pollution, *(ii)* Deforestation, *(iii)* Land slides and earthquakes, and *(iv)* Loss of agricultural land for nonagricultural purposes.

6.4.1.1 Soil degradation

It has been reported from many parts of the country that agricultural intensification, faulty irrigation techniques, deforestation and excessive use of pesticides and herbicides have lead to widespread soil degradation. According to one estimate, about 300 million ha of land is now highly degraded and about 1.2 billion ha or 10% of the earth's vegetated surface, is described as moderately degraded. In India, the worst examples of soil degradation have been observed in Gujrat, Rajasthan and Bihar. The desertification of soil in Rajasthan is attributed to deforestation and cattle grazing. Agricultural land is also highly susceptible to degradation in arid and semiarid regions. Both overgrazing and excessive cultivation expose the soil directly to erosion by the wind. In the process of desertization, arid and semiarid land that has supported subsistence or nomadic agriculture gives way to desert. The process may be slowed by irrigating the land, which gives a temporary remission but lowers the water table, and salts accumulate in the topsoil (*salinization*). Once salts have started to accumulate, the process of salinization tends to spread and leads to sterile white salt deserts. This has been observed in many areas of India and Pakistan. However, the degradation of soil by intensive agriculture can be prevented or slowed by (*a*) returning the land to grazed pasture or rangeland; (*b*) use of farmyard manure, crop residues and animal wastes and (*c*) alternating years under cultivation with years of fallow.

6.4.1.2 Soil erosion

Soil erosion is a process in which useful land is washed off through excessive rains and soil runoff. The Rio conference,1992 also adopted Agenda 21 for soil conservation and recommended measures, now coordinated by the United Nations Commission on Sustainable Development, to prevent soil erosion, in particular minimizing soil runoff and sedimentation. The main factors or forces responsible for removal or displacement of soil from one place to another are water and wind. Water erodes and removes soil mainly in the following four ways:

(*i*) *Sheet erosion*: The soil is removed in small but uniform amounts from all over and therefore does not leave a mark behind. The evidence of this type of erosion can be seen in the heavy quantity of silt that is deposited elsewhere.

(*ii*) *Rill erosion*: The runoff water moves rapidly and cuts several stream-like structures. As the water moves very rapidly, its cutting effect is quite visible.

(*iii*) *Gully erosion*: In this process, several rills converge towards the steep slopes and join to form broad channels of water called gullies. When rills and gullies are formed in abundance, ploughing becomes impossible due to uneven topography and cultivation of crops cannot be done.

(*iv*) *Riparian erosion*: It takes place on the banks of fast running rivers. The fast flowing waters cut the margin of the bank laterally. Once much of the soil from beneath is cut away, the top soil of the river banks tumbles down into the water with big splash.

6.4.1.2.1 Soil Erosion due to wind

During some parts of the year soil erosion due to wind is very common. In a duststorm, huge quantity of dust is raised high and transported to great distances. The wind lifts finer particles high up whereas the large particles roll along the surface. The rolling particles rub the ground and due to abrasive action help in loosening the soil. The process continues and more dust particles gather as the storm advances with speed.

6.4.1.2.2 Measures to check Soil Erosion

The principal methods to check soil erosion aim at reducing the physical and biological forces that cause erosion and protecting the soil by plant cover. However, the plant cover should be such as to yield food, fodder and other resources. A cost-effective technology used in reducing soil erosion is *contour - based cultivation* (contour-farming). In India, contour ditches have helped to increase four times the survival chances of tree seeding. Deeply rooted, hedge-forming vetiver grass, planted in contour strips across hill slopes, slows down water runoff efficiently, reduces erosion, and increases moisture content for crop growth. Therefore, currently

90 per cent of soil conservation efforts in our country are based on such biological systems. Ramakrishnan (2001) stated that in the north-eastern parts of India shifting agriculture (Jhum) helps in sustainable production as the soil fertility is recovered during its vegetational fallow phase. Simple technology involving rock embankments (bunds) constructed along contour lines for water and soil conservation greatly succeeded. Such bunded fields yield average of 10 per cent more than traditional fields in a normal year and almost 50 per cent more in dry years (UN,1998). Such terracing provides a very high quality of soil conservation, but is possible at a place where labour is cheap.

Contour farming is recommended on lands with gentle slopes. But on somewhat greater slopes, *strip cropping* is done where strips of land are ploughed at right angles of the direction of the slope and different crops are raised in adjacent strips. On less steep slopes, by ploughing and cultivating in strips along the contour, soil runoff can be reduced significantly. Singh, Misra and Ambasht (1980) have shown that plants play a very significant role in soil conservation. They have quantified their soil binding efficiency, which was regarded above 95 per cent for dominant grasses.

Plant cover is also the most efficient method for checking soil erosion due to wind. Tree plantations in short blocks are called *wind breaks.* Extensive tree plantations are called *shelter belts.* These are planted in rows at right angles to the direction of the wind. The wind breaks and shelter belts reduce the wind velocity on wind-ward side upto about the distance equivalent to 5 times the height of trees and on lee-ward side to about 30 times the tree height (Ambasht, 1990).

It may be concluded that proper soil conservation strategy may help not only in achieving sustainable agriculture but also in checking soil degradation and erosion.

6.4.1.3 Soil pollution

Almost all the sources that pollute inland water bodies also pollute agricultural land and other soils. These are:

(*a*) **Sewage and domestic wastes:** In villages, only few houses have septic tanks for treatment of sewage and domestic wastes. Many people go to open fields for easing out; such faecal matter of human origin pollutes both soil and air.

(*b*) **Industrial wastes and effluents:** Industrial wastes and effluents that are discharged to some water bodies also pollute the adjoining lands. The toxicity of these industrial pollutants to crop plants and soil organisms is variable but all of them contaminate the soil environment.

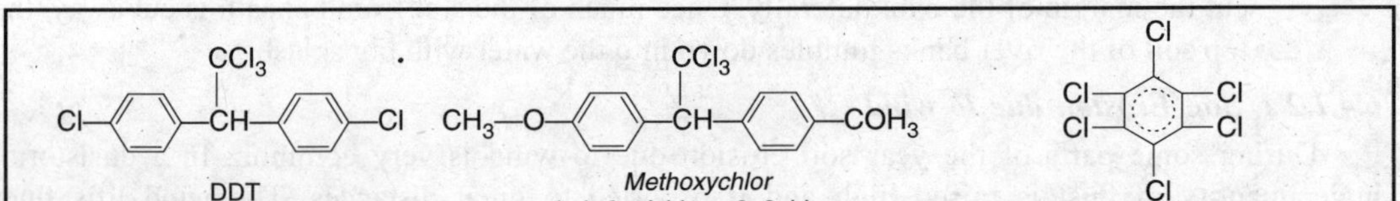

Fig. 6.5 (*a*): Formulae of some insecticides (Chlorinated hydrocarbons)

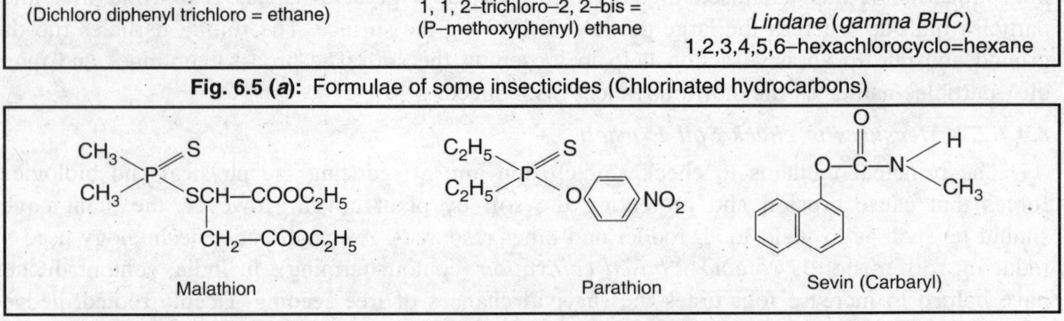

Fig. 6.5 (*b*): Formulae of some organophosphorous insecticides (Malathion and Parathion) and carbamate (Carbary)

(*c*) **Pesticides:** Pesticides are synthetic chemicals used for pest control. They are classified as *insecticides* (organochlorines such as DDT, organophosphates such as Malathion and Carbamates such as Carbaryl), (Fig. 6.5), *rodenticides* (such as Sodium fluoroacetate), molluscicides (such as Sodium pentachlorophenate), and *piscicides* (such as Toxaphene) Rachael Carson for the first time highlighted the dangers of pesticide pollution in her famous book "Silent Spring" (1963).

Besides pesticides, a variety of other synthetic chemicals called *fungicides* (such as Phenylmercury, Carboxin,) and *herbicides* (such as Paraquat, 2, 4-D) (Fig. 6.6) are also used in modern agriculture for killing fungi and herbs and weeds respectively. Pesticides may also be grouped on the basis of their mode of action as fumigants, contact poisons, repellents, systemic poisons and growth regulators.

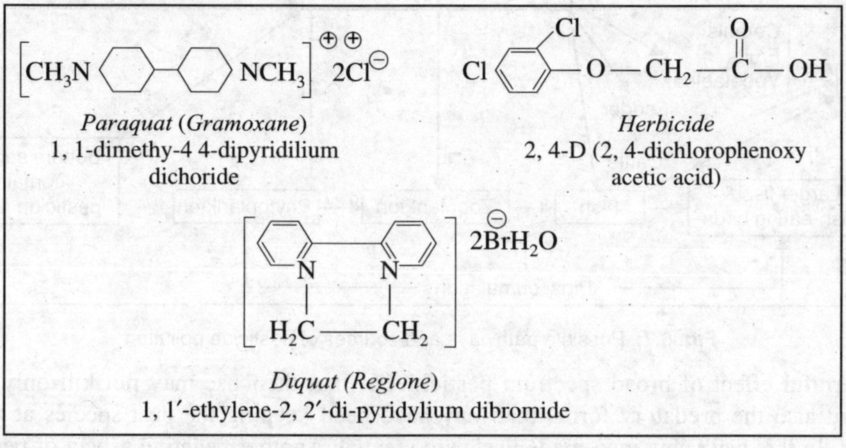

Fig. 6.6: Formulae of some herbicide/molluscicide

Of all the insecticides, the mode of action of organophosphorus compounds is better known. They function by blocking acetylcholine - esterase. They effectively poison the enzyme by phosphorylation and thus block hydrolysis of acetylcholine into choline. The inhibition results in accumulation of acetylcholine at the postsynaptic membrane, which is then unable to return to its original state and takes sometime to regain its activity.

Pesticides reach soil through spraying, runoff, rain, atmospheric fall out, and industrial effluents. Pesticides used on the crops or agricultural land may reach the aquatic systems and also the atmosphere. The biogeochemical cycle of a pesticide (Fig. 6.7) shows that after entering the environment, a pesticide cycles like a mobile substance. The hazards of pesticide pollution are alarming since many pesticides are highly persistent. They remain in the environment through *bioconcentration* and *biomagnification* (Flint and Van den Bosch, 1981). However, the use of synthetic pesticides and fertilizers is a necessary evil of modern agriculture.

Effects of pesticides on the living forms are too many. Though pesticides have made enormous contribution to the green revolution and economic development, their indiscriminate and widespread use has caused several life threatening health hazards to humans and other animals. Their use is directed against economic pests, but they also affect a wide range of nontarget organisms, including desirable animals and plants. The chlorinated hydrocarbons may reach such high levels in animals at the summit of the aquatic food web that they are unable to reproduce or they may be poisoned outright. Many pesticides affect the reproduction of fish-eating birds and other higher animals. DDT affects the steroid sex hormones in birds. The herbicides 2,4-D and 2,4,5-T at relatively high doses lower reproductive activity in chickens. Pesticides can change the growth rate and also produce some behavioural changes in nontarget organisms. The most widely

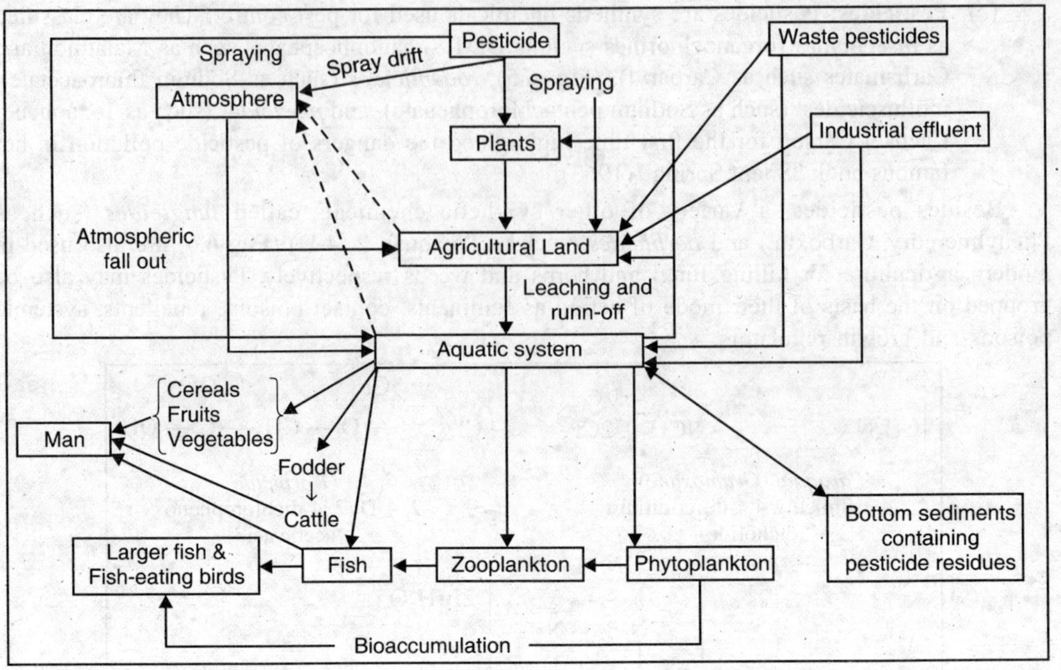

Fig. 6.7: Possible pathways and sources of pesticide pollution

known harmful effect of broad spectrum pesticides is that their use may not kill only the target insects, but also the predatory forms that may have been keeping the pest species at reasonably low levels. In this way, new man-made pests are created. Another negative aspect of pesticide use is that repeated applications of a given pesticide may result in the development of resistant strains of target pests. The development of such genetic resistance in pest populations is well-known. It is documented that by 1980s more than 60 species of malaria mosquitoes had developed resistance to DDT. There are examples of multiple - pesticide failure in some cases of mosquito pest.

Instances of direct killing of humans through consumption of heavy doses of pesticides such as DDT, Gamaxine or other poisonous chemicals used in the manufacturing of pesticides are not uncommon. The memory of the so-called 'Bhopal Tragedy' is still fresh. The leakage of *methyl isocynate* (MIC) from the Union Carbide Factory, Bhopal in 1984 caused over 2500 deaths and other ailments in thousands of people. MIC is an intermediate chemical used in the production of carbamate pesticides such as Carbaryl and Aldicrab. Many cases of pesticide poisioning do occur almost every day in the country and many such cases go unreported. More than 5 million of deadly pesticides, most of which already banned in Europe and USA, are used by farmers in Asia. These are believed to cause an estimated one million deaths yearly, especially among pesticide handlers (FAO,1995). In India, several women who experienced still births were found to have high levels of pesticides such as aldrin and DDT in their maternal blood and umbilical cords. The observed DDT residues in the fat of Indian population are very high, ranging from 12.5 ppm for Gujrat to about 30.0 ppm for Delhi. High infant mortalities have been reported from places where high residues of DDT were found in human milk. A study from Punjab revealed high levels of DDT and BHC in breast milk samples.

A recent study conducted by Greenpace International has reported on health-effects of chlorinated pesticides such as DDT, BHC and Endosulphan. Indian men exposed to these pesticides while working in cotton fields were found to have decreased fertility. Accumulation of these pesticides in vital organs such as liver, kidney, gonads, lungs, adrenal glands, and blood causes histopathological changes and carcinogenic, mutagenic and teratogenic effects. However, the management of pesticide risks in non-human species is even less studied excepting in veterinary medicine. In pesticide threatened ecosystems, its reversal and subsequent rehabilitation needs a better understanding of pesticide risk management. The same applies to environmental friendly disposal of pesticides including land fill and incineration (Misra *et al*, 1998).

6.4.1.4 Forests and related problems

During past 50 years forests have suffered serious depletion. At present forests constitute over 21% of the total geographic area of the country as against the target of 60% as laid down by the National Forest Policy 1952 (revised in 1988, re-revised in 2011). The preamble of National Forest Policy (NFP), 1988 and 2011 says that forest depletion is attributable to relentless pressures arising from ever-increasing demand for fuelwood, fodder and timber, inadequacy of protection measures, diversion of forest lands to non- forest uses without ensuring compensatory afforestation and the tendency to consider forests as revenue earning resource. Therefore, NFP 2011 has suggested a new strategy of forest conservation.

Table 6.4: State-wise details of the Protected Area Network of the country

S. No.	*State/UT*	*No. of National Parks*	*No. of Wildlife Sanctuaries*	*No. of Conservation Reserves*	*No. of Community Reserves*
1.	Andhra Pradesh	6	21	0	0
2.	Arunachal Pradesh	2	11	0	0
3.	Assam	5	18	0	0
4.	Bihar	1	12	0	0
5.	Chhatisgarh	3	11	0	0
6.	Goa	1	6	0	0
7.	Gujarat	4	23	1	0
8.	Haryana	2	8	2	0
9.	Himachal Pradesh	5	32	0	0
10.	Jammu & Kashmir	4	15	34	0
11.	Jharkhand	1	11	0	0
12.	Karnataka	5	22	2	1
13.	Kerala	6	16	0	1
14.	Madhya Pradesh	9	25	0	0
15.	Maharashtra	6	35	1	0
16.	Manipur	1	1	0	0
17.	Meghalaya	2	3	0	0
18.	Mizoram	2	8	0	0
19.	Nagaland	1	3	0	0
20.	Orissa	2	18	0	0
21.	Punjab	0	12	1	2
22.	Rajasthan	5	25	3	0
23.	Sikkim	1	7	0	0

(Contd.)

S. No.	State/UT	No. of National Parks	No. of Wildlife Sanctuaries	No. of Conservation Reserves	No. of Community Reserves
24.	Tamil Nadu	5	21	1	0
25.	Tripura	2	4	0	0
26.	Uttar Pradesh	**1**	**23**	**0**	**0**
27.	Uttaranchal	**6**	**6**	**2**	**0**
28.	West Bengal	5	15	0	0
29.	Andaman & Nicobar	9	96	0	0
30.	Chandigarh	0	2	0	0
31.	Dadar & Nagar Haweli	0	1	0	0
32.	Lakshadweep	0	1	0	0
33.	Daman & Diu	0	1	0	0
34.	Delhi	0	1	0	0
35.	Pondicherry	0	1	0	0
	Total	**102**	**515**	**47**	**4**

Note: Names and numbers of National Parks and Sanctuaries in U.P. and Uttarakhand are given separately.

According to the *State of Forest Report, 2011,* the forest cover of the country has been estimated to be 6,92027 km^2, which is 21.05% of the geographic area of the country. This includes very dense forest (2.54%), moderately dense forest (9.76%) and open forest (8.75%). The scrub (1.28%) and non-forest are (77.67%) is also there. (Table 6.6)

6.4.1.4.1 Protected Areas

The PAs covers about 5% of total geographic area of the country. In includes 102 national parks, 515 wildlife sanctuaries, 47 conservation reserves and 4 community reserves (Table 6.4). These 4 categories of PAs are a means of *in-situ conservation* of biodiversity. A list of National Parks of U.P. and Uttarakhand is given separately (Table 6.5).

1. **National Park:** A national Parks is an area having adequate ecological, faunal, floral, geomorphological, natural or zoological significance. The national park is declared for the purpose of protecting, propagating or developing wildlife or its environment. The Corbett National Park is the oldest national park of the country.
2. **Sanctuary:** A sanctuary is an area which is of adequate ecological, faunal, floral, geomorphological, natural or zoological significance. The sanctuary is declared for the purpose of protecting, propagating or developing wildlife or its environment. Certain rights of people living inside the sanctuary could be permitted.
3. **Conservation Reserves:** These can be declared by the State Government in any area owned by the Govt. However, such declaration should be made after consulting the local communities. The conservation reserves are declared for the purpose of protecting landscapes, seascapes, flora and fauna and their habitat. The rights of people living inside a conservation reserve are not affected.
4. **Community Reserves:** These can be declared by the State Govt. in any private or community land, not comprised within the above named PAs, where an individual or community has volunteered to conserve wildlife and its habitat. Community reserves are declared for

the purpose of protecting fauna, flora and traditional or cultural conservation values and practices. For example, Vishnois of Rajasthan protect black buck as part of their culture. As in the case of a conservation reserve, the rights of people living inside a community reserve are not affected.

Differences between a national park and a sanctuary: The difference between a national park and a sanctuary lies in the vesting of rights of people living inside. Unlike a sanctuary, where certain rights can be allowed, in a national park, no rights are allowed. No grazing of any livestock shall also be permitted inside a national park; while in a sanctuary, the chief wild life warden may regulate, control or prohibit it. In addition, while any removal or exploitation of wildlife or forest produce from a sanctuary requires the recommendation of State Board for Wildlife, removal from a national park requires the recommendation of the National Board for Wildlife. However, a sanctuary is always meant for protecting a particular part of wild life, for example, muskdeer sanctuary or bird sanctuary; in a national parks are efforts is made to protect the entire wildlife and the habitat. In other words, we could say that a sanctuary is species-specific or species-oriented.

Names and numbers of national parks and sanctuaries of some states is given in Table 6.4.

Table 6.5: Name and No. of National Parks and Sanctuaries in U.P. and Uttarakhand

	UTTAR PRADESH		
	Name of the PA	*Year or Notification*	*Total Area*
A.	**National Parks**		
1.	Dudhwa NP	1977	490
B.	**Wildlife Sanctuary**		
1.	Bakhira WLS	1990	28.94
2.	Chandraprabha WLS	1957	78
3.	Hastinapur WLS	1986	2,073.00
4.	Kaimur WLS	1982	500.73
5.	Katerniaghat WLS	1976	400.09
6.	Kishanpur WLS	1972	227
7.	Lakh Bahosi WLS	1988	80.24
8.	Mahavir Swami WLS	1977	5.41
9.	National Chambal WLS	1977	6.35
10.	Nawabganj WLS	1984	2.25
11.	Okhla WLS	1990	4
12.	Parvati Aranga WLS	1990	10.84
13.	Patna WLS	1990	1.09
14.	Ranipur WLS	1977	230.31
15.	Saman Bird WLS	1990	5.26
16.	Samaspur WLS	1984	7.99
17.	Sandi WLS	1990	3.09
18.	Sohagibarwa WLS	1987	428.2
19.	Sohelwa WLS	1988	452.47

(Contd.)

UTTAR PRADESH			
	Name of the PA	*Year or Notification*	*Total Area*
20.	Sur Sarovar WLS	1991	4.03
21.	Surha Tal WLS	1991	34.22
22.	Turtle WLS	1989	7
23.	Vijai Sagar WLS	1990	2.62
	Total		**5221.88**
	Grand Total		**5711.88**

UTTARAKHAND			
	Name of the PA	*Year of Notification*	*Total area*
A.	**National Parks**		
1.	Corbett NP	1936	520.82
2.	Gangotri NP	1989	2390.02
3.	Govind NP	1990	472.08
4.	Nanda Devi NP	1982	624.6
5.	Rajaji NP	1983	820.42
6.	Valley of Flowers NP	1982	87.5
	Total		**4915.44**
B.	**Wildlife Sanctuary**		
1.	Askot Musk Deer WLS	1986	599.93
2.	Binsar WLS	1988	45.59
3.	Govind Pashu Vihar WLS	1955	485.89
4.	Kedarnath WLS	1972	975.2
5.	Mussoorie WLS	1993	10.82
6.	Sonanadi WLS	1987	301.18
	Total		**2418.01**
C.	**Conservation Reserve**		
1.	Asan Barage Wetland CR (Bird)	2005	4.44
2.	Jilmil Jheel CR (Bird)	2005	37.84
	Total		**42.28**
	Grand Total		**7376.33**

Table 6.6: Forest and Tree cover of India in 2011

	Class	*Area (km²)*	*% of geographic area*
	Forest Cover		
(*a*)	Very Dense Forest	83,471	2.54
(*b*)	Moderately Dense Forest	3,20,736	9.76
(*c*)	Open Forest	2,87,820	8.75
	Total Forest Cover*	**6,92,027**	**21.05**
	Scrub	42.177	1.28

(*Contd.*)

Class	*Area (km²)*	*% of geographic area*
Non forest	2,553,059	77.67
Total Geographical Area	**3,287,263**	**100**

(*includes 4662 km^2 area under mangroves)

6.4.1.4.2 Biosphere Reserves

Biosphere reserves, a concept originated in 1974, are a set of unique ecosystems identified on the basis of their bio-diversity, naturalness and effectiveness as a conservation unit. There are 18 Biosphere Reserves (Table 6.7) with a geographical area of 89219.70 sq km. A large part of this area spreads even beyond protected areas and they are distinct geographic entities.

Wetlands and Ramsor Sites

Six internationally important wetlands of India have been declared as "Ramsar Sites" under Ramsar Convention (1971). These are *Keoladeo National Park* (Rajasthan), *Lake Chilika* (Orissa), *Lake Sambhar* (Rajasthan), *Lake Loktak* (Manipur), *Lake Harike* (Punjab) and *Lake Wular* (J & K). For intensive conservation and management eleven more wetlands have been identified in the country. Wetlands are increasingly threatened by anthropogenic activities.

Ramsar Convention promotes the concept of "wise use", which refers to "sustainable utilization for the benefit of humankind in a way compatible with the maintenance of the natural properties of the ecosystem". The Ministry of Environment and Forests, the nodal agency for wetland conservation in India, constituted in 1993, a National Wetland Committee (National Committee for Wetlands, Mangroves and Coral Reefs, since 1992), which advises the Government on research, conservation and management of wetlands in the country.

Table 6.7: List of Designated Biosphere Reserves (2012)

S. No.	*Name of the BR & total geographical area (km²)*	*Date of designation*
1.	**Nilgiri (5520)**	01.08.1986
2.	**Nanda Devi (5860.69)**	18.01.1988
3.	**Nokrek (820)**	01.09.1988
4.	Manas (2837)	14.03.1989
5.	**Sunderban (9630)**	29.03.1989
6.	**Gulf of Mannar (10500)**	18.02.1989
7.	Great Nicobar (885)	06.01.1989
8.	**Similipal (4374)**	21.06.1994
9.	Dibru-Saikhowa (765)	28.07.1997
10.	Dehang Debang (5111.5)	02.09.1998
11.	**Pachmarhi (4981.72)**	03.03.1999
12.	Khangchendzonga (2619.92)	07.02.2000
13.	Agasthyamalai (3500.36)	12.11.2001
14.	Achanakmar-Amarkantak (3835.51)	30.03.2005
15.	Kachchh (12,454)	29.01.2008
16.	Cold Desert (7770)	28.08.2009
17.	Sashachalam (4755.997)	20.09.2010
18.	Panna (2998.98)	25.08.2011

*Sites with bold letters have been included in the World Network of BRs of UNESCO

6.4.1.4.3 Deforestation

Deforestation means felling of tress or cutting of forests without regard to their vital role in the life support system and ecological balance and environmental stability. Deforestation is a worldwide phenomenon. Forest losses may be attributed to ever-increasing demand for timber, fuelwood and fodder and pressures on forest land for creating space for agriculture and industries. Forest areas in the proximity of the villages are reported to be degrading faster due to collection of fuelwood and cattle grazing as compared to forests located in inaccessible areas. For centuries, forests have been looked upon as revenue earning resource and this concept has promoted deforestations all over the world.

Adverse Effect of Deforestation

It has been observed that deforestation, while changing a mountain belt from stable to unstable, creates a number of problems. We can actually see the different stages of such a change—stability, vulnerability, fragility and finally instability (Messerli, 1985). The term '*stability*' implies the long-term sustainability in the use and exploitation of land and natural resources in each ecological belt, and also a sustained long - term interaction between the different belts. The term '*vulnerability*' is used to characterize a system for which stability can be maintained only by careful management and by a high input of energy. The term '*fragile*' is used to imply that irreversible change or damage can be inflicted easily. Finally, the term '*instability*' is used in situations where damage or change is occuring not only in terms of the resource and land use of each ecological belt, but also through the interaction between belts, or between highland and lowland systems. For example, soil erosion in one system produces destabilizing effect on adjacent ecosystems. The process of destabilization of one ecological belt through spread of instability into a neighbouring belt is especially dangerous in mountain regions because of the high energy environment or downslope effect that is characteristic of them (Messerli, 1985). Thus, deforestation effects not only the biota of the forests and neighbouring ecosystems, but soil is eroded, land is degraded, ground water channels are altered and water becomes polluted and scarce.

Forestry and agriculture are two important land uses, the latter competing with the former under relentless pressure of an ever-increasing population. Human population has grown from 361 million in 1951 to 955 million in 1997 and 7.139 billion in 2014. To meet the requirement of food production, the area under agriculture has increased from 118 million ha in 1951 to 142 million ha at present. It is therefore unlikely that the agricultural land will be available for expansion of forest cover of the country. It is only the "culturable wastelands, covering about 14 million ha and part of the "fallow land and other than current fallows", covering about 10 million ha, which seem to be potential areas for expanding forest cover through afforestation. In addition, efforts will have to be made to raise trees outside conventional forest areas through innovative agro - forestry programmes.

6.4.1.4.4 Joint Forest Management

The National Forest Policy, 2011 emphasizes on creating massive people's movement through involvement of village communities living near the forest in protection and development of forests. Pursuant to this policy, the Government of India issued a notification requesting the State Governments to involve local communities in the conservation and management of forests. It is envisaged that the communities, in lieu of their participation in protection and development of forest areas, will be entitled to sharing of usufructs in a manner specified by the concerned State Forest Departments. This has led to the development of Joint Forest Management (JFM) programme and many State Governments have issued resolutions in this regard. The State of M.P. accounts for more than 50% of the area under JFM. About 36, 130 forest Protection Committees, in which the participation of women is also ensured, are managing a total of 10.25 million ha of forest area in different parts of the country.

6.4.1.4.5 Other Suggestions to Increase Forests

The National Forest Policy, 2011 has suggested a number of steps for forest management. Some of the steps are—regulated grazing, careful planning of forest based industries, forest extension programmes, forestry education and research, forest survey and database, legal and financial support for forestry. The role of tribal people and women in forest management is very important. Tribal people should be closely associated in the protection, regeneration and development of forests.

One of the major causes for degradation of forest is illegal cutting and removal by contractors and their labour. The new policy suggests that in order to put an end to this practice, contractors should be replaced by institutions such as tribal cooperatives, labour cooperatives, and government corporations as early as possible.

6.5 HYDROSPHERE

Water covers about 73 per cent of the earth's surface entirely, and it is a major constituent of the lithosphere and the atmosphere. Water is also the most abundant component of the protoplasm, and therefore it is the major requirement of all living organisms. In metabolism, water is the only source of hydrogen and one of the several sources of oxygen. The major pools of water occur in the oceans – 97.3% of the total for the biosphere (Bener and Berner, 1987), the ice of the polar ice - caps and glaciers (2.06%), groundwater (0.67%) and in rivers and lakes (0.01%).

6.5.1 Hydrological Cycle

Water or Hydrological cycle (Fig. 6.8) is defined as the movement of water from ocean, by evaporation, to atmosphere and by precipitation to land and back via river flow, to ocean (Begon *et al*, 1996). It is a balanced continuous process of evaporation, transpiration, precipitation, surface runoff and ground water movements.

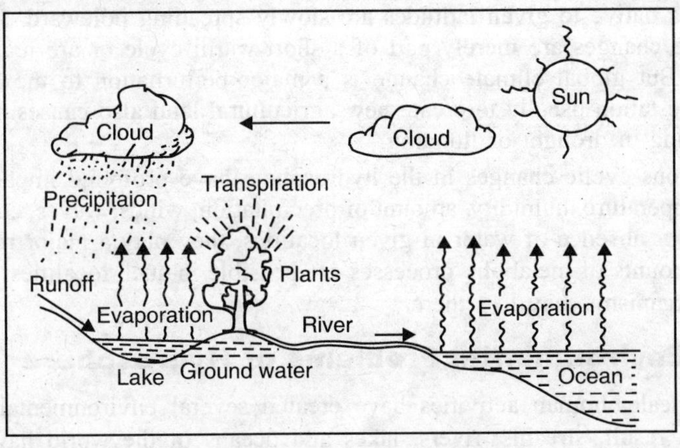

Fig. 6.8: The hydrologic cycle

Thus, through the natural process of hydrological cycle, water is exchanged between the atmosphere (troposphere), land, sea, all living plants and animals, and industrial plants.

6.5.1.1 Role of water in metabolism

In using water as a metabolic raw material, organisms withdraw it mainly from the hydrosphere. Aquatic animals absorb water directly from their liquid environment. They excrete some of it back while alive and after death the remaining water is returned through decay. Terrestrial animals absorb water from the reservoir present in soil and in bodies of fresh water. Animals and plants move such water through their bodies and in the process retain required amounts. The rest is excreted, partly as liquid water but more particularly as water vapour, which raises the moisture content of the atmosphere. Thus, a given amount of environmental water goes from hydrosphere to atmosphere more faster through the metabolic processes of living organisms than if that water were simply allowed to evaporate directly from the hydrosphere. In other words, the metabolism of terrestrial organisms actively contributes toward the maintenance of the global water cycle. Sometimes this may have an effect on the climate. For example, the tropical forest trees release so much water vapour that the surrounding air remains permanently saturated with moisture. After terrestrial organisms die, any liquid water in their bodies again returns to the hydrosphere through decay.

Water influences metabolism not only as the most important nutrient but also through its effect on all aspects of climate and weather, both in the sea and on land. In the sea, water warmed in the tropics becomes light and rises to the surface, whereas cool polar water sinks. These up-down displacements bring about massive horizontal shifts of water between equator and poles. The rotation of the earth introduces east- west displacements. These effects, reinforced substantially by similar wind-producing air movements result in *oceanic currents*. The latter influence climatic conditions not only within the seas, but also in the air and on land.

Another climatic effect is the result of the thermal properties of water. Among all liquids, water is one the slowest to heat or cool, it stores a very large amount of thermal energy. The oceans thus become huge reservoirs of solar heat. The result is that sea air chilled by night becomes less cold because of heat radiation from water warmed by day. Conversely, sea air warmed by day becomes less hot because of heat absorption by water cooled by night. Warm or cool onshore winds then moderate the inland climate in daily patterns.

Thirdly, global climate over long period of time is determined by the relative amount of water locked into *polar ice*. As polar ice is melting, water levels are now rising and coast lines are gradually being submerged. If trends during the past fifty years are reliable indications, the earth appears to be warming up generally. Deserts are presently expanding, snow-lines on mountains are receding to higher altitudes. In given localities more days of the year are snow-free, and the flora and fauna native to given latitudes are slowly spreading poleward. It is difficult to be sure whether these changes are merely part of a short warm cycle or are really indicative of a long-range trend. But global climate change is a major perturbation to the hydrologial cycle. Large- scale deforestation usually to create new agricultural land also causes perturbation of the water cycle resulting in drought or flooding.

All these various cyclic changes in the hydrosphere have profound impact on metabolism. By influencing temperature, humidity, amount of precipitation, winds, waves, currents, and indeed the very presence or absence of water in given localities, they play a major role in determining what kinds and amounts of metabolic processes are possible in such localities, hence what kinds and numbers of organisms may live there.

6.5.2 Major Environmental Problems of Hydrosphere

On a global scale, human activities have created several environmental problems of the hydrosphere. As a result, streams, rivers, lakes and oceans of the world have been adversely affected. Some of the major problems are:

(*i*) Water pollution
(*ii*) Eutrophication
(*iii*) Loss of aquatic biodiversity
(*iv*) Floods and related problems

6.5.2.1 Water pollution

Water pollution is a major environmental problem. The main desirable uses of water are: (*a*) Municipal water supply (drinking), (*b*) Industrial water supply, (*c*) Recreational uses such as swimming, boating, aesthetics, (*d*) Agriculture and fisheries, and (*e*) Ecological balance.

For each of these uses there are certain water quality requirements. Major water quality problems arise through interference with various uses of the water. These problems are confirmed through water quality monitoring and analysis. For example, the problem of low dissolved oxygen (DO) content in a stream interferes with the fish life of that stream besides promoting growth of pollution tolerant organisms and disappearance of clean water species, and the manifestation of that interference is with foul odours.

Water pollution sources are generally described as *point sources* and *non-point sources.* A point source is any single identifiable source of pollution from which pollutants are discharged, *e.g.*, the outlet pipe of a municipal wastewater treatment plant or industrial discharge. A nonpoint source is more diffuse and includes a variety of wastes from streets, construction sites, agriculture and food processing and mining wastes.

6.5.2.1.1 Water Pollutants

Water is polluted by: (*i*) Sewage and domestic wastes, (*ii*) Pesticides, (*iii*) Detergents and synthetic fertilizers, (*iv*) Toxic metals and pollutants, (*v*) Oil, grease and other chemicals, (*vi*) Acid deposition (increased pH), (*vii*) Suspended solids, (*viii*) Thermal pollution, and (*ix*) Pathogenic microorganisms.

6.5.2.1.2 Harmful effects of water pollutants

(*i*) *Effect of sewage and BOD problem.* The sewage and other organic matter discharged to a water body are degraded by oxygen-requiring micro-organisms. The amount of oxygen consumed by the microbes is the *Biochemical Oxygen Demand* (BOD). BOD is an indication of the presence of sewage and other organic wastes. After the addition of organic wastes, several zones may be distinguished in a flowing stream. These are zone of degradation, zone of active decomposition, zone of recovery and unaffected zone. If the volume of incompletely digested sewage is small in relation to the volume of water into which it is released, the oxygen demand may be easily satisfied. But when the volume of sewage is greater than the carrying capacity of the water, the oxygen demand may remain unsatisfied for a long period of time creating low-oxygen (hypoxic) conditions. Thus high levels of BOD can deplete the oxygen in water. Near the point of organic discharge, the bacterial population rapidly increases with active growth of sewage fungus (Fig. 6.9). Protozoa species that feed on bacteria increase in numbers, but the oxygen depletion causes a decrease of algae and clean water fauna. Fish being sensitive to DO concentration are often eliminated through mass fish deaths and only a few macroinvertebrates such as sludgeworms (*Tubifex*) and bloodworms (*Chironomus*) can exist in hypoxic water.

Effects of organic pollution in a lentic water body, as opposed to the flowing water environment, are modified by the morphometric features of the receiving water. Zonal changes that have been described for flowing water do exist, but may become compressed in great measure either laterally or vertically when the discharge is to a lake or estuary. Such compression may tend to decrease the severity of pollution that is often observed in lotic water body and on the other hand, may increase substantially the development of algal blooms or rooted aquatic plants that may develop from the nutrients released with the introduced organic waste and decomposed flora.

(*ii*) *Effects of Pesticides, Detergents and Synthetic Fertilizers.* Pesticides are synthetic chemicals used for pest control. A major source of pesticides in water bodies is the runoff from agricultural fields. Some pesticides also enter inland waters from industries, which use pesticides in their manufacturing processes, or from the manufacturing of pesticides themselves. Pesticides adversely affect a wide range of organisms including insects and fish. Pesticides accumulate in the tissues of aquatic organisms through bioconcentration and biological magnification and adversely affect their metabolic processes including reproduction.

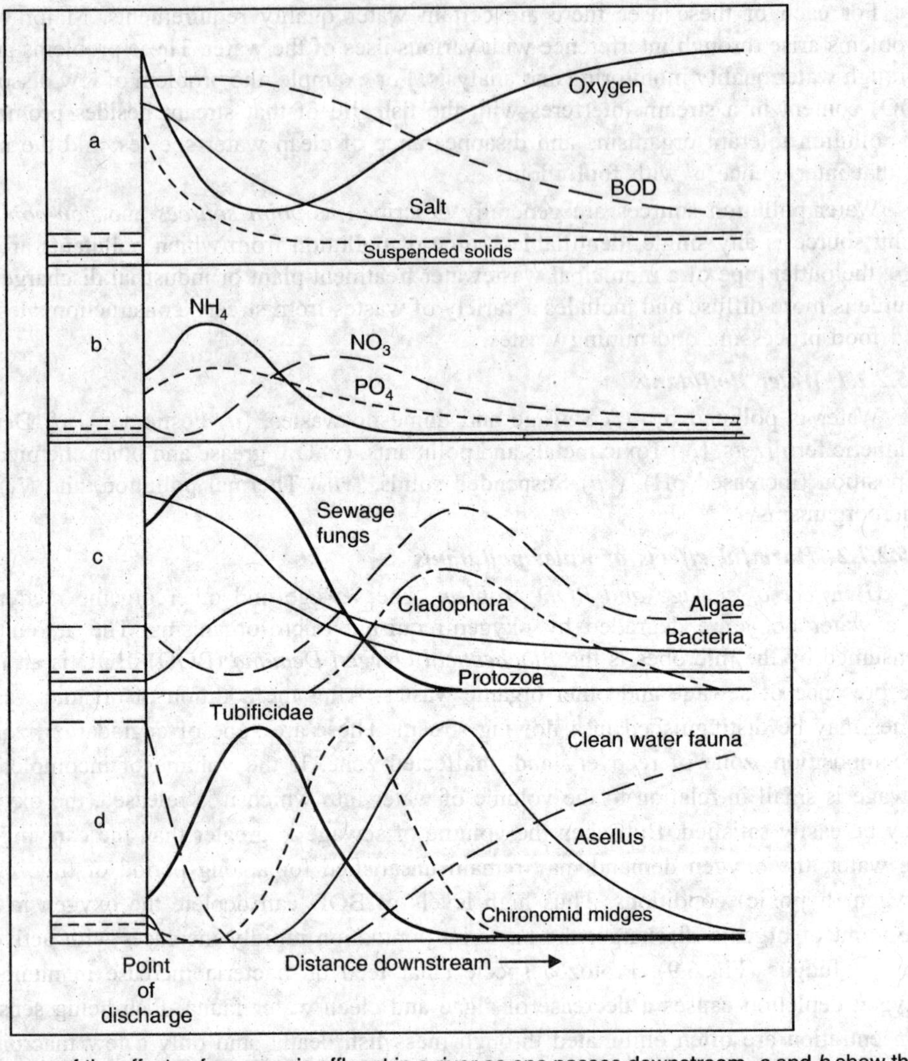

Fig. 6.9: Diagram of the effects of an organic effluent in a river as one passes downstream. *a* and *b* show the physical and chemical changes; *c* shows the changes in the algae and microorganisms; *d* summarizes the changes in the large animals (after Hynes).

Some pesticides kill not only the target organisms but they also adversely affect many nontarget organisms. Hence the term 'biocide' is often used for them. In the environment, a biocide may be detoxified or its toxicity is changed in some way. A good example of the second possibility is DDT, which is quite toxic to many insects and relatively nontoxic to birds. DDT is metabolized by detritivores along two pathways, under unaerobic condition to TDE (also called DDD) and under aerobic condition to DDE, as shown below (Fig. 6.10)

These metabolites, DDD and DDE both are toxic. However, the biocidal activity of the metabolites is different from DDT. DDE is relatively nontoxic to insects, but it seriously affects female birds. It disrupts their calcium metabolism. As a result, the eggs laid by affected birds may have very thin shells (Peakall, 1974). The birds most affected by DDT poisoning are the raptors, the carnivorous birds, since they are at the end of the food chain. Pesticides show the phenomenon of 'food chain magnification', by which species at the end of the food chain ingest more amount of the chemical than those at the beginning. Many other insecticides show similar patterns of biodegradation. By bacterial action, Aldrin is changed to Dieldrin, which is more toxic than the parent molecule.

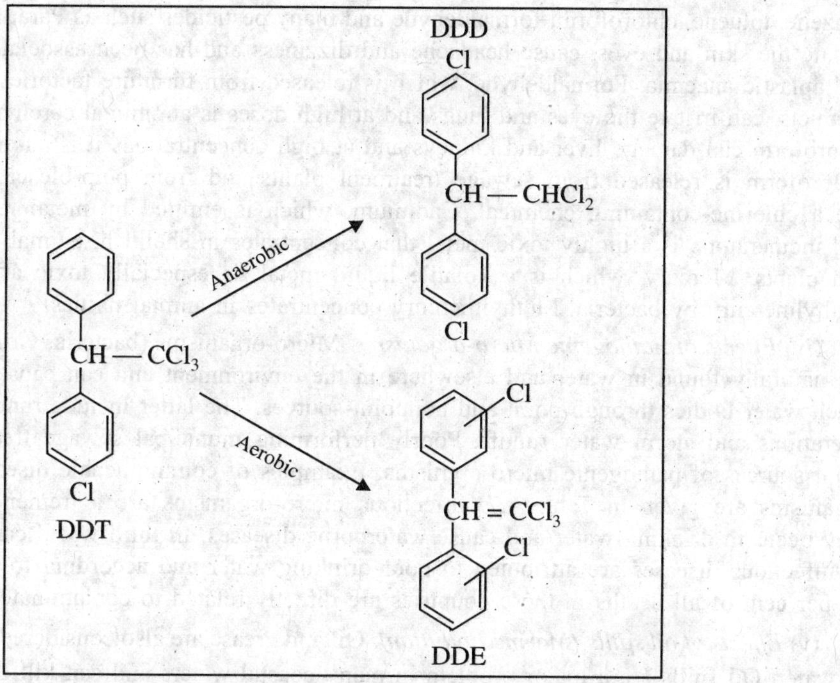

Fig. 6.10: Breakdown paths of DDT into DDD (TDE) and DDE

Pesticides also affect nontarget organisms. These effects may be mainly physiological and behavioural. *Physiological effects* may be manifested by death, sterility or some other dysfunction. The worst harm is done when the nontarget organism is a beneficial insect. Pesticides enter the human body by a number of routes. Some insecticides remain in the body, they are stored in certain tissues (especially in adipose tissue) and cause pathological effects. *Behavioural changes* caused by pesticides may be too many. For example, sublethal doses of DDT cause trout to forget most of their learned avoidance responses. Salmon exposed to sublethal dose of DDT become increasingly sensitive to cold water and may even be led to lay their eggs in abnormally warm water (Ogilive and Anderson, 1965). Mosquito fish exposed to relatively low concentrations of DDT tend to prefer waters that are more saline than usual for the species. Such behavioural changes might have quite serious impacts on the populations of desirable species.

Detergents are washing powders and materials. They are composed of complex phosphates which eventually breakdown into phosphates usable by aquatic plants. The use of detergents has been responsible for increase in the phosphorus in sewage effluents. Phosphate pollution of rivers and lakes causes extensive growth of algae, which depletes the DO content of water and disrupts the natural food chains.

Fertilizers containing nitrates and phosphates are used in modern agriculture. Some of these are washed off the agricultural fields through irrigation, rainfall and drainage into nearby rivers and ponds where they seriously disturb the aquatic ecosystem. Excessive use of fertilizers often leads to accumulation of nitrates in water. When such waters are used by animals, these nitrates are reduced to toxic nitrates by intestinal bacteria. Nitrates can cause a serious disease called *methaeglobinemia,* which can damage respiratory and vascular systems and even cause suffocation.

(*iii*) *Effects of Hazardous Pollutants (Toxic Metals and Other Chemicals*). As is the case with toxic air pollutants, there are many potentially toxic water pollutants. Of these, inorganic pollutants include arsenic and the metals cadmium, mercury, lead, nickel, zinc and copper. These can kill or sicken fish and other aquatic animals. Organic pollutants under this category are benzene, toluene, chloroform, formaldehyde and many pesticides such as Parathion. Benzene can irritate the skin and eyes, cause headache and dizziness and has been associated with leukemia and aplastic anaemia. Formaldehyde, which is released from furniture factories or pressed wood products, can irritate the eyes and lungs and at high doses is an animal carcinogen. Exposure to chloroform can damage liver and kidneys and at high concentrations it is an animal carcinogen. Chloroform is released from sewage treatment plants and from pulp-bleaching facilities that use a chlorine-containing chemical. Cadmium, which is emitted by metal refining operations and incinerators, is a highly toxic metal that concentrates in shellfish, animal kidneys and liver and plants. Mercury, which is a volatile liquid metal, is especially toxic after conversion to methylmercury by bacteria. Methylmercury concentrates in animal tissues.

(*iv*) *Effects of Pathogenic Micro-organisms.* Micro-organisms (bacteria, viruses and protozoa) are naturally found in water and elsewhere in the environment and can cause infections. They reach water bodies through point and nonpoint sources. The latter include runoff from livestock operations and storm water runoff. Poorly performing municipal sewage treatment plants are point sources of pathogenic micro-orgnisms. Examples of communicable disease indicators and organisms are given in Table 6.7. Infectious micro-organisms are a tremendous threat when they occur in drinking water and cause waterborne diseases. In third world countries, upto 80% of infectious diseases are attributed to poor drinking water and according to WHO, more than 35 per cent of all deaths in these countries are directly related to contaminated drinking water.

(*v*) *Effects of oil spills (Marine Pollution).* Oil and grease are also considered serious pollutants of water. Oil spills are a major problem in many coastal waters and can kill or adversely affect fish, phytoplankton and zooplankton, and birds and mammals. Oil spills have been reported from various parts of the world. Steiner (1994) has recently described the dreadful effects of the Alaskan oil spill, the *Exxon Valdez* that affected the entire biotic community including river otters, bald eagles, seals, sea lions, whales, and diving birds.

According to Hill (1997), oil spills can kill or reduce populations of organisms living in coastal sands and rocks, and may kill the insects and worms that serve as food to many birds and other animals. When oil moves into coastal wetlands, it kills fish, shrimps, birds and other animals. An oil spill may also foul beaches used for swimming. However, spills are not the only source of oil in water. Oil leaking from vehicles or released during accidents is washed off roads with rain water. Improper disposal of used oil from cars is another source. This oil adversely affects soil and water organisms and green plants.

Table 6.8: Examples of communicable disease indicators and organisms

Indicators	***Sources***
Total Coliform (TC)	Faeces of humans and other warm-blooded animals
Faecal Coliform (FC)	Intestinal tract of humans and other animals,
Faecal streptococci (FS)	Intestines of humans and animals
Pathogenic bacteria	**Disease caused**
Vibrio cholerae	Cholera
Salmonella spp	Typhoid and Paratyphoid
Shigella spp	Dysentery
Viruses	**Disease Caused**
Hepatitis A	Hepatitis

Enteroviruses	Polio
Polioviruses	Diseases of the central nervous system (CNS)
Echoviruses	Diseases of CNS
Coxsackieviruses	
Retroverus	AIDS
Pathogenic Protozoa and Helminths	**Disease caused**
Entamoeba histolytica	Amoebic dysentery
Giardia lambia	Giardiasis
Whipworm, hookworm, dwarf tapeworm, nematodes	Several

(Note: all diseases are not shown)

(*vi*) *Effects of Acids deposition.* Some water bodies may be naturally acidic, but others are made so by acidic deposition or acid runoff from mines. The sulphur in coal produces acids in the water, which is sometimes responsible for lowering the pH of streams. The lowered pH usually kills all the organisms of streams. Acid rain or acid deposition, which lowers the pH of lakes and streams to 4 or 5, adversely affects fish, algae and other sensitive organisms. Acid rain also affects green plants; there is reduced rate of photosynthesis and growth and increased sensitivity to drought and disease. Acid deposition weakens trees, so they could be easily killed by drought and pathogens. Nutrients like nitrates may be leached from the soil by acid runoff waters. The activity of nitrogen-fixing bacteria is inhibited, thus reducing soil fertility.

(*vii*) *Effects of Thermal Pollution*. Thermal pollution may be defined as "the warming up of an aquatic ecosystem to the point where desirable organisms are adversely affected" (Owen, 1985). A large number of industrial plants use cold water from the rivers and discharge it hot. Effect of thermal pollution depends on the temperature of the hot water released at a particular place (or point). In extreme cases it may cause direct mortality of fish and other desirable organisms. The lethal temperature for trout is above 25°C and for yellow perch 31°C. The trout eggs will not develop in water having temperature above 14°C. The thermal pollution causes interference in spawning and reproduction of fish since all fish require an optimum temperature for breeding. It is apparent that because thermal pollution may raise stream temperature to about 32°C during summer, any coldwater fish will have to leave the area to survive. It is well known that most freshwater fauna populations decline with rising temperatures and few species can exist in a water temperature of over 40°C. Above 30°C green algae and diatoms are reduced in numbers, but there is increased growth of blue-green algae and sewage fungus.

Thermal pollution causes reduction in DO content of water. For example, at a temperature of 0°C, water has DO content of about 14.5 ppm, but at 18°C it contains only 6.5 ppm. It is apparent that coldwater fish, which require about 6 ppm to survive, could not tolerate the high water temperatures resulting from thermal pollution. If they remained in the area, they would die from oxygen starvation. Besides these effects thermal pollution causes increased vulnerability to disease and permits invasion of organisms that are tolerant to warm water and disturb ecological balance.

(*viii*) *Effect of Eutrophication*. Eutrophication means the excessive growth of aquatic plants, both rooted and planktonic, to levels that are considered to be an interference with desirable water uses. Claphem Jr (1981) defines eutrophication as a natural phenomenon whereby the productivity of a lake increases from initial *oligotrophic* stage, through a moderately productive or *mesotrophic* stage, to a very fertile or *eutrophic* stage. This phenomenon is called *natural eutrophication* and leads to lake-aging in due course of time. When the process of eutrophication is speeded up by anthropogenic activity, it is called *cultural eutrophication.* The increased production of aquatic plants has the following consequences regarding water uses:

(*a*) Large diurnal variations in DO can result in low levels of DO at night, which in turn can result in the death of desirable fish species.

(*b*) Phytoplankton and weeds may settle to the bottom of the water body and create a sediment oxygen demand (SOD), which results in low levels of DO in the hypolimnion of lakes and reservoirs as well as in the bottom waters of deeper estuaries.

(*c*) Large diatoms and filamentous algae can clog water treatment plant filters.

(*d*) Toxic algae have sometimes been associated with eutrophication in coastal regions and have been a key factor in the occurrence of "red tide", which may result in paralytic shell-fish poisoning.

(*e*) Excessive growths of rooted aquatic macrophytes interfere with navigation, aeration and channel-carrying capacity.

(*f*) The algal blooms impair water quality by giving it a bad taste and odour.

The eutrophication of a given water body may vary depending on the degree of penetration of the solar radiation to different depths, the amount and type of nutrient inputs (Phosphorus, Nitrogen, Silica), the particulars of the water movement through flow transport and dispersion, and the geographical location of the water body. The principal external sources of nutrient inputs are municipal sewage and wastes, industrial wastes, agricultural runoff, forest runoff, and urban and suburban runoff. The nutrients are present in several forms. However, all the forms are not readily available for uptake of the phytoplankton. For example, total phosphorus is present in two forms – dissolved and particulate. The dissolved form, in turn, is composed of several forms, one of which is the dissolved reactive phosphorus. This form of phosphorus is used for phytoplankton growth. Total nitrogen is composed of four major components – organic, ammonia, nitrite and nitrate forms. The latter three forms make up the total inorganic nitrogen, which is utilized for phytoplanktonic growth. The organic form of nitrogen represents both the dissolved and the particulate component. The particulate form, in turn, is composed of organic detritus particles and the phytoplankton. The direct input of nutrients may be reduced by several means.

In many cases the process of eutrophication is speeded up by human activity; this is called *cultural eutrophication* and affects recreation, fisheries, water supply and public health. It is well known that about 80% of the nitrogen and 75% of the phosphorus added to the water bodies has its source in human activities. This human-generated nutrient input is derived from several sources, including domestic sewage, agricultural fertilizers, detergents, industrial wastes and livestock wastes. When the average concentration of soluble inorganic nitrogen exceeds 0.30 ppm and the soluble inorganic phosphorus exceeds 0.01 ppm, algal blooms may appear. During the summer, the algal bloom problem usually becomes more acute with adverse effects on the whole biota of the lake. Such cultural eutrophication has caused lake Erie to age 15000 years in only 25 years between 1950 and 1975 (Owen, 1985). Lake Mendota and lake Washington have also undergone rapid eutrophication as a result of human activities. In India, the evergoing problem of cultural eutrophication has already reduced the recreational value of the Kashmir lakes. The Nainital lake is also undergoing rapid eutrophication, especially due to inputs of sewage and domestic wastes and detergent discharges (Nagdali and Gupta, 2002). Several studies have indicated that cultural eutrophication can be controlled if phosphate inputs are reduced.

6.5.2.1.3 Control of water pollution

There are several measures by which the water quality in a system can be controlled (Thomann and Mueller, 1987). Some of these are:

1. *Reducing the effluent concentration of the waste input by* :
   (*a*) Wastewater and sewage treatment
   (*b*) Industrial in-plant process control
   (*c*) Eliminating effluent constituents by pretreatment prior to discharge to sewer systems or by different product manufacturing for an industry.
2. *Reducing the upstream concentration by upstream point and non – point source controls*
3. *Reducing the effluent volume by*:
   (*a*) Reduction of direct industrial discharge volumes into the municipal sewer system.
   (*b*) Reduction in infiltration into municipal sewer systems.
   (*c*) Reduction of waste volumes through process modifications in industries.
4. *Increasing the upstream flow by low flow augmentation, i.e.*, releases from upstream reservoir storage or from diversion from nearby water bodies.
5. *Increase the environmental, instream degradation rate of the substance*: This can be accomplished by a redesign of the chemical to result in a more rapid breakdown of the chemical by the natural heterotrophic bacteria in the stream. For example, redesign of synthetic detergents to reduce foaming and downstream transport through increased biodegradation rate. In manufacturing of potentially toxic chemicals also an attempt may be made to increase biodegradation rates so that a chemical buildup does not occur.
6. *Enviromental Controls*: The three broad categories, for control of pathogenic bacteria, viruses and parasites are – control at the input source of the micro-organism, control at the area of water use, and control of the product that is affected by contamination. Municipal wastes causing point pollution are the usual principal inputs of communicable disease organisms. Such inputs can be reduced by treatment of wastes without disinfection and disinfection by chlorination, ozonation, chlorine dioxide and ultraviolet radiation. Controls of the area of water use would include bathing restrictions on a transient basis, construction of dikes and diversion structures to protect a given area. Controls of the product would include chlorination or other methods of disinfection of water used for drinking water supply, treatment plants for contaminated shellfish to allow depuration of bacteria prior to marketing, and distribution of high-quality bottled water during emergency.
7. *Harvesting of Biomass*: Water hyacinth *(Eichhornia crassipes)* and other aquatic weeds are used to upgrade wastewater treatment lagoons and treat chemical wastewaters. The water hyacinth has been used for purifying not only domestic wastewater but also industrial wastewater. It can readily absorb, accumulate and concentrate heavy metals such as lead, cadmium, mercury and nickel. Other studies have evaluated the potential of water hyacinth to absorb various organic chemicals like phenols, toxaphene, etc. Water hyacinth can remove even radioactive metals from effluents, which is very difficult otherwise. As the metals are extracted from the effluents, it causes considerable cleaning of the effluent. The pH of effluent after treatment has been found to be between 6.8 and 7.8, which is an ideal range for freshwater. Further, it does not allow the algal bloom to develop and the BOD and COD are lowered to the level at which the effluent can be allowed to mix with freshwater without causing any bad effect. The controlled use of water hyacinth in conjunction with waste stabilization ponds not only increases the BOD removal capacity of these systems, but also reduces the high total suspended solids normally associated with sewage lagoons (Table 6.7). The harvested plant material is also a potential source of high quality protein, energy (biogas), fertilizer and other products.
8. *Bioremediation*: In bioremediation, living organisms are used to remediate chemicals at a hazardous waste site, at an oil spill, or in contaminated water or sediments. These micro-

organisms are able to degrade a chemical because they use it as a source of energy to promote their growth. The waste site is first investigated to identify the pollutants present there and the characteristics of the water and soil and to know the micro-organisms that are already naturally present. To speed up the action of microbes, nitrogen or other nutrients may be added to the soil and aeration may be used, if required.

An example of one of the best known uses of biomediation was in case of Exxon Valdez Alaskan oil spill in 1989. Naturally present microbes had already begun to degrade the oil hydrocarbons and tidal influx provided needed oxygen. A nitrogen fertilizer was added along the coast to speed up microbial action. Micro-organisms that received fertilizer degraded both surface and subsurface oil 3 to 5 times faster than did microbes in unfertilized areas. The oil that could have taken 10 years or more to naturally degrade was broken down in 2 to 3 years.

Bioremediation can also be useful in metal contaminated sites. Microbes can convert metals to forms that are more easily dealt with. They can also carry out reactions, that indirectly aid metal stabilization. For example, microbes can convert sulphate into hydrogen sulphide. For bioremediation, bioengineered organisms are also developed. Bioengineered bacteria can decompose PCBs and dioxins. Plants can also be used in bioremediation (*phytoremediation*). One aquatic plant reduced the TNT in contaminted water from a concentration of 128 ppm to 10 ppm. A recent study demonstrated that sunflower roots submerged in water concentrated radioactive wastes from water.

9. *Wastewater reclamation* : The sewage treatment yields irrigation water that contains a number of essential nutrients like nitrogen, phosphorus and potassium to make it a fertilizer. In West Bengal, the practice of irrigating fish ponds with sewage for raising fish is common. Another interesting aspect is the reuse or reclamation of sewage effluents for industrial purposes. This holds good for coping with ever increasing demand for water by industrial establishments in big cities. A few industries in Mumbai have already taken steps in this direction and the treated sewage is used for air-conditioning and other purposes.

6.5.2.1.4 Wastewater Treatment

Wastewater (or sewage) treatment is done in three stages – primary, secondary and advanced. Primary treatment first removes large objects through screening, and then smaller objects such as sand and small stones. Suspended solids remain in the wastewater, which are settled out in a sedimentation tank. After removal of solid materials, the wastewater is given secondary treatment (Fig. 6.11), where micro-organisms digest soluble contaminants. However, phosphours and nitrogen are not removed by this process and we require advanced or tertiary treatment for this purpose. The major function of tertiary treatment is to reduce the load of phosphorus and nitrogen through precipitation, nitrogen stripping, carbon absorption and chlorination. Two most common additional methods of tertiary treatment are physico-chemical and advanced biological treatment. The physico-chemical treatment includes the addition of reagents to wastewater to precipitate various entities, as well as carbon adsorption columns, ion exchange and reverse osmosis. It is especially suited to industrial application such as those involving heavy metals, and to wastes with very low or very high pH. In advanced biological treatment, wastewater is treated by biological communities, such as water hyacinths, after secondary treatment and before it is discharged into receiving waterways. However, due to cost factor advanced treatment is not always used.

Usually the cost of treatment of wastewater doubles from primary to secondary. The advanced treatment plants are twice as expensive to build as secondary treatment plants and probably four times as expensive to operate. Obviously, the cost of membrane separation (reverse osmosis), though quite useful for industrial wastewater treatment, is exorbitantly high. In areas where sewer lines are not available, the traditional *septic tanks* are used for sewage and wastewater treatment.

These systems produce a concentrated *septage,* which must be periodically pumped out and, if possible, taken to a municipal plant for treatment.

Table 6.9: Reduction of pollutants (mg/l) from effluent in a canal covered with water hyacinth

	Total *dissolved solids*	Organic *carbon*	BOD	Nitrogen	Phosphorus	Metals
Inflow	380	75	33	2.36	0.48	0.99
Outflow	212	13	3.6	0.43	0.08	0.001

Sewage/Wastewater (400 : 300)

↓

Primary treatment **SCREENING → Grit**

↓

SEDIMENTATION

Suspended solids allowed to settle → Primary sludge

↓

Secondary treatment **Effluent (150:200)**

↓

BIOLOGICAL TREATMENT

Uses micro-organisms to digest soluble substances

↓

SEDIMENTATION ⟶ Activated studge (secondary sludge)

Advanced treatment ↓

Effluent (30 : 20)

↓

SPECIALIZED TREATMENT

To remove nitrogen, phosphorus and other specific contaminants

↓

DISINFECTION

↓

Effluent (10:10)

Fig. 6.11: Domestic sewage/wastewater treatment methods. Sewage with suspended solids 400 mg/l and BOD 300 mg/l (400:300) gives an effluent of 10:10 strength as a result of treatment

MODEL QUESTIONS

1. Write an essay on man-biosphere relationship.
2. Define the term 'biosphere'. What is the contribution of biosphere toward maintenance of life?
3. What is the role of atmospheric ozone ? Which factors are responsible for ozone depletion?
4. Define 'Greenhouse effect'. What is the impact of global warming on climate and living organisms ?

5. Give a list of criteria of air pollutants. Write a note on the impact of air pollution on living organisms.
6. What are major environmental problems of lithosphere ? How soil pollution affects living forms ?
7. Describe the forest related problems of India. Write a note on National Forest Policy.
8. What are the major environmental problems of hydrosphere ? Describe the sources and harmful effects of water pollution.
9. What is marine pollution ? Describe the effects of oil spills on marine biotic communities.
10. Define eutrophication. How eutrophication affects our lakes? Write a note on cultural eutrophication of the Kashmir lakes.
11. Write notes on the following:
    (*a*) Ozone depletion, (*b*) Global warming, (*c*) Photochemical smog,
    (*d*) Soil erosion, (*e*) Protected Areas, (*f*) Biosphere Reserves,
    (*g*) Deforestation, (*h*) Eutrophication, (*i*) Marine pollution,
    (*j*) Thermal pollution, (*k*) Acid rain, (*l*) Wastewater treatment

MULTIPLE CHOICE QUESTIONS

12. Which of the following is not included in criteria air pollutants?
    (A) Carbon dioxide (B) Lead
    (C) Carbon monoxide (D) Nitrogen oxides
13. Which of the following is considered a pollutant?
    (A) Ozone in stratoshphere (B) Excess ozone in troposphere
    (C) Carbon dioxide (D) All the above
14. Which of the following is not a greenhouse gas?
    (A) Nitrous oxide (B) Carbon dioxide
    (C) Oxygen (D) Methane
15. The Bhopal gas tragedy was caused by leakage of:
    (A) Methyl isocyanate (MIC) (B) Carbon monoxide (CO)
    (C) Nitrogen monoxide (NO) (D) Some other poisonous gas
16. Which of the following parametres is most affected by water temperature?
    (A) Total dissolved solids (TDS) (B) Conductivity
    (C) Dissolved oxygen (DO) (D) All the above
17. The most effective way to fight oil spills in the ocean is:
    (A) Phytoremediation (B) Bioremediation
    (C) Microbial degradation (D) Microbial degradation and addition of nutrients
18. Of the following wetlands all except one are included among six Ramsar sites, point out which is that one ?
    (A) Chilka lake (B) Loktak lake
    (C) Nainital lake (D) Koeladeo National Park
19. Who of the following is considered the leader of chipko movement ?
    (A) Baba Amte (B) Sunderlal Bahuguna and C.P. Bhatt
    (C) Medha Patkar (D) Kiran Bedi

20. Who gave the concept that all States of our country should have a State animal and a State plant ?
    (A) M.S.Swaminathan (B) Menka Gandhi
    (C) S.Z.Qasim (D) T.N.Khoshoo
21. Photochemical smog does not contain:
    (A) Ozone (B) Peroxyacyl Nitrate
    (C) Particulates (D) Mercury
22. Which of the following are precursors of acid rain ?
    (A) Sulphur dioxide (B) Nitric oxide
    (C) Nitrogen dioxide (D) All of them
23. Ozone depletion is caused by:
    (A) CFCs (B) HFCs
    (C) HCFCs (D) All of them
24. "Blue baby" syndrome is caused by the presence in the blood of:
    (A) Lead (B) Nitrate
    (C) Methemoglobin (D) Haemoglobin
25. Which of the following is a nonchemical way to disinfect water ?
    (A) UV light (B) Chlorine dioxide
    (C) Chloramine (D) Ozone
26. Which of the following is not included in biological variables of water quality ?
    (A) MPN (B) TPC
    (C) BOD (D) Species diversity
27. Who first adopted the well-known BOD test for assessment of water quality ?
    (A) National Research Council (NRC)
    (B) US Food and Drug Adminstration (FDA)
    (C) US Environmental Protection Agency (EPA)
    (D) Royal Commission on Sewage Disposal (RC SD)
28. Sewage fungus community occurs in which of the categories of pollution ?
    (A) Oligosaprobic (B) Beta-mesosaprobic
    (C) Alpha-mesosaprobic (D) Polysaprobic
29. Which of the following is not included in particulate matter ?
    (A) Grit (B) Fly ash
    (C) Aerosols (D) Formaldehyde
30. Which of the following gases is not an air pollutant?
    (A) SO_2 (B) NO_x
    (C) O_2 (D) O_3

CHAPTER 7 Biodiversity

7.1 INTRODUCTION

Ever since the dawn of human civilization, biodiversity has been the backbone of human food, health and livelihood providing systems. The importance of biodiversity is economic, environmental, scientific and medical. The Global Convention on Biodiversity (CBD) was adopted at the UN Conference on Environment and Development in 1992. The CBD officially came into force in December 1993 and by January 2000, over 170 nations had ratified the CBD, which is a unique international agreement for conservation and sustainable use of biodiversity.

The term "biodiversity" is a concise form of "Biological Diversity" and was coined by *Walter G. Rosen* in 1986 (Wilson, 1994). The definition of biodiversity, as provided in Article 2 of Convention on Biodiversity (CBD), is as follows. "Biological diversity means the variability among living organisms from all sources including, *inter alia*, terrestrial, marine and other ecosystems and the ecological complexes of which they are part; this includes diversity within species, between species and of ecosystems".

Biodiversity is not just the numbers of organisms present in any natural ecosystem. Biodiversity as such is not a resource but a property of living systems. CBD is one of the main global attempts to set an agenda for maintaining biodiversity. According to Harvey B. Lillywhite (2002), the term refers to "the variety and variability among living organisms and the ecological complexes in which they occur".

These days we find lot of focus and public concern for biodiversity. As example, the Academic Press has recently announced the publication of a five-volume '*Encyclopedia of Biodiversity*' containing more than 300 separate articles and covering 20 different subject areas related to this topic. Numerous aspects of research in the field of "conservation biology" are related to issues of biodiversity and biological researches at community or ecosystem level are biodiversity dependent. The subject also is important to many developments in biotechnology, for example the genetic engineering for control of exotic species. To illustrate the level of concern for biodiversity issues that has emerged internationally, the Organization for Economic Cooperation recently approved plans for creating the world's largest biodiversity databank.

The Global Biodiversity Information Facility (GBIF) will contain not only a catalogue of all the known species, but will also include scores of databases related to molecular and genetic collections, geospatial informations and natural history data (MacCabe, 1999). During recent years our predominant approaches to understanding biodiversity related issues have come from the fields of ecology, population biology and genetics. However, we have ignored physiological approch to understand issues and paradigms related to biodiversity.

India is one of the 12 mega biodiversity countries in the world. The country is divided into 10 biogeographic regions. The wide variety in physical features and climatic conditions have resulted in a diversity of ecological habitats like forests, grasslands, wetlands, coastal and marine ecosystems and deserts which harbour and sustain immense biodiversity. Biogeographically, India is situated at the trijunction of three realms, Afrotropical, Indo-Malayan and Palaeo-Arctic, and therefore, makes the country rich and unique in biological diversity. The country is also one of the 12 primary centres of origin of cultivated plants and domesticated animals. It is considered to be the homeland of 167 important plant species of cereals, millets, fruits, condiments, vegetables, pulses, fibre crops, and oilseeds and 114 breed domesticated animals. *Endemism* of Indian biodiversity is significant. About 4900 species of flowering plants are endemic to the country. These are distributed over 141 genera belonging to 47 families. These are concentrated in the floristically rich areas of North-East India, Western Ghats, North-West Himalayas and the Andaman and Nicobar islands. These areas constitute 2 of the 25 hot spots identified in the world. It is estimated that 62% of the known amphibian species are endemic to India of which a majority is found in Western Ghats. Thus, endomism is an important characteristics of India's biodiversity. The number and percentage of species of bacteria, fungi, plants and animals are shown in Table 7.1.

Biodiversity hotspot

The concept of biodiversity hotspot was originated by *Norman Myers* in 1988. The hotspots idea was also promoted by *Russell Mittermeier* in his popular book 'Hotspots Revisited' (2004).

To qualify as a biodiversity hotspot a region must meet two strict criteria– (*a*) it must contain at least 0.5% or 15000 species of vascular plants as *endemics* and (*b*) it has to have lost at least 70% of its primary vegetation. Around the world at least 25 areas qualify under this definition, with nine other possible candidates. These sites support nearly 60% of the world's plant and animal biodiversity with a high share of endemic species. The present distribution of hotspots is as follows:

(*i*) Africa - 8
(*ii*) Asia Pacific - 13
(*iii*) Europe and Central Asia - 4
(*iv*) North and Central America - 4
(*v*) South America - 5

Two hotspots of India are– (*a*) The Himalayas and (*b*) Indo-Burma. Other important hotspots of the world are Barrier, Reefs, Borneo, Philippines and Madagascar, etc. The highest concentration of species in the western ghats is believed to be the Agasthyamalai hills in the extreme south. This region harbours over 450 bird species, about 140 mammalian species, 260 reptiles and 175 amphibians. Over 60% of the reptiles and amphibians are completely endomic to the hotspot. However, this remarkable diversity is severly threatened today.

Table 7.1. Number and Percentage of species of Bacteria, Fungi, Plants and Animals (Source B.S.I and Z.S.I., 1994)

Sl.No.	*Taxon*	*No. of species*	*Percentage*
1.	Bacteria	85	0.08
2.	Fungi	23000	21.2
3.	Algae	2500	2.3
4.	Bryophyta	2564	2.4
5.	Pteridophyta	1022	0.9

(*Contd.*)

6.	Gymnosperms	64	0.06
7.	Angiosperms	15000	13.9
8.	Insecta	53430	49.3
9.	Mollusca	5050	4.7
10.	Pisces	2546	2.4
11.	Amphibia	204	0.2
12.	Reptilia	446	0.4
13.	Aves	1228	1.1
14.	Mammalia	372	0.3
	Total	108276	100.0

7.2 BIODIVERSITY DISTRIBUTION

Biodiversity is not uniformly distributed. Several theories have been proposed to explain uneven distribution of biodiversity. Some important theories are – Evolutionary time theory, climatic stability theory, spatial heterogeneity theory, and productivity theory. These theories attempt to explain that biodiversity depends on resource gradient; biodiversity is a function of frequency of disturbance and biodiversity is highest at intermediate level disturbance. Neighbourhood recruitment limitations and herbivores and pathogens also play an important role in biodiversity distribution.

7.3 GEOLOGICAL HISTORY OF BIODIVERSITY

The geological history of biodiversity is about 3.5 to 4 billion years old (Table 7.2). The first appearance of multicellular organisms was perhaps a milestone in the history of biodiversity which, did not diversify until about 600 *million years ago* (Mya). In fact, by the end of the Cambrian we find that almost all of today's major animal phyla are present in the fossil record. Broadly speaking there were relatively few species during the Palaeozoic and Mesozoic eras. But about 100 Mya there was a tremendous increase in biodiversity that culminated at the end of the Tertiary and beginning of the Quaternary (Pleistocene). During these periods we had more living species and higher taxa of plants and animals than at anytime before or afterwards.

During the entire history of biodiversity we find dramatic increases as well as mass extinctions in certain periods (Seposke 1992). Thus, we find clear patterns of *adaptive radiation*, stabilization and extinction for both the terrestrial and marine groups. We find an overall increase in diversity through time, but we do not know the answer as to why the temporal diversification took place. We have also noticed several differential patterns of diversification—most animals are insects (> 80%), more species of mammals are rodents (> 40%) and most species of plants are angiosperms (> 75%). Such disproportionate diversifications are not easy to explain. Some scientists believe that *phytophagy* (plant eating) led to disproportionate diversification in insect groups (Mitter *et al* 1988). We are living in the Quaternary (Holocene) in a time of decreasing biodiversity, which is correlated with changes in climate and large - scale human activities.

Table 7.2. Geological Eras and Periods and the major events recorded (After Schopf, 1992)

Sl.No.	*Era*	*Period*	*Mya*	*Major Events*
1.	Precambrian	P∈	4500	The Origin of life, first multicelluler organisms
2.	Palaeozoic	Cambrian (∈)	550	All of the major phyla present in fossil record

(Contd.)

		Ordovician(O)	550	Jawless fish (First Vertebrates)
		Silurian (S)	440	Colonization of land by plants and arthropods
		Devonian (D)	410	Diversification of teleosts, first amphibians and insects
		Carboniferous(C)	360	Extensive spread of vascular plants, origin of reptiles, Amphibians dominant
		Permian (P)	290	Mass extinction of marine invertebrates,.
3.	Mesozoic			Origin of mammal-like reptiles and 'modern insects'
		Triassic (TR)	250	Origin and diversification of Ruling reptiles, Origin of mammals, Gymnosperms dominant
		Jurrasic (J)	210	Dominance of ruling reptiles (Dinosaurs) and gymnosperns, origin of birds
		Cretaceous (K)	140	Origin of flowering plants (angiosperms), Dinosaurs and many invertebrate groups became extinct toward the end of period
4.	Cenozoic			Diversification of mammals, birds, pollinating insects and angiosperms
		Tertiary (T)	65	Late Tertiary/early Quaternary, the zenith of biodiversity
		Quaternary (Q)	1.8	The origin of humankind

The present scenario of biodiversity of different groups possibly indicates that the numbers of described species may not be accurate. The reliability of accuracy is good for plants and chordates, it is moderate for insects and poor or very poor for some other groups like nematodes, bacteria and viruses. It is possible that many species became extinct even before they were described by taxonomists and the present state is what we discribe as biodiversity crisis.

7.4 THREE MAIN ELEMENTS/GROUPS OF BIODIVERSITY

The key elements of biodiversity are: ecological diversity, genetic diversity and organismal diversity (Heywood and Baste, 1995). These elements (Table 7.3) are the basic building blocks of biodiversity. These groups are intimately linked and may have even common elements. For example, populations appear in all the three elements. According to Gaston and Spicer (1998), the elements of biodiversity are not independent, the elements can be viewed as forming nested hierarchies. Within genetic diversity we have populations consisting of individuals, each individual having a complement of chromosomes. These chromosomes have genes which have nucleotides as their building blocks. Similarly, within organismal diversity, kingdoms, phyla, families, genera, species, subspecies, populations and individuals form a nested sequence, in which all elements at lower levels belong to one example of each of the elements at higher levels. Thus, elements of genetic diversity are more easily defined than others. Genetic diversity of course, fuels the engine of evolution (Ricklefs and Miller, 2000). But when we consider organismal diversity, there may be debate over the number of kingdoms or over the definition of species. Likewise, it is not easy to define the elements of ecological diversity. A large ecosystem may have several habitats, which in turn may have many microhabitats or niches.

Table 7.3. Three main elements of biodiversity (after Heywood and Baste, 1995)

Ecological Diversity	*Genetic Diversity*	*Organismal Diversity*
Biomas		Kingdoms
Bioregions		Phyla
Landscapes		Families
Ecosystems		Genera
Habitats		Species
Niches		Sub species
Populations	Populations	Populations
	Individuals	Individuals
	Chromosomes	
	Genes	
	Nucleotides	

7.4.1 Varieties of Ecological Diversity

The following varieties of ecological diversity are recognised (Singh, 2002, Singh and Kumar 2003).

7.4.1.1 Species Diversity

The number of species that occurs in a particular area is called its species richness (Donovan and Welden, 2002). It means the species richness in any habitat and is common currency of the study of biodiversity. Species richness index is essentially a measure of the number of species in a defined sampling unit (Magurran, 1988). Species richness is a function of sample size. However, it should not be confused with species abundance. Each natural habitat has a variety of species, which differ in their relative abundence. No community consists of species of equal abundance; some species are rare, others are common and still others may be abundant.

Species diversity measures are often more informative than species counts alone. According to Harper, (1977), there is "importance of taking an organism's eye view of community diversity". This comment is relevant to structural diversity as it is to species composition.

7.4.1.2 Resource Diversity

It means the diversity of resources that an organism (species) utilizes. For example, some fish species in the hillstreams have a wide trophic niche and depend on zooplankton, insects, and algae and diatoms for their food(Singh and Bahuguna 1983). In many cases food resources consumed by an organism differ during different stages of the life cycle, such as fry, fingerling and adult stages in case of fish. Thus, niche width is the measure of the diversity of resources utilized by a species. The usual approach is to use the Simpson index or the Shannon index to calculate the niche width. The number of resource types observed (*e.g.*, types of food items eaten, varieties of habitat utilized, kinds of behaviour employed) replace number of species in the equation. A separate value must be calculated for each type of resource and measures of abundance will depend on the way in which the index is being used. If the niche width of a particular species is under consideration then abundance may be measured as the number of individuals either eating each type of food, living in each sort of habitat, or adopting each kind of behaviour. However, if we wish to measure the niche width of an individual, then abundance can be taken as the amount of each food type eaten, the time spent in each habitat or the frequency with which each behaviour is performed (Magurran, 1988).

7.4.1.3 Habitat Diversity

It is the number of habitat types in a defined geographical area. This is an index, which measures the structural complexity of the habitat. This structural complexity of environment, in turn, is responsible for the presence of a wide variety of spatial and trophic niches. This means that if any habitat supports more microhabitats its biological diversity will be more as compared to a habitat which has less number of microhabitats. More studies on habitat diversity have been made for terrestrial environments. The number of substrate types has been related to species diversity for aquatic insects, molluscs and benthic macroinvertebrates. Gorman and Karr(1978) have taken bottom type, depth and current into account to investigate the link between habitat diversity in streams and fish species diversity. The author found that habitat diversity was more in small hillstreams and some tributaries than large, snow-fed rivers in the Garhwal Himalayas (Dobriyal and Singh 1988, Kumar 1992).

7.4.1.4 Differentiation Diversity

It is also called beta diversity. It means degree of change in species composition between sites or communities or along gradients. A number of studies on faunal diversity of fish and insects have clearly indicated that their dirstribution and abundance is governed by gradient and altitude, among other factors (Singh *et al.* 1994, Singh and Nautiyal, 1990). For example, stoneflies and rheophilic fish species in the rhithron parts of hillstreams are characteristic and are absent from their potamon parts. One study, indicated that there is zonation of animals within a river; *Simulium monticole* occurs from source to 12 km, *Simulium variegatum* from 12 to 35 km., and *Simulium equinum* from 20 to 50 km. All these studies indicate that greater diversity of species and habitats means greater ecological quality.

7.4.2 Types of Species Diversity

Each community has a variety of species, some of which are rare, others are common and still others are abundant. Thus, no community consists of species of equal abundance. Diversity tends to increase with increase in the area sampled, since the sampling progressively brings in a wider spectrum of habitats, ecosystems and biomes for consideration. The following types of species diversity are described. These are whittaker's (1977) four levels of inventory diversity.

7.4.2.1 Point Diversity

This is the diversity on the smallest scale, the diversity of a micro-habitat or sample taken from within a homogeneous habitat. It is but natural to expect differences in point diversity of the two sites (or micro-habitats) of a habitat.

7.4.2.2 Alpha Diversity

This is the second category of diversity. It is defined as within-habitat diversity (Mac Arthur 1965). It is also defined as the variety of organisms occurring in a particular habitat (often called local diversity).

7.4.2.3 Gamma Diversity

It is the third category of inventory diversity. It is defined as the diversity of a larger unit such as an island or landscape. It is considered to be the overall diversity of a group of areas of alpha diversity. Gamma diversity equals alpha diversity × beta diversity. Thus, it is the inclusive diversity of all the habitat types within an area (region).

7.4.2.4 Epsilon Diversity

This is the fourth category of inventory diversity. The epsilon or regional diversity is defined as the total diversity of a group of areas of gamma diversity.

According to Magurran (1988), a single plant may be considered as a unit of alpha diversity; a leaf as an area of point diversity; a group of plants occurring together as an area of gamma diversity; and the forest within which the plants are located as an area of epsilon diversity.

7.4.2.5 Other Categories

7.4.2.5.1 Beta diversity

The term β-diversity was coined by Whittaker (1960, 1977). It is essentially the same as MacArthur's (1965) between habitat diverity. Beta diversity is a measure of how different or similar a range of habitats are in terms of the variety of species found in them. It tells us how species diversity changes along gradient. Another way to view beta diversity is to compare the species composition of different communities. For example, forests have more species than grasslands; or more fish species occur in the potamon part of a river than in the rhithron part. Beta diversity is also called differentiation diversity or turnover diversity.

According to Ricklefs and Miller (2000), it is not practical to measure beta diversity directly because the habitat distributions of species overlap. But we can calculate the number of unique habitats recognised by species within a region from the following relationship: beta diversity = gamma diversity/alpha diversity (Fig. 7.1).

7.4.2.5.2 Delta diversity

It is defined as the change in species composition and abundance between areas of gamma diversity, which occur within an area of epsilon diversity. It represents differentiation diversity over wide geographic areas.

7.5 MEASURES OF SPECIES DIVERSITY

Any measure of species diversity, by itself, does not convey much information; we appreciate its significance only when we compare with any other measure. Measures of species diversity can be divided into three categories (Magurran,1988). These are:

(*i*) Species richness indices,

(*ii*) Species abundance models, and

(*iii*) Species proportional abundance based indices

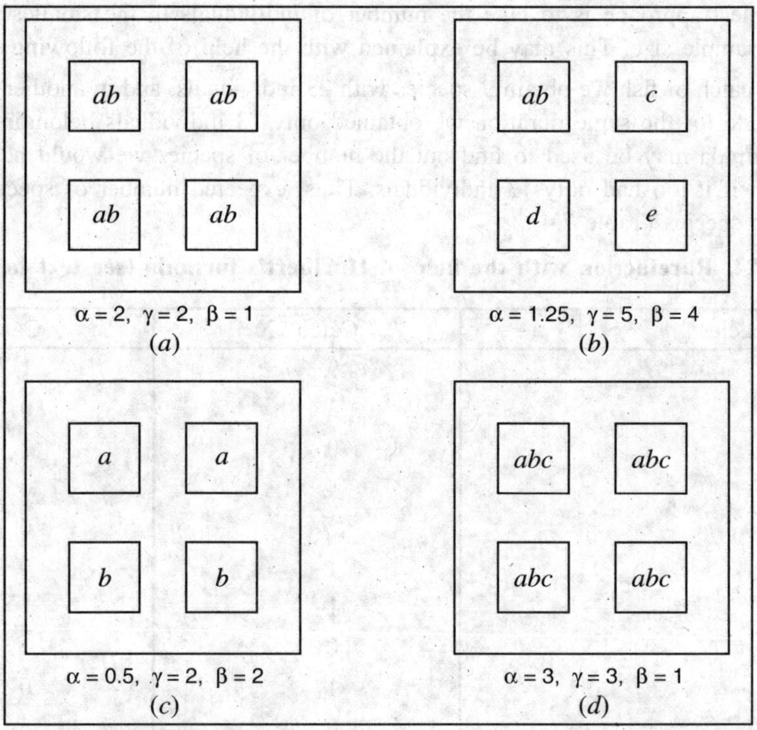

Fig. 7.1 Relationship between alpha, gamma and beta diversity (after Ricklefs and Miller, 2002). Each large box (region) has four small boxes (habitats). In (*a*), the diversity of each habitat (alpha diversity) is the same for all 4 habitats, each contains species a and b, species richness of 2. The regional diversity (gamma) is 2. The beta diversity is gamma/alpha, 2/2 = 1. In (*b*), alpha diversity is 2 for one habitat (species a and b) and 1 for the other three (species c, d and e occur alone in a habitat), yielding an average a diversity of 1.25; gamma diversity is 5 (No. of habitats × average alpha diversity), so beta diversity is gamma/alpha, 5/1.25 = 4. In (*c*) average alpha diversity is 2/4 = 0.5, gamma diversity 2, beta diversity 2/1 = 2. In (*d*) alpha diversity is 12/4 = 3, gamma diversity 3, beta diversity is gamma/alpha, 3/3 = 1. Regions (*a*) and (*d*) have different gamma diversities but the same beta diversity, indicating little species turnover in those areas.

7.5.1 Species Richness Indices

Species richness, as measure of diversity, has been used by ecologists. Species density or the number of species per m^2 is most commonly used to measure species richness. However, species richness increases with sample size. The smallest sample size may be 1 km^2 and the largest may be the entire region or country.

7.5.1.1 Rarefaction

As the sample sizes are always unequal, Sanders technique called Rarefaction is used to cope with this difficulty. Sanders's formula, as modified by Hurlbert (1971) is as follows:

$$E(S) = \sum \left\{ 1 - \left[\binom{N - N_i}{n} \Big/ \binom{N}{n} \right] \right\}$$

where $E(S)$ = expected number of species in the rarefied sample

n = standardized sample size

N = the total number of individuals recorded in the sample to be rarefied

N_i = the number of individuals in the *ith* species in the sample to be rarefied

The simplest approach is to take the number of individuals in the smallest sample as the standardized sample size. This may be explained with the help of the following example :

If in one catch of fish we obtain 9 species with 23 individuals, and in another catch from the same area made for the same duration we obtained only 13 individuals belonging to 6 species, Hurlberts' formula may be used to find out the number of species we would have expected in the first catch if it too had only 13 individuals. Thus, expected number of species for the first catch x is 6.6 species (Table 7.4).

Table 7.4. Rarefaction with the help of Hurlbert's formula (see text for details)

Species	Catch X	Catch Y
A	9	1
B	3	0
C	0	1
D	4	0
E	2	0
F	1	0
G	1	1
H	0	2
I	1	0
J	0	5
K	1	3
L	1	0
Total No. of species (S)	9	6
Total No. individuals (N)	23	13

1. The term $\binom{x}{y}$ is a 'combination' which is calculated as follows

$$\binom{x}{y} = \frac{x!}{y!(x\ \ y)!}$$

x ! is a factorial. For example 5! = 5 × 4 × 3 × 2 × 1 = 120

With these points in mind the computations can proceed

2.The first step is to take each species abundance from catch X and insert it in the formula

$$\left\{1-\left[\left(\frac{N-N_i}{n}\Big/\frac{N}{n}\right)\right]\right\}$$

Thus, for the species A which was represented by 9 individuals, the calculations are :

$$\left\{1-\left[\left(\frac{14!}{13!\times 1!}\right)\Big/\left(\frac{23!}{13!\times 10!}\right)\right]\right\} = \{1 - [14/1144066]\}$$

$$= 1 - 0.00 = 1.00$$

The result for each species is listed and summed to give the expected species number for catch X.

Ni	
9	1.00
3	0.93
4	0.98
2	0.82
1	0.57
1	0.57
1	0.57
1	0.57
1	0.57
Expected No. of Species for catch X	E(S) = 6.58 or 6.6

7.5.1.2 Menhinick's Index ($I\ M_n$)

This index is based on the ratio of number of species (S) and the square root of the total number of individuals(N)

$$I_{Mn} = S/\sqrt{N} \quad \text{or} \quad D_{Mn} = S/\sqrt{N}$$

It is claimed that this index may be used to compare samples of different sizes and that the effect of the number of individuals is reduced. However, some authors have shown that this index is not independent of sample size.

Using the data given in Table 7.4, the value of IM_n for catch *x* and catch *y* will be 1.88 and 1.66 respectively.

7.5.1.3 Margalef's Index (I_{Mg})

This index also relates the number of species to the number of individuals.

$$I_{Mg} = S - 1/\log_e N \quad \text{or} \quad D_{Mg} = (S - 1)/\ln N$$

The index is influenced by sample size. However, some authors have demonstrated that both this and Manhinick's index are insensitive to changes in community structure.

Using the data given in Table 7.4, the value of I_{Mg} for sample *x* and sample *y* will be 2.55 and 1.95 respectively.

7.5.2 Species Abundance Models

No community has species of equal abundance. Some species are very abundant, others may have medium abundance and still others may be rare or represented by only a few individuals. This observation led to the development of species abundance models. Species diversity data is frequently described by one or more patterns of distribution (Piclou, 1975), diversity is usually examined in relation to the following four models:

(*a*) the geometric series

(*b*) the log normal distribution

(*c*) the logarithmic series

(*d*) the broken stick model (the random niche boundary hypothesis)

when plotted on a rank abundance graph, the four models represent a progression ranging from the geometric series where a few species are dominant with the remaining fairly uncommon, through the log series and log normal distributions where species of intermediate abundance become more common and ending in the conditions represented by the broken stick model in which species are equally abundant as may be hardly observed.

7.5.3 Species Proportional Abundance Based Indices

These indices provide an alternative approach to the measurement of diversity. These indices are called *heterogeneity indices* (Peet 1974) as they take both species richness and evenness into consideration. Southwood (1978) called them *nonparametric indices* in view of the fact that no assumptions are made about the shape of the underlying species abundance distribution. The following indices are used.

7.5.3.1 Simpson's Index

This index relates the contribution made by each species to the total number of individuals present.

$$I = \Sigma p_i^2$$

Where pi is the proportion of individuals in the *ith* species. The equation given by Wilhm (1967) is the following :

$$I = \Sigma(ni(ni - 1)/N(N - 1)$$

Where ni = the number of individuals in the *ith* species and N = the total number of individuals. The values of Simpson's index range from zero to 1 (unity) and are inversely proportional to the wealth of species (As I increases, diversity decreases). Pielou (1969) has given the following form of equation.

$$I = 1 - \Sigma(ni(ni - 1)/N(N - 1)$$

Therefore, index is usually expressed as 1 – I or 1/I. The reciprocal form of Simpson's index ensures that the value of the index increases with diversity.

7.5.3.2 Shannon Index

The index independently derived by Shannon and Wiener from the application of information theory is known as the Shannon index of diversity. It is sometimes incorrectly referred to as the Shannon - weaver index (Krebs, 1985). The index assumes that:

(*a*) all species are represented in the sample, and

(*b*) individuals are randomly sampled from an 'indefinitely large' population (Pielou, 1975). It is calculated from the equation :

$$H' = -\Sigma \text{ pi ln pi}$$

Where pi is the proportion of individuals found in the ith species. It is estimated as (ni/N). N is total number of individuals in S species. The value of Shannon index usually varies between 1.5 and 3.5 and rarely exceeds 4.5. The value of H′ is related to species richness but is also influenced by the underlying species abundance distribution. May (1975) has shown that if the underlying distribution is log normal, 10^5 species will be required to give a value of $H' < 5.0$. Log_2 is often used to calculate Shannon index. Usually the index is obtained from the series.

$$H' = -\Sigma \text{pi ln pi} - \frac{S-1}{N} + \frac{1-\Sigma pi^{-1}}{12N^2} + \frac{\Sigma\left(pi^{-1} - pi^{-2}\right)}{12N^3}$$

7.6 ECOLOGICAL MECHANISMS CONTRIBUTING TO DIVERSITY

A number of simple and complex ecological mechanisms contribute to biodiversity. According to one view, the so-called local or deterministic view, the biodiversity is determined principally by biological interactions such as *competition and predation*. The other view considers the importance of other environmental factors such as soil type, moisture, temperature, gradient, productivity, niche and habitat diversity, and some other factors like stability, disturbance, immigration, extinction and species differentiation and movement at the regional level and the interaction between local and regional processes. Thus, there are a variety of determinants of species richness, some of which are discussed below.

7.6.1 Competition

The ecological effects of interspecific competition are too many. These effects, however, depend on the scale of competition. On a small scale we can observe the co-occurrence of only a few species with complementary ecological niches; but on a broad scale the community will contain more species, occurring within a patchwork, with each patch supporting only a few. In nature, competition is often avoided by differential resource utilization, *e.g.*, different species of fish feeding at different depths. When exotic animal or plant species are introduced to a new habitat, they sometimes prove to be better competitors and many indigenous species suffer. For example, introduction of common carp to some reservoirs of Punjab and elsewhere has adversely affected the native populations of major carps.

In brief, the significance of interspecific competition depends on how widespread are its evolutionary and ecological consequences. Interspecific competition tends to affect communities and their biodiversity in many ways some of which are less understood even today.

7.6.2 Predation

Predation affects both prey populations and whole ecological communities. When predation promotes the coexistence of species among which there would otherwise be competitive exclusion, this is called *predator- mediated coexistence.* The effect of predation on a group of competing species depends on which species suffer most. If it is subordinate species, then these may be driven to extinction and therefore the total number of species in the community will decrease. However, if dominant species suffer most, heavy predation will create space and resources for other species, consequently species numbers may then increase. The number of species in a community are usually maximum at intermediate levels of predation.

The effects of predation on biodiversity have been extensively studied in aquatic systems, in which the introduction of a predatory fish, starfish or salamender can greatly change the community structure of primary producers and consumers (carpenter *et al* 1987, 1988). The effect of herbivory on plant species diversity has also been studied. The *pest pressure hypothesis* suggests that seedlings are most dense close to the parent tree, but their survival is maximum at a distance from the parent, because herbivores will be more common among the dense seedling adjacent to parent tree. Several authors suggested that herbivory promotes more diversity in tropical forests (Clark and Clark, 1984).

7.6.3 Productivity

Nutrient input and productivity affect biodiversity. Plant productivity often depends on the nutrient or condition which is most limiting to growth. Animal productivity also depends on resource levels and some other key factors such as temperature and moisture (for terrestrial environments) and temperature, dissolved oxygen and depth for aquatic systems. However, these are not the only factors that affect productivity.

It was suggested by Connell and Orias(1964) that biodiversity should be highest in relatively stable habitats having high productivity.This is called *productivity-stability hypothesis* (Tilman,1982). Tilman and Pacala (1993) suggested that biodiversity does not increase monotonically with productivity for any group of species, but that species richness varies depending on what environmental factor is used as a measure of productivity and which species are being taken into consideration. It has been also suggested that predator-prey ratios increase with the increase in productivity. However, at high productivity levels, predators consume a disproportionate share of the available production, thereby causing a decline in community biodiversity. The *intertaxon competition hypothesis* (Rosenweig and Abramsk 1993) holds that the peaks of species diversity for different multispecies taxa should occur in areas showing different productivity levels. But, Tilman's (1982) hypothesis suggested that habitat heterogeneity increases with productivity to a certain point only after which it decreases. However, increase in biodiversity with productivity is not a universal phenomenon.

7.6.4 Spatial Heterogeneity

Spatially heterogeneous habitats offer a wide spectrum of resources and food chains. Accordingly, they are expected to support more species as they provide a greater variety of microhabitats (spatial niches), a greater range of microclimates, more types of places to hide from predators and a greater variety of trophic niches. Gould and Walker (1997) have shown a positive relationship between number of vascular plant species and index of spatial heterogeneity (ranging from 0 to 1), based on a number of things including soil pH, slope, drainage pattern and substrate types. There are several studies indicating a postive relationship between animal species richness and plant spatial heterogeneity.

7.6.5 Climate

The effects of climatic variation on species richness depend on whether the variation is unpredictable or predictable. In seasonally changing environment, different species may occur at different times of the year. Therefore, more species may be expected living together in a seasonal environment than a completely constant one. For example, in temperate regions different annual plants germinate, grow, flower and produce seeds at different times during a seasonal cycle. However, there is no firm relationship between species richness and climatic instability. Stable climate is likely to support more species. Tropics often showing better climatic regulation are richer in species than temperate regions.

7.6.6 Harshness of Environment

An environment may be called *harsh* or *extreme* if organisms are unable to live there. However, some organisms do occur in very cold and very hot environments and grossly polluted rivers and lakes. But the distribution pattern of organisms generally indicates that species richness is quite lower in harsh environments. Many studies have indicated that diversity of benthic macroinvertebrates and fish was quite low in streams and lakes having low pH. Most caves and hot spring also exhibit low biodiversity. The deepest parts of the oceans (200 m to 8000 m) also have few species of fish such as tripod fish and lizard fish, adapted to living in complete darkness, low temperature and very high pressures, some of them showing interesting specialisation of eyes and luminous organs.

7.6.7 Disturbance

Many communities experience periodic physical disturbance. Anthropogenic activities that alter habitat charactristics also affect local species diversity. The *intermediate disturbance hypothesis* (Connell, 1978) suggested that communities are expected to have more species when the frequency of disturbance is neither too high nor too low. This hypothesis was proposed to account for patterns of species richness in tropical rain forests and coral reefs.

In upland streams disturbances are created by a number of factors including water diversion for fishing or other purposes and quarrying. These activities affect indigenous fish species, benthic invertebrates and riparian vegetation. Sometimes flash floods with enormous gushing waters carrying tons of silt load play havoc with the natural communities as the entire river bed is destroyed and boulders and fish species are washed off by speedily flowing water (Singh and Badola, 1980, Sharma and Singh 1980, Kumar and Singh 2013). Disturbance by municipal sewage also affects biological diversity (Nautiyal *et al* 1996,2000). Townsend *et al* (1997) observed that the pattern of richness of macroinvertibrate species conformed with the intermediate disturbance hypothesis. Disturbances often keep the community in early stages of succession and therefore poor diversity.

7.6.8 Other Factors

Other factors and mechanisms that affect biodiversity are community succession, latitudinal gradient, altitudinal gradient and depth, immigration, emigration, extinction, and evolutionary age (time) and evolutionary adaptations.

The communities may differ in species diversity because some are closer to ecological or evolutionary equilibrium and others are still evolving. For example, tropics are richer in species than temperate regions because, among other things, tropics have existed over long periods of evolutionary time, whereas the temperate regions are still recovering from the Pleistocene glaciations.

Latitudinal gradients of diversity are quite obvious, species of plants, and animals increasing toward the equator. For example, within a small region at 60° north latitude one might find 10 species of ants; at 40°, there may be 50 to 100 species; and in a similar area within 20° of the equator, between 100 to 200 species (Ricklefs and Miller, 2000). Diversity in marine environments follows a similar trend. This increase in diversity from the poles to the tropics has been attributed to a number of factors including greater predation, productivity, light, temperature and water regimes.

Altitudinal gradients (Singh and Nautiyal 1990) and depth also affect species richness. In hillstreams, low diversity of macroinvertebrates was observed at higher altitudes. In terrestrial environments, a decrease in species richness with altitude is a widespread phenomenon. Bird, mammal and vascular plant species richness declined with the increase in altitude in Himalayan mountains of Nepal (Hunter and Yonzon, 1992; Whittaker,1977). Singh and Kumar (2003) found no fish species in Garhwal hillstreams at high altitudes (2400 to 3600 m). Number of species of phytoplankton, zooplankton and fish also tend to decline with increasing depth in the ocean.

Isolation and extinctions have also played their role in affecting species richness in different parts of this earth. The extinctions of many large animals in the Pleistocene may reflect the role of human migration. It is wellknown that over the past 30,000 to 40,000 years, a major loss of animal biodiversity has occurred over Australia, North America, New Zealand and Madagascar. All over the globe, human factor still continues to be the worst enemy of natural biodiversity and has created the so-called biodiversity crisis, discussed later in this chapter.

7.7 DIVERSITY- STABILITY RELATIONSHIP

A.R. Wallace had suggested that diversity in the tropics was greater than in temperate regions, because tropical regions have enjoyed longer periods of *stability* and, thus, have had more time for species differentiation. This idea is now sometimes referred to as the *time hypothesis* of species diversity (Ricklefs and Miller, 2000). A number of evidences are available to support the time hypothesis. Evidence from the fossil record that diversity has increased over-time, also supports the time hypothesis.

Four hypotheses have been proposed to explain how much biodiversity is enough and how ecosystem processes may respond to reduction in biodiversity. The *diversity-stability hypothesis* proposes that ecological communities will decrease in ability to recover from disturbance and in productvity if there is reduction in species richness (Johnson *et al* 1996). According to *rivet hypothesis* (Ehrlich and Ehrlich 1981), the relationship between species richness and ecosystem functioning is non-linear, but may follow a variety of possible trajectories. The loss of a few species (or rivets holding together an aeroplane) will create no problem, but beyond a certain point losses will cause catastrophic effect. The *redundant species hypothesis* predicts that a minimal diversity is necessary for ecosystem processes, but beyond this most species are redundant in their roles (Walker 1992). And finally, the *idiosyncratic response hypothesis* (Lawton, 1994) proposes that with changes in biodiversity ecosystem processes also change, but the magnitude and directiion of change is not predictable because each individual species has complex and varied roles.

Howover, it is difficult to answer the question – how much biodiversity is needed? But diversity-stability hypothesis is supported by many studies (spring-fed hillstreams in the Garhwal region support more biological diversity of aquatic insects (Kumar,1992) than large snow-fed rivers which are more disturbed) and we may like to have some 'functionally redundant'(?) species because they are 'beautiful', if not 'useful'.

7.8 BIODIVERSITY CRISIS

The present and projected losses of biodiversity have been termed a 'crisis' (Wilson, 1985). The crisis is particularly more serious in developing countries like India. Serious losses of biodiversity are presently taking place all over the world. It follows that the 'bottom line' is how much of biodiversity can we alter, or loose, before we threaten our own existence (Gaston and Spicer, 1998). The erosion of biodiversity may be attributed to the following factors.

Table 7.5. Animal Species in major taxonomic groups that are listed as extinct, extinct in the wild or threatened (after IUCN, 1996)

Sl.No.	*Group*	*Extinct*	*Extinct in the wild*	*Threatened*
1.	Mammalia	86	03	1096
2.	Aves	104	04	1107
3.	Reptilia	20	01	253
4.	Amphibia	05	00	124
5.	Cephalaspidomorphi	01	00	03
6.	Elasmobranchii	00	00	15
7.	Actinopterygii	80	11	715
8.	Sarcopterygii	00	00	01
9.	Arachnida	00	00	10
10.	Chilopoda	00	00	01
11.	Crustacea	09	01	407
12.	Insecta	72	01	537
13.	Onychophora	03	00	06
14.	Oligochaeta	00	00	05
15.	Polychaeta	00	00	01
16.	Bivalvia	12	00	114
17.	Gastropoda	216	09	806
18.	Enopla	00	00	02
19.	Turbellaria	01	100	00
20.	Anthozoa	00	00	02
	Total	**609**	**30**	**5205**

7.8.1 Extinction of Species

The list of animal and plants species that have become extinct or extinct in the wild (Table 7.5) is quite alarming. Since 1600 there have been over 1000 recorded extinctions of plant and animal species and about half of these extinctions took place between 1900 and 2000 (Smith *et al* 1993). Probably early humans were directly responsible for the extinction of many large mammals during the late Pleistocene and also indirectly responsible for the loss of many smaller ones (Martin 1984, Owen-Smith 1987). Colonization of tropical Pacific islands by prehistoric peoples led to large number of avian extinctions (Pimm *et al,* 1995). However, our studies and data on extinctions are biased toward birds and mammals and also toward islands and some developed nations. Regarding invertebrate and plant extinctions we have little information. It is sometimes advocated that the current extinction rates are possibly 4 or 5 times more than the rates in the fossil record. The Red List of 2003 has more than 12 thousand species and the current extinction phase is effecting large mammals and birds and freshwater invertebrates. These changes call for Rapid Biodiversity Assessment (RBA).

7.8.2 Human Population Growth and Urbanization

According to World Resources Institute the world's human population is estimated to about 7.13a billon and to be increasing by more than 86 million per annum. Howover, between 1800 and 2000 the rate has increased dramatically. This upprecedented human population growth led to increase in the number and sizes of the megacities in the world. A large portion of the world's total population resides in about 25 megacities. Though the world's cities take up only 2% of the earth's surface, they account for approximately 78% of carbon emission from human activities, 76% of industrial wood use and 60% of the water tapped for use by people (UNEP-IETC,2000). Space created for expansion of cities or for conurbation have been through large-scale deforestations and use of agricultural land for human settlements. These practices have created serious threats to biodiversity. It has been estimated that with the present rate of destruction of tropical forests, 20 to 25% of the world's plant species would have been lost by the year 2000. According to another estimate, 90% of tropical forest area containing more than 500 varieties of plant species will be destroyed during the next 20 years.

7.8.3 Energy Consumption and Emissions

It is estimated that from before the agricultural revolution to the present time,the total energy consumption by humans has multiplied about 7000 to 13000 fold (Ehrlich, 1995). The global commercial energy production in 1993 was 338 EJ (1EJ = 10^{18} J), or about 163 million barrels of oil, 40% greater than in 1973. As this trend of energy consumption is further increasing year after year, we have more and more emissions of harmful gases to the troposphere. Global emissions of CO_2 during 1992 were estimated at 26.4 Gt (1Gt = 10^9 t), of which about 85% was from industrial activity (World Resources Institute). The power generation activities and emissions of gases have adversely affected biodiversity both directly and indirectly.

7.8.4 Changes in Land use Patterns

According to Richards (1990), changes in land use patterns during the past 300 years have caused about 19% reduction in forest cover, 8% reduction in grasslands and pastures and 500% increase in croplands. Expansion of modern agriculture and extensive and intensive aquaculture practices have also led to serious changes in land use over the past 50 years. These changes have in turn caused severe losses of biodiversity in agricultural ecosystems . During green revolution a large number of wild crop varieties were replaced with a few hybrid species. This resulted in the small disappearance of our genetic resources of crop plants especially for wheat and rice.

With the disappearance of the plants, the associated microorgnisms and fauna were also lost. Besides this, indiscriminate use of synthetic fertilizers and other chemicals reduced the microbial species diversity.

7.8.5 Habitat Degradation and Loss

Habitat degradation and loss due to human activities and other disturbances are well known factors all over the world.The pattern of human disturbance may vary but it is causing unforeseen changes in all types of ecosystems and thus threatening the biodiversity. As a result of habitat degradation wild populations become more vulnerable to predators and diseases and fight for their existence. This is especially true for wildlife, which suffer due to habitat loss and fragmentation.

7.8.6 Pollution

Some studies have shown a direct connection between economic growth and pollution and also between economic development and pollution. Pollution of air, water and soil and the presence of toxic and hazardous pollutants in different components of biosphere have created serious problems. As a result, our freshwater resources have suffered irreparably and many species of fish and aquatic birds and mammals have been threatened. Fish and birds are also killed by unregulated pesticide use. In case of an electric power plant, only about 40% of the coal's energy value is used to run turbines to produce electricity; the remaining 60% is low grade energy which ordinarily becomes thermal pollution and harms all the aquatic communities and their natural food chains. Thus, environmental pollution is considered as one of the immediate causes affecting current levels of biodiversity (Mc Neely *et al* 1995).

7.8.7 Global Climate Change

As a result of global warming and other factors, a climate change is experienced. The potential environmental implications of climate change are too many. As a result of climate change snow caps are receding, summer temperatures are rising and many coastal areas are going under water, besides many unnoticed changes in populations and ecosystem functioning. It is estimated that the greenhouse effect is expected to threaten 400 species of mammals, 460 species of reptiles, 650 species of birds, 660 species of freshwater fish, and the tens of thousands species of invertebrates and plants in the United States (Sweeny). An increase in extinction rate of some of these species and changes in their migration patterns are predicted.

The global climate change is also going to affect crops fields, forests and coastal wetlands which are very rich in biodiversity. Due to pressure on account of environmental changes, 11046 species of plants and animals are facing high risk of extinction today (IUCN, 2000). In Oct.,2002 in Delhi 169 countries participated in UN conference and prepared Delhi declaration on climate change (Ddcc), which emphasizes on proper use of energy and ratification of Kyoto protocol by all the countries.

7.8.8 Introduction of Exotic Species

Foreign species introductions are also considered one of the immediate causes affecting current levels of biodiversity. Although man has moved many species of plants and animals from one part of the globe to another and many of such species have established themselves in new environments but few studies have been made to assess the impact of exotic species on the local fauna and flora. At a regional scale, new introductions may tend to increase biodiversity, but they also tend to homogenize biodiversity between regions (Gaston and Spicer, 1998). In many cases exotic species outcompate or prey on the pre-existing biota. One of the most dramatic examples is of the brown tree snake *Bioga irregularis,* which since its introduction has caused high decline in the avifauna of the island of Guam (Pimm 1991). The indigenous fish fauna of India has suffered due to exotic species like common carp and the African cat fish, *Clarias gariepinus,*which has faster growth rate (NBFGR, AR 1999-2000).

7.8.9 Other Factors

A number of other factors such as global trade and economic systems and policies that do not fully appreciate the value of natural resources have detrimental impacts on biodiversity. Illegal trade and poaching of wildlife products have reduced the populations of many wild animals such as African elephant, Asiatic lion, one horned rhinoceros and musk deer. Stuart Chapman recently commented 'the Asian rhinos' future hangs in the balance unless we reduce the persistent demand for its horn used in traditional Asian medicine as well as protect them from poachings'. Trade in medicinal plants has also played an adverse role.

It may be concluded that biodiversity is being destroyed mainly by human activities at alarming rates almost in all parts of the earth. If the present rate of loss of biodiversity continues, it is not surprising that many valuable species will become extinct even before they are described. The biodiversity loss results in ecological and environmental deteriorations.

7.9 CONSERVATION OF BIODIVERSITY

In view of the importance of biodiversity (economic, environmental, scientific and medical) the urgent need for conservation of biodiversity was felt in Rio conference (1992). The World Summit on sustainable development held in Johannesburg in August 2002 reiterated that the conservation of biodiversity was necessary for the survival of human race on this earth. In the Convention on Biodiversity (CBD), which has 42 Articles, Article 8 and 9 are about in- situ conservation and ex-situ conservation respectively. The objectives of CBD clearly stated the conservation of biological diversity, the sustainable use of its components, and the fair and equitable sharing of the benefits arising from the utilization of genetic resources.

7.9.1 *In-situ* Conservation

In-situ conservation is possible through protected area systems or networks. A large number of protected areas in different countries of the world are recognised by IUCN *Commission on Parks and Protected Areas,* although they vary in their effectiveness in conserving biological diversity. The Article 8 of CBD also mentions to ensure that development in areas adjacent to protected areas does not undermine the capacity of those protected areas to conserve biodiversity.

The Protected areas cover about 5% of total geographic area of the country. In India, the protected area network includes National Park (102), Wildlife sanctuaries (515), Conservation reserves (47), Community reserves (04) Biosphere reserves (18) Wetlands and Ramsar sites (Table 7.6) Of them, some National Parks have tiger reserves (Table 7.7)

7.9.2 *Ex-situ* Conservation

Ex-situ conservation measures include seed banks, sperm and ova banks, plant tissue cultures, captive breeding of animals and artificial propagation of plants. Article 9 of CBD deals with ex - situ conservation. It is mentioned in Article 9 that ex-situ measures are necessary for complementing in -situ measures and each nation shall (*a*) "adopt measures for the ex-situ conservation of components of biological diversity, preferably in the country of origin of such components; (*b*) estiblish and maintain facilities for ex -situ conservation of and research on plants, animals and microorgnisms, preferably in the country of origin of genetic resources; (*c*) adopt measures for the recovery and rehabilitation of threatened species and for their reintroduction into their natural habitats under appropriate conditions; (*d*) regulate and manage collection of biological resources from natural habitats for ex-situ conservation purposes so as not to threaten ecosystems and in-situ populations of species, except where special temporary ex-situ measures are required under subparagraph (*c*) above ; and (*d*) cooperate in providing financial and other support for ex-situ conservation outlined in subparagraphs (*a*) to (*d*) above and in the establishment and maintenance of ex-situ conservation facilities in the developing countries."

The CBD is currently one of our 'best hopes' and a number of countries have begun to develop and follow action plans in accordance with the articles of the Convention.

7.9.3 Biodiversity Conservation in India

In India, several Bills and Action Plans have been passed by the Parliament for the conservation of biodiversity. Some policy and legal frameworks recently developed to protect the biodiversity are The Forest (Conservation) Act 1980, The Wildlife (Protection) Act 1972, The Environment (Protection) Act 1986, The National Biodiversity Strategy and Action Plan (NBSAP) 1999 and Biological Diversity Bill 2002. The Bill provides for the constitution of a separate high-power National Biodiversity Authority (NBA) at Chennai. The Bill also envisages the setting up of State Biodiversity Boards, which would advise the State Governments on issues relating to conservation of biodiversity and sustainable use of its components.

In India, commendable efforts have been made for in-situ conservation of biodiversity. Under Protected Area network, now we have 18 Biosphere Reserves set up in different biogeographic regions of the country, besides 515 wildlife Sanctuaries and 102 National Parks. Some special projects were also started for providing protection to wildlife in their natural habitats. *The Project Tiger* was started with the financial assistance of WWF in 1973 with 9 Tiger Reserves. Now we have 39 tiger reserves covering a large area. Besides preserving the tiger and its varied habitats, the wild fauna and flora of these habitats have also benefited from the project.

Other wild animals which have been protected and rehabilitated through such projects are the Asiatic lion, the Black buck, the Rhinoceros, the Musk deer, the Hangul, the Ghariyal. India is also a party to CITES (Convention on International Trade in Endangered Species). In India, specific programmes for conservation of wetlands, mangroves and coral reef systems are also started.

Major wetlands of India have already been listed under the Ramsar Convention (1981), six internationally significant wetlands of India have been declared as "Ramsar Sites".

The Project Tiger: Realising the need and importance of the tiger conservation, the project is the result of the recommendation of a Task Force of the Indian Board for Wildlife. The Project Tiger was so named because the tiger is an animal which symbolises the biological diversity of the Indian subcontinent.

Tiger Reserves: It is an area declared by notification (as a tiger reserve) by the State Government on recommendation of the Tiger Conservation Authority to promote the conservation of the tiger and its habitat under the management regime of a 'Tiger Conservation Plan' prepared by the State Government.

The tiger reserves are the especially constituted reserves representing different habitat types with the aim of maintenance of available tiger population in India in its natural environment. A core zone of minimum of 300 sq. km has been identified in each Reserve along with a suitable area of buffer. While the core areas of tiger reserves are to be kept inviolate without people—the buffer areas are to be managed as safe cohabitation space for people and the wild cat.

Tiger Census Methods: The 'pug mark' count was the usual method in the past for tiger census. It used to be performed every fourth year with the help of voluntary organizations along with representations of Wildlife Institute of India (WII), Dehradun. The method was considered scientific and approved by government.

Another method was non-invasive DNA testing of stool droppings in forest areas. This method was evolved by the Centre for Cellular and Molecular Biology (CCMB), Hyderabad. According to some experts, it is now increasing. Faeces samples are carefully collected and preserved in absolute alcohol. It is first screened for species identification of tigers or leopard. After DNA extraction individual tigers are identified and result matched with the findings of camera traps in the relevant area.

The new method is "*Camera trap method*". This method has been used in the latest census of tigers. The camera trap head, count is done in selected areas and according to some experts, thee is every possibility that certain other areas have been left out. This method has shown a sharp decline in their numbers. The number of tigers in 2001 was 3642 but in 2008, the number is 1411 only. In Uttar Pradesh, from 273 in 2005, the number of tigers in the state, as per the latest census announced in Feb. 2008, has come down to 91 minimum and 127 maximum; which denotes a sharp decline. This includes 95 in Dudhwa Tiger Reserve. The total population of tigers is about 1500.

Performance of Project Tiger: The performance of the project during the past years has been quite satisfactory. As a result, not only the habitat conditions for wildlife have improved, the number of wild animals is also on the increase. For example, the Kanha had only 39 tigers when it came under Project Tiger in 1973, but in 1979 their number rose to 62 and in 1984 there were about 700 tigers in M.P.

Under the project, the number of Royal Bengal Tigers rose from 1800 in 1973 to 4100 in 1988. The official 1993 tiger census estimated a total of 3750 tigers. Of these only 1266 (34% of the total) were found within the boundaries of the then 19 Tiger Reserves. The 1993 census in M.P. estimated a population of 910 tigers, the highest density of tigers anywhere in the world.

According to wildlife experts, 'Project tiger', has proved beneficial to both tigers as well as tribals. Forty tribal villages have been rehabilitated with central assistance under the project and are now reaping the fruits of development. The Union Government has spent nearly 19 million U.S. dollars on the project since its inception. Some 6000 villages has to be removed out of the core areas. These communities were provided with new houses, schools, biogas plants to replace wood and fuel and give assistance to improve their farming.

Besides preserving the tiger and the habitats, the wild fauna of the latter have also benefited from the project. Now rare animals like the snow leopard, caracal hispid hare, pigmy hog and large sized animals like the India rhino and wild elephants are living more comfortably in the relatively disturbance-free environments of the different Tiger Reserves. Now it is proposed to set up a field research station in all the Tiger Reserves so that more attention may be paid in the field of wildlife management under the project.

Poaching for commercial gains threatens Royal Bengal Tigers. According to a report of Wildlife Protection Society of India (WPSI), 63 tigers were killed in 1994 and 64 in 1995. Intensive poaching activity occurs in all areas, which have a large number of tigers.

Table 7.6. List of Wetlands of International Importance from India (under Ramsar Convention)

Sl.No.	*Name of Wetland*	*State*	*Date of declaration*
1.	Chandratal	H.P.	08/11/05
2.	Renuka	H.P.	08/11/05
3.	Rudrasagar	Tripura	08/11/05
4.	Uppar Ganga	U.P.	08/11/05
5.	Hokarsar	J & K	08/11/05

(Contd.)

6.	Surinsar-Mansar	J & K	08/11/05
7.	Bhitarkanika Mangroves	Orissa	19/08/02
8.	Bhoj Wetland	Madhaya Pradesh	19/08/02
9.	Deepor Beel	Assam	19/08/02
10.	East Calcutta Wetlands	West Bengal	19/08/02
11.	Kolleru Lake	Andhra Pradesh	19/08/02
12.	Sasthamkotta Lake	Kerala	19/08/02
13.	Ashtamudi	Kerala	19/08/02
14.	Tsomorari	Jammu & Kashmir	19/08/02
15.	Vembanad-Kol Wetland	Kerala	19/08/02
16.	Point Calimere Sanctuary	Tamil Nadu	19/08/02
17.	Pong Dam Lake	Himachal Pradesh	19/08/02
18.	Kanjli	Punjab	22/01/02
19.	Ropar	Punjab	22/01/02
20.	Harike Lake	Punjab	23/03/90
21.	Loktak Lake	Manipur	23/03/90
22.	Sambhar Lake	Rajasthan	23/03/90
23.	Wular Lake	Jammu & Kashmir	23/03/90
24.	Chilika Lake	Orissa	01/10/81
25.	Keoladeo National Park	Rajasthan	01/10/81

Table 7.7. List of the Tiger Reserves in India

	Year of Creation	*Name of Tiger Reserve*	*State*
1.	1973-1974	Bandipur	Karnataka
2.	1973-1974	Corbett	Uttarakhand
3.	1973-1974	Kanha	Madhya Pradesh
4.	1973-1974	Manas	Assam
5.	1973-1974	Melghat	Maharashtra
6.	1973-1974	Palamau	Jharkhand
7.	1973-1974	Ranthambhore	Rajasthan
8.	1973-1974	Similipal	Orissa
9.	1973-1974	Sunderbans	West Bengal
10.	1978-1979	Periyar	Kerala
11.	1978-1979	Sariska	Rajasthan
12.	1982-1983	Buxa	West Bengal
13.	1982-1983	Indravati	Chhattisgarh
14.	1982-1983	Nagarjunsagar	Andhra Pradesh
15.	1982-1983	Namdapha	Arunachal Pradesh
16.	1987-1988	Dudhwa	Uttar Pradesh
17.	1988-1989	Kalakad Mundathurai	Tamil Nadu

(*Contd.*)

18.	1989-1990	Valmiki	Bihar
19.	1992-1993	Pench	Madhya Pradesh
20.	1993-1994	Tadoba-Andheri	Maharashtra
21.	1993-1994	Bandhavagarh	Madhya Pradesh
22.	1994-1995	Panna	Madhya Pradesh
23.	1994-1995	Dampa	Mizoram
24.	1998-1999	Bhadra	Karnataka
25.	1998-1999	Pench	Maharashtra
26.	1999-2000	Pakke	Arunachal Pradesh
27.	1999-2000	Nameri	Assam
28.	1999-2000	Satpura	Madhya Pradesh
29.	2008-2009	Anamalai	Tamil Nadu
30.	2008-2009	Udanti-Sitanadi	Chhattisgarh
31.	2008-2009	Satkosia	Orissa
32.	2008-2009	Kaziranga	Assam
33.	2008-2009	Achanakmar	Chhattisgarh
34.	2008-2009	Dandeli-Anshi	Karnataka
35.	2008-2009	Sanjay-Dubri	Madhya Pradesh
36.	2008-2009	Mudumalai	Tamil Nadu
37.	2008-2009	Nagarhole	Karnataka
38.	2008-2009	Parambikulam	Kerala
39.	2009-2010	Sahyadri	Maharashtra

Efforts have also been made for *ex-situ conservation*. There are 33 Botanical Gardens, and about 275 Zoos, Deer Parks and Aquaria. The Central Zoo Authority manages all the large Zoos and provides financial assistance to all types of Zoos in the country.

In spite of these efforts during recent years, the erosion of biodiversity is continued. M.Gadgil (1998) stated 'the ongoing erosion of India's biological diversity is also prompted by the narrow focus of the official management as well as conservation efforts on a handful of animal and plant species'. It is expected that with the increasing awareness toward the values and importance of diversity and stringent measures we shall be able to protect our biodiversity. We need to conserve biodiversity keeping in view the direct use values, indirect use values, and intrinsic value of biodiversity.

In fact, the very existence of human species and sustainable development depend on biodiversity conservation. Both in-situ and ex-situ conservation have been part of the cultural traditions in India. By recognising the role of community in such conservation, these traditions can be revitalised (Swaminathan, 2000). It is expected that NBSAP will help us in this regard.

MODEL QUESTIONS

1. Define the term 'Biodiversity'. What is importance of biodiversity for humankind ? Describe briefly the theories that explain the distribution of biodiversity ?
2. Give an account of geological history of biodiversity. How will you explain diversifications of biodiversity ?

3. Define 'ecological diversity'. Give an account of varieties of ecological diversity.
4. What do you mean by species richness ? How measures of species diversity are helpful?
5. Describe the ecological mechanisms that contribute to and affect biodiversity.
6. Define 'biodiversity crisis'. Which factors are responsible for losses of biodiversity ?
7. What is the need to conserve biodiversity? Review the efforts made in India for conservation of biodiversity.
8. Write note on the following:

   (*a*) Genetic diversity, (*b*) Gamma diversity, (*c*) Shannon index, (*d*) Diversity- stability relationship, (*e*) Biosphere reserve, (*f*) Endemism of Indian diversity, (*g*) Project Tiger, (*h*) Ramsar sites

MULTIPLE CHOICE QUESTIONS

9. Who coined the term biodiversity ?
   (A) M.K.Hill (B) W.G.Rosen
   (C) C.R.Townsend (D) M.S.Swaminathan
10. The highest biodiversity was observed in :
    (A) Tertiary period (B) Late Tertiary and early Quaternary period
    (C) Jurassic period (D) Devonian period
11. Differentiation diversity is also called :
    (A) Beta diversity (B) Point diversity
    (C) Alpha diversity (D) Delta diversity
12. The value of Shannon index usually varies between :
    (A) 1 and 2 (B) 2 and 5
    (C) 1.5 and 4.5 (D) 3 and 5
13. Diversity may be more at :
    (A) Constant disturbance stage (B) No disturbance stage
    (C) Highest disturbance stage (D) Intermediate disturbance stage
14. The term 'biodiversity crisis' was proposed by :
    (A) E.O. Wilson (B) R.H.Whittaker
    (C) A.E.Magurran (D) D.M.Raup
15. Which of the following is an exotic species ?
    (A) *Moschus chrysogaster* (B) *Panthera tigris*
    (C) *Clarias gariepinus* (D) *Pavo cristatus*
16. Who has written the famous essay "The Land Ethic" ?
    (A) Aldo Leopold (B) Lester Brown
    (C) J. R. Kribs (D) Mark Kat
17. Who authored the book "The Closing Circle, Nature Man and Technology"?
    (A) Rachael Carson (B) Barry Commonor
    (C) E. P. Odum (D) A. H. Strahler
18. Which of the following is not an indigenous species?
    (A) *Labeo rohita* (B) *Clarias gariepinus*
    (C) *Clarias batrachus* (D) *Tor tor*

19. The Present number of biodiversity hotspots is as follows–
    (A) North and Central America - 4 (B) South America - 5
    (C) Europe and Central Asia - 4 (D) Asia Pacific - (give the number)
20. Which of the following countries has at present eight hotspots?
    (A) Africa (B) Australia
    (C) India (D) South America
21. India has ____________ of the 25 hot spots.
22. The concept of biodiversity hot spots was originated by ____________.
23. Namdapha flying squirrel is described from ____________.
24. The national aquatic animal of India is ____________.
25. The *Saola* was discovered in Vietnam in 1992. It is ____________.

CHAPTER 8 Environmental Toxicology

8.1 INTRODUCTION

The term 'environment' is defined as physical surroundings and conditions, especially those affecting the living organisms in the biosphere. Living organisms interact with their physical and biotic environment and sometimes are able to partly modify their environment. However, all the activities of humans are not environmentally friendly and our synthetic methodologies generate substances that are highly toxic to human health and the environment.

Any substance or chemical that is poisonous, harmful or deadly is called *toxic* and the scientific study of poisons is called *toxicology*. According to Lu and Kacew (2002), "toxicology is the study of the nature and mechanism of toxic effects of substances on living organisms and other biologic systems. It also deals with quantitative assessment of the toxic effects in relation to the level, duration and frequency of exposure of the organisms." *Environmental toxicology* deals with the sources, transport, degradation, bioconcentration in the environment and adverse effects on humans and other desirable species of a variety of toxicants that reach their vital tissues via air, water, food additives, pesticides and other chemicals often at low levels but may have long-term adverse effects.

Thus, toxicology is one of the several multidisciplinary areas of environmental science. Some of these areas overlap and share common things. The presence of toxic substances in the environment at levels described as lethal or causing harmful effects to the life is considered a serious problem. The scientific study of environmental toxicants, including their chemical nature, sources and status, environmental fate and harmful effects to living organisms may be called as *Environmental Toxicology.*

Now the efforts have been started in the developed and some developing countries that the chemical products should be designed to preserve efficacy of function while reducing toxicity. In U.S. a programme was started in 1998 by the American Chemical Society (ACS) and the U.S. Environmental Protection Agency (E.P.A.) to replace more toxic and environmentally harmful toxic substances, wastes and insecticides. The environmentally benign chemistry that is focused on products and processes that eleminate or reduce the use and generation of hazardous substances and chemicals is called *Green Chemistry* (Cann and Connelly, 2000). Environmental toxicants are present in all the three components of biosphere and many of them contaminate our food, food products and food organisms.

The problem of toxic substances in the environment, in general, is tackled from several angles and perspectives. Accordingly, we use the terms like analytical toxicology, occupational toxicology, economic toxicology, clinical toxicology and forensic toxicology. *Analytical toxicology* deals with the identification of toxicants through analysis of gut contents and body fluids. Occupational exposure to toxicants may cause intoxication leading to acute or chronic adverse effects. Such cases fall in the domain of *occupational toxicology.*

The *Economic toxicology* deals with the use and risks involved in the use of food additives in food preservation and processing, synthetic fertilisers and pesticides in modern agriculture. Poisoning, accidental or occupational, which is so common in the use of synthetic pesticides and the contamination of food stuffs by these toxic chemicals are areas of interest in economic toxicology.

The *Clinical Toxicology* is concerned with the diagnosis and treatment of harmful effects caused by *xenobiotics*, *i.e.*, toxic substances of foreign origin. It also involves administration of *antidotes* to counter the specific toxicity.

Forensic toxicology deals with the medico – legal aspects of the harmful effects of toxic chemicals on humans, often in postmortem material pertaining to cases of suicide by poisoning or attempted murder or murder after giving poison in milk or food. The determination of alcohol concentration in blood and identification of suspicious substances in connection with trade in narcotics, as well as investigations on doping in sports by consuming steroids are included in forensic toxicology. Thus, toxicology has a vast scope and toxicity tests are conducted to evaluate the adverse effects of a chemical on living organisms under laboratory conditions.

A toxic chemical can exist in the environment in several different forms. For example, in aquatic environment, a chemical can exist in three different forms, which in turn affect its bioavailability. These are dissolved, adsorbed to an abiotic or biotic component and suspended in the aquatic system or deposited on the bottom material, and incorporated or accumulated in tissues of animals and plants. Such chemicals can be concentrated and accumulated in organisms (bioconcentration and bioaccumulation) or may be magnified through food chain (biomagnification). Chemicals may also be excreted back into the environment or metabolised (biotransformation) into other forms, which may be more or less toxic than the original chemical.

In the beginning more studies were conducted in *mammalian toxicology* using laboratory rat as the test species. Of late, *aquatic toxicology* has developed as an important branch obviously due to the importance of water as an indispensable resource and the importance of aquatic animals in aquaculture and research. However, the two differ in several aspects. In mammalian toxicology, test organisms are homoiothermic (warm-blooded); their body temperature is relatively uniform and almost independent of environmental temperature; hence toxicity is rather predicatable. On the other hand, in aquatic toxicology, test organisms are poikilothermic, *i.e.*, they are cold-blooded; their body temperature is variable and usually dependent on environmental temperature; thus toxicity may not be sufficiently predictable. In mammalian toxicology, the 'dose' of a test chemical can be measured directly and accurately, and administered by a number of routes. But in aquatic toxicology, the 'dose' is known only in terms of the concentration of the chemical in water and the length of exposure to it; the actual 'absorbed dose' is sometimes determined experimentally using bioconcentration and metabolic studies. In mammlian toxicology, extensive 'basic research' has been conducted and mechanisms of toxic action are relatively better understood. However, in aquatic toxicology, little basic research has been conducted and emphasis is given on measuring toxic effects with an toward regulatory needs. The field of aquatic toxicology is developing fast and a number of organisms are used throughout the world. (Table 8.1)

8.2 NOMENCLATURE OF TOXINS

Toxins are produced by bacteria, viruses, fungi and some other organisms. Some fungi contain naturally toxic metabolites called *mycotoxins* and can lead to *toxicosis*. The *endotoxin* is normally released from gram-negative bacterial cells in the bloodstream. The word *extoxin* may be used to describe any extracellular protein produced by bacteria which induces a toxic effect on host cells. Exotoxins may be divided into three groups based on their sites of action. These are (*a*) membrane damaging toxins, (*b*) toxins that act on sites inside the cell, and (*c*) superantigens, toxins that over-induce the immune system.

Some toxins, especially those that are the clear cause of disease symptoms, are named after the bacteria that produces them such as *Cholera toxin* (produced by *Vibrio cholerae*) or *dephtheria toxin* (*Corynebacterium diphtheriae*). Others are named after their sites of action such as *neurotoxins* (acting on nerve cells), *leukotoxins* (acting on white blood cells). Some toxins are just given a letter or number such as *Pseudomonas aeruginosa* toxin *A*, etc. Some toxins are named on the basis of their activities such as *phospholipase C* produced by *Listeria monocytogenes* and *adenylate cyclase* by *Bordatella pertussis*. Some gram-positive bacteria may produce a number of different toxins.

8.3 ANTIBIOTICS AND VACCINES

However, vaccines are not available for all *bacterial toxins*. Antibiotics are molecules, produced by microbes, which inhibit the growth or kill other microbes. The antibiotics normally used to treat bacterial infections are those that are selectively toxic to bacterial cells and do not harm the host. The effectiveness of *antibiotic therapy* is endangered by bacterial antibiotic resistance and antibiotic-resistance genes between bacteria.

Vaccination is the use of a vaccine to stimulate the immune response to protect against the wild-type virus. Vaccines are either live (attenuated) virus or inactivated (killed yirus). Some DNA vaccines are being developed and most of them are at an experimental stage. A DNA vaccine consists of injecting a host with a *plasmid* containing DNA, which encodes antigenic portions of the virus. Some currently available live attenuated viral vaccines are *oral polio*, *Measles* and *Mumps*. The currently available inactivated viral vaccines include *inactivated polio*, *Hepatitis B*, *influenza*, *Rabies*, *Hepatitis A*, etc.

Immunization: The earlier experiments of Jenner with cowpox formed the basis for to day's vaccination programmes. The major types of white cells are *lymphocytes (T-cells or B-cells)*. T-cells (cytotoxic)–CD_8 or helper cells–CD_4 cells pass through the *thymus* during their maturation. B-cells give rise to antibody-producing plasma cells. T_4–helper cells help B-cells react to highly specific antigenic stimuli (*e.g.* viral proteins). The host also produces non-specific responses such as cytokines, natural killer cells and macrophages.

Active immunization is the artificial introduction of antigens from a microbe into an individual (in a controlled way) leading to the stimulation of the immune system without the symptoms of full-blown diseases. This results in the production of *memory cells* within the host, so that after a second encounter with the microbe, the immune system can generate a rapid antibody response, thereby preventing infection. However, the bacteria must be attenuated. Live vaccines have distinct advantage over killed. The WHO has an ambitious vaccination programme which seeks to get vaccines to eradicate many viruses in the next decade.

Immunity resulting from injections of readymade antibodies is immediate and is called *passive artificial immunity*. Its duration is 6 weeks or less. Passive artificial immunity is used in the prevention as well as treatment of disease. Diseases against which equine serum is occasionally used for prophylaxis (means to guard against) are *diphtheria* and *tetanus*. Effective human sera are also available for prophylaxis against rabies and pertussis. However, the most common forms of vaccines are live *attenuated* bacteria or viruses which can evoke a protective immune response without causing pathology.

8.4 FACTORS AFFECTING TOXICITY

In the environment, the concentration, transport, transformation and fate of a chemical are affected by several factors. Factors which affect overall toxicity may be described under four headings –

8.4.1 Factors Related to Chemical

Table 8.1. Species commonly used in aquatic toxicology tests

Freshwater species	*Saltwater species*
Invertebrates	**Invertebrates**
Water flea (*Daphnia magna, D. pulex*)	Sprimp (*Penaeus sp.*)
Midge (*Chironomus sp.*)	Grass Shrimp (*Palaemonetes vulgaris*)
Scud (*Gammarus fasciatus, G.pulex*)	Mysid shrimp (*Mysidopsis bahia*)
Crayfish (*Orconectes sp. Astacus sp.*)	Eastern oyster (*Crassostrea virginica*)
Grass shrimp (*Palaeomonetes kadiakensis*)	Blue mussel (*Mytilus edulis*)
Myflies (Ephemeroptera)	
Caddisflies (Trichoptera)	
Planarian (Tricladida)	
Oligochaeta (*Tubifex*), *Pheretima*	
Vertebrates	**Vertebrates**
Brown trout (*Salmo trutta*)	Striped mullet, (*Mugil cephalus*)
Rainbow trout (*Salmo gairdneri*)	Sheephead minnow (*Cyprinodon varie*gatus)
Fathead minnow (*Pimephales promelas*)	Spot, (*Leiostomus xanthurus*)
Goldfish (Carassius auratus)	Mummichog, (*Fundulus heteroclitus*)
Bluegill sunfish (*Lepomis macrochirus*)	Mugil, (*Mugil corsula*)
Channel catfish (*Ictalurus punctatus*)	
Brooktrout (*Savelinus fontinalis*)	
Common guppy (*Poecilia reticulata*)	
Mosquito fish (*Gambusia affinis*)	
Ticto (*Puntius ticto*)	
Rasbora (*Rasbora daniconius*)	
Mystus (*Mystus cavasius*)	
Singhi (*Heteropneustes fossilis*)	
Magur (*Clarias batrachus*)	
Gachua (*Channa gachua*)	
Barilius (*Barilius bendelisis*)	
Plants	**Plants**
Algae, *Chlorella sp.*	Algae, *Skeletonema sp.*
Algae, *Microcystes sp.*	Algae, *Dunaliella sp.*
Algae, *Selenastrum sp.*	Algae, *Thalassiosira sp.*
Algae (diatom), *Navicula sp.*	
Duckweed (*Lemna minor*)	

8.4.1.1 Physicochemical properties

The physico-chemical properties of a compund that are considered important include its molecular structure, solubility, ionization and vapour pressure. All these characteristics affect the toxicological and pharmacological properties of the compound. The nature of chemical reaction is determined by the functional groups of the substance and its toxicity is a response of organisms consequent upon reaction between the compund and certain part of the organism. The nonpolar (lipophilic) compounds are highly soluble in lipids and other organic solvents. Therefore, they can easily penetrate the membrane's lipoprotein layers and produce their potential effects. However, polar (hydrophilic) compounds are not soluble in lipids; therefore, they cannot cross the membrane barrier to reach the target site. Other properties of chemicals such as rate constant for hydrolysis, photolysis, evaporation, sorption and partition coefficients are also important and affect environmental concentration of toxicants.

8.4.1.2 Concentration or dose

A toxic substance produces its toxic effects after its interaction with appropriate receptors of the organism. The effect is directly dependent on the concentration of the chemical at the target site and concentration at the target site is related to the amount (dose) of the chemical administered. Paracelsus stated "No substance is a poison by itself. It is the *dose* (the amount of the exposure) that makes a substance a poison and the right dose differentiates a poison and a remedy". This statement became the basis for "*dose-response relation*". The lower doses of the chemical cause mild effects, whereas high doses may cause serious and long lasting effects. If the dose is administred through mouth, or applied on the skin, or into the respiratory tract, transport across the membranes may be identical with the dose administered. In environmental exposures, an estimate of the dose can be made from the measurement of environmental and food concentrations as a function of time, and involves the assessment of food intake, rate of inhalation, and the deposition and retention factors. The doses in the tissues and organs may be estimated from (*a*) intake or administered dose, (*b*) assessment of concentrations in samples of tissues and organs, and (*c*) measurement of concentration in excreta or exhaled air.

However, estimation of dose by these methods depends on time taken for transport, absorption, distribution, retention, biotransformation and excretion of the chemical or its metabolites. When the site of action is very near to the site of application, then the time concentration estimate may be quite reliable. But when the site of action is located away, as the liver cells, then estimates of dose may not be reliable. The presence of a chemical in low or high concentration in the blood indicates aborption. However, the blood concentration of a toxic chemical depends on several other factors such as rate of absorption, distribution, tissue storage, metabolic transformation and excretion.

8.4.1.3 Selective toxicity

Certain toxic substances are nonselective in their mode of action *i.e.*, they exert their adverse effect on a large number of organ tissues of the body of an organism. As such, they may be effective in low concentrations. On the other hand, a selective chemical adversely affects only one type of tissue. For example, snake poison either affects blood tissue (haemotoxic) or nerve cells (neurotoxic). The affected and unaffected cells and tissues may be in the same or different organisms. In the latter case, the chemical is called species- specific in its selective toxicity. Selective toxicity results from biological diversity and variations in the response of living cells and tissues to different chemicals. There appear to be involved two important mechanisms for the selective action of a chemical. These are presence or absence of specific receptor or target sites and the factors that may be responsible for the distribution and change in the concentration of the chemical at specific sites.This results from processes such as selective absorption, translocation, biotransformation and excretion.

8.4.1.4 Chemical interaction

In nature, sometimes organisms are exposed not only to one chemical, but to two or three chemicals at the same time. These chemicals may interact with each other and such chemical interactions may have great toxicological significance. In the laboratory, organisms may be exposed to chemical mixtures containing two or more chemicals. These chemicals often interact and affect the toxicity of each constituent. The combined effect of two chemicals may be equal to the sum of the effect of each chemical when given alone. This type of effect is called *additive*. For example, effect of combination of two organophosphorus pesticides on AcheE activity. The combined effect of two chemicals may be more than the sum, then this type of interaction is called *synergistic*. For example, effect of ethanol and carbon tetrachloride on liver, or effect of cigarette smoking and asbestos exposure on the lung tissue, or the effect of malathion and EPN when given simultaneously to test animals are far more toxic than the sum of their toxicities when given separately. A term used to express this condition is sensitization or *potentiation*, which means that when two chemicals are acting on the same or different organs or systems, one chemical is made more effective in the presence of other. A third type of interaction between two chemicals is called *antagonistic*. In such cases the combined effect of two chemicals may be less than the sum. In other words, there is no potentiation of activity between similarly acting chemicals, rather one chemical may antagonise the toxicity of another and we observe negative summation or antagonism. For example, chelation of heavy metals by dimercapol. Thus, according to Sprague's interpretation, the toxicity or effects of chemical mixtures are described as simply additive, more than additive (synergistic) or less than additive (antagonistic). In other words, the interaction of one chemical with another may have no effect on their toxicity, or increase their toxicity, or decrease their toxicity.

8.4.2 Factors Related to Exposure

Any toxic substance cannot show its effect unless it comes in direct contact of the organism living in natural environment or maintained in the laboratory for bioassay. This contact reaction between the organism and chemical is called exposure. In toxicity studies the most significant factors related to exposure are –

(*i*) the concentration of the chemical

(*ii*) the kind, duration and frequency of exposure, and

(*iii*) exposure routes

Aquatic animals are naturally exposed to chemicals and wastes present in water, sediment and food items. Water soluble toxicants are more readily available to organisms than water insoluble chemicals that may be tightly absorbed, or otherwise bound to suspended particles or organic matter. Water soluble chemicals may enter an organism through integument, mouth or gills. Chemicals concentrated in food items may be ingested and absorbed through the alimentary canal. Adsorbed chemicals may also enter the body through the general body surface and gills as they gradually dissociate from particles to water in immediate contact with these areas. The chemical produces more rapid and immediate effect when given by intravenous route (injection). Besides the routes mentioned above a toxicant may also gain access to the body of an organism by inhalation or by intraperitoneal, intramusculer and subcutaneous injections.

Toxic effects can be produced in the natural environment or in the laboratory by short-term (*acute*) or long-term (*chronic*) exposure to toxic chemicals (*bioassay*). Exposures of intermediate duration are called *subchronic* esposures. In routine bioassay, organisms are exposed for 24, 48, 96 hrs or for more days depending upon the type of chemical, organism and test conditions. The concentration of the chemical required to kill 50% test organisms is called LC 50 or LD 50 and expressed as 24 hr- LC 50, 48 hr-LC 50 or 96 hr-LC 50. For example, 96 hr-LC 50 is the concentration of a chemical that is lethal to 50% of the test organisms in 96 hr. Obviously, the safe concentration, which permits successful reproduction, growth and all other normal life processes in the organisms natural environment is usually much lower that LC 50. The LC 50 values of toxicants decrease with increase in duration of exposure. The LC 50 value obtained for long-term exposure is much less than those determined for short-term exposure.

A variety of exposure systems are used in the laboratories. The choice of the system depends upon the test organisms and facilities available. In the *Static System*, a toxicant is mixed in the water and the organisms are exposed to the toxicant in still water. In *Recirculatory System*, a toxicant is recirculated through certain pumps. In *Renewal System*, a toxicant solution is renewed after certain interval, say 24 hrs or more. In *Flow through System*, the toxicant solution flows into and out of the test chamber continuously or intermittently.

8.4.3 Factor Related to Organism

Toxicity of a chemical is generally evaluated against a particular organism or a group of organisms. The toxicity of a chemical varies according to the size, life history stage, age, sex and health and nutritional status of the organism. Acclimation of the organisms to laboratory conditions prior to bioassay also plays an important role. The same species or organisms collected from different habitats may exhibit variations in their limits of tolerance to the chemicals. Certain individuals of a species may be susceptible to a chemical while others may be resistant. Therefore, it is often suggested that to have a reasonably clear picture of the toxicity of a chemical, it must be tested against one plant such as an alga, one or two nonchordates (an insect or a crustacean) and at least one chordate such as a fish. In general, invertebrates are more sensitive to toxicants than the vertebrate species. Not all the species of a group are equally sensitive to toxicants. Among various fish species, the toxicity of a chemical may vary considerably. Generally, trouts and minor carps are more sensitive than catfish and snakeheads.

The most important considerations in the selection of test organisms for bioassays are: (*i*) their economic, ecological and recreational importance; (*ii*) their sensitivity to the material or environmental factors under consideration; (*iii*) their geographical distribution, abundance and availability throughout the year; (*iv*) knowledge of their environmental requirements and culture methods for rearing in the laboratory; (*v*) their general physical conditions and freedom from disease; and (*vi*) their suitability for bioassay tests. Enold (1984) has stated that healthy animals are of paramount importance for reliable toxicity studies.

In most bioassays usually the most sensitive locally important species may be used. For anyone series of bioassays, the test species should be obtained from a common source and collected at one time. Test organisms that are nearly uniform in size and belong to the same age group or life history stage are considered ideal. A variety of organisms including algae, protozoa, cladoceran, copepods, annelids, crustaceans, molluscs, aquatic insects, fishes and macrophytes have been used in different bioassays. But all these organisms are not suitable for all types of bioassays. For a long-time it was felt that only fish bioassays were necessary when estimating the probability of harm to aquatic life from industrial wastes. However, fish are not invariably the most sensitive organisms in the aquatic ecosystem. Therefore, to estimate harm to aquatic life, toxicity tests are designed to include algae, invertebrate and fish together. Generally, the most desirable species for toxicity testing are those that have relatively short life cycles, mature quickly, thrive in laboratory aquaria, can be easily handled and are small as adults (Wayne and Nimmo, 1984).

8.4.4 Factors Related to Environment

There are many abiotic and biotic factors of the environment that may modify the toxicity of chemicals. The factors of the environment affecting the toxicity of chemicals have been studied extensively for the aquatic medium and 'aquatic toxiology' is now considered an important branch of science (Wayne and Nimmo, 1984). Important factors are mentioned below.

Of all the environmental factors of the aquatic medium, the *water temperature* greatly affects the toxicity of xenobiotic chemicals. An increase in water temperature increases the solubility of many substances, alters the chemical structure of some toxicants and also lowers the dissolved oxygen content of water. However, some pesticides have been reported to be more toxic at higher temperatures while some others show stronger lethal action at low temperatures. Some studies reported that DDT was three times more toxic at 10ºC than at 27ºC. Omkar (1983) in his toxicity studies of pesticides, using freshwater prawns as test organisms, observed that increase in water temperature increases the toxicity of pesticides. The pH of the medium also affects toxicity of chemicals. The pH of water may have greater effects on the toxicity of chemicals that can ionize under the influence of pH. Generally, an undissociated form of a chemical is more toxic to organisms because it can penetrate the cell membrane. The unionized from of ammonia (NH_3) is highly toxic to fish and quite low concentrations, 0.2 to 0.7 mg./l, adversely affect salmonids. However, the ionized from of ammonia, ammonium ion (NH_4^+) is little or no toxic. An increase of pH by one unit within the usual middle ranges increases the proportion of NH_3 by about six folds, hence six folds increase in the toxicity of the chemical. The pH has profound effect on metal toxicity, as metals ionize under the influence of pH.

The *salinity* and *hardness* of water also affect toxicity. Generally, euryhaline organisms are more resistant in water containing about 30-35% salinity. An appreciable decrease in salinity of water often renders the marine organisms less tolerant. Therefore, some studies concluded that the toxicity of xenobiotics may be increased with an appreciable decrease in the salinity of the surrounding medium. Herbert *et al* (1965) observed that the toxicity of zinc and ammonium chloride for rainbow trout (*Salmo gairdneri*) decreases with an increase in salinity. However, the *total hardness* of water has little effect on the toxicity of most of the chemicals except metals. A study on the toxicity of the petro chemicals in soft and hard water revealed significant differences for only 4 of them and they were generally two times more toxic to fathead minnows (*Pimephales promelas*) in soft water than in hard water (Pickering and Henderson, 1966). Of the many pesticides tested, only two carbamate compounds showed three times increase in toxicity with decrease in water hardness (Macek *et al* 1977). Several studies indicated that many heavy metals are more toxic to aquatic biota, including fish, in very soft water than hard water. The increase in toxicity of metals with respect to decrease in hardness of water has been attributed to alteration in fish gill permeability caused by the calcium content of water. The *suspended and dissolved solids*, which affect conductivity of water, also affect the toxicity of chemicals. They may partly detoxify some of the xenobiotic chemicals such as metals as a result of sorption or binding. The addition of suspension of clay affects the toxicity of an insecticide, endrin to fathead minnows, while the addition of soil suspension to endrin solution moderates its toxicity.

The *DO* and *BOD* content of water also affect the toxicity of chemicals. The sewage and other organic matter discharged to a water body are degraded by oxygen requiring microbes. The amount of DO consumed by the microorganisms is biochemical oxygen demand (BOD). If the BOD value is more, the water contains less amount of DO. Freshwater can dissolve 14.6 mg/l of oxygen at OºC, which gradually decreases to 9.1 mg/l at 20ºC and further decreases to 7.5 mg/l at 30ºC. It is, therefore, obvious that rise in water temperature decreases its DO content. Since oxygen is essentially required for respiration, any physiological change affecting the rate of oxygen flow to the blood of the respiratery organ (gills), will obviously affect the concentration of toxicant at the gill surface and thus the overall toxicity of the chemical.

In the presence of low oxygen content of water fish become more susceptible to disease and attack of other organisms, which release toxins. Such toxins cause wounds or ulcers on the body of fish (Hill, 1997). Thus, it might be expected that reduction in DO content of water imposes stress on the aquatic organisms, which may greatly increase the toxicity of a chemical present in water.

8.5 MECHANISM OF TOXICITY

Mechanism of action of toxicants varies from chemical to chemical. Some chemicals cause *cytotoxicity*, *i.e.*, they act by causing cell death. A number of chemicals are able to cause *genotoxicity*, *i.e.* they are able to produce their effect by altering gene expcession. Generally, toxicants produce their effects by interference with physiological or biochemical homeostatic (regulatory) mechanisms. Some of them can modify the regulation of cell division with harmful effects. Many toxic chemicals cause proliferative lesions on the skin of the organism. A dinoflagellate toxin causes wounds (ulcers) on the body of fish (ichthyotoxin). When the toxic algal bloom die, they are decomposed by oxygen-requiring microorganisms, thereby causing hypoxic condition in water,which may lead to massive fish kills. Humans who eat shellfish that have fed on red tide organisms (blooms of dinoflagallates) may suffer from paralytic shellfish poisoning caused by the toxin that they produce (Hill, 1997).

Certain environmental agents called *carcinogens* cause mutations that can lead to cancer. These mutations activate oncogenes (or inactivate tumor-suppressor genes). An *oncogene* (Gk. onco = tumor) is a cancer-causing gene. A *tumor-suppressor gene* is so called because it prevents cancer from occurring. Certain pathogens produce toxins. The most dangerous food poisoning, called *botulism*, is attributed to *Clostridium botulinum*. This is a widely distributed soil bacterium. It is developed in incompletely sterilised meat products. It derives its name from its occurrence in sausages (Latin, *botulus* = sausage). It produces a toxin that on injection of the food, causes paralysis of the nervous system (neurotoxin) and is lethal by virtue of respiratory paralysis. Botulinus toxin A, which is excreted by *C.botulinum*, is a heat-labile protein. It has LD 50 value of one nanogram per kg body weight. Thus, this neurotoxin is the most effective of all lethal poisons for humans (Schlegel, 1995). Table 8.2 shows some examples of how toxicants exert their adverse effects.

Table 8.2. Some examples of how toxicants exert their harmful effects

S.No.	*Name of Toxicant*	*Mechanism of Action*
1.	Botulinum toxin	Blocks release of the neurotransmitter Acetylcholine at neuromuscular junction and causes death by paralysis of respiratory muscles
2.	Carbon monoxide	It binds to blood haemoglobin, thereby blocking the formation of oxyhaemoglobin.If enough haemoglobin is blocked, death results.
3.	Nerve gases	They prevent acetylcholine from being degraded.
4.	Cyanide and azide	Inhibit cytochrome oxidase, thereby producing cytotoxic anoxia
5.	Methyl mercury	By causing damage of neurons
6.	Doxorubicin (Adriamycin)	By affecting neurons
7.	Organophosphate	By causing inhibition of acetylcholine esterase (AchE) activity resulting in accumulation of acetylcholine
8.	DDT	Affects nervous system by targetting nerve membrane Na^+ channels
9.	Lead	Affects various parts of nervous system; also effects Blood vessels.

8.6 BIOCONCENTRATION

Bioconcentration may be defined as a process by which the concentration of toxic substance gradually increases in the tissues of the organism and becomes many times more than the concentration available in the surrounding medium. By the process of direct uptake and bioconcentration an organism constantly receives the chemical directly from the medium. Most of the organochlorine pesticides are absorbed into the tissues of animals and plants and often reach at concentrations many thousand times greater in the biota than in the water or soil. For example, *Daphnia* can concentrate DDT 100,000 times even if the DDT concentration in the water is only 0.5 ppb. Thus, potential for uptake and bioconcentration into aquatic organisms is an important characteristic, which helps in determining the effects of a particular pesticide or any other toxic chemical on an aquatic system. (Fig. 8.1)

8.7 BIOACCUMULATION

The term bioaccumulation is used to refer to storage of a pollutant at levels higher than found in the environment (Table 8.3). Many chemicals including dioxins, polycyclic aromatic hydrocarbons (PAHs), polychlorinated biphenyls (PCBs) and organometallic forms of metals bioaccumulate in animal fat (Hill, 1997) and are difficult to degrade. Lead and fluoride bioaccumulate in bones. Chemicals bound to proteins and cadmium can bioaccumulate in the liver, kidney and other tissues. The route of almost 90% of human exposure to PAHs is from food consumption, especially the leafy vegetables and unrefined grains.

Table 8.3. Some examples of Bioaccumulation

Sl. No.	*Chemical or toxic metal*	*Ttissues where the chemical bioaccumulates*
1.	Lead	Bones
2.	Fluoride	Bones
3.	Cadmium	Kidney, liver
4.	Cadmium	Plants (tobacoo plants)
5.	Organochlorine pesticides	Animal fat
6.	PAHs	Animal fat
7.	PCBs	Animal fat
8.	Dioxins	Animal fat

Breathed into the lung as fine particulates, PAHs can cause respiratory problems and a number of them are *carcinogens*.Cadmium bioaccumulates in the kidney and the amount of this toxic metal stored in this organ increases with age. Cadmium can also accumulate to high levels in liver. Organochlorine pesticides have low solubility in water; they are highly soluble in lipid materials, including animal fat, in which they bioaccumulate to high levels.

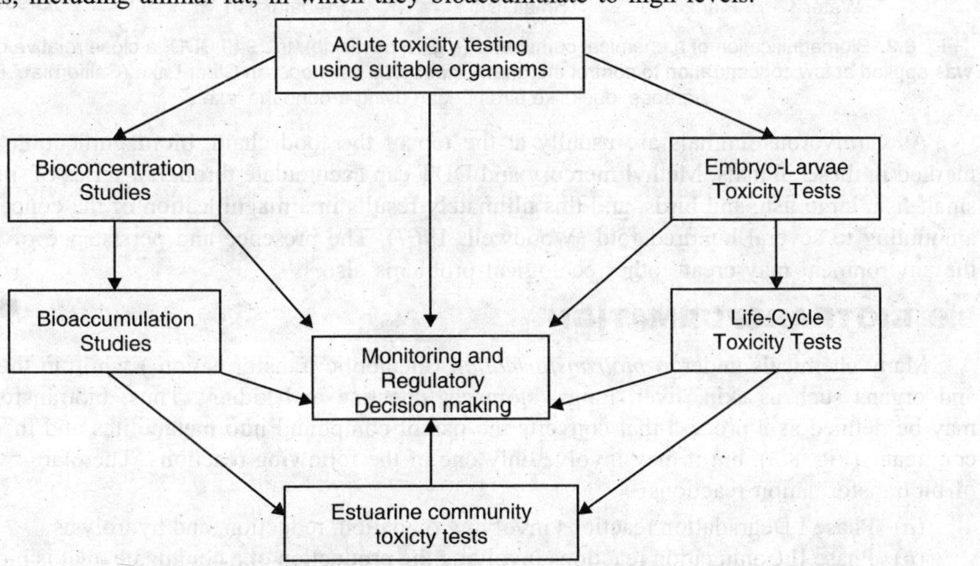

Fig. 8.1. Some aspects and linkages of toxicity tests in aquatic toxicology

The term bioaccumulation is also used to describe increase in quantity of a chemical through the food chain. For example, in a food chain—phytoplankton → zooplankton → small fish → large fish → fish eating birds – each component of the food chain receives some chemical from the environment (bioconcentration) but the amount of chemical becomes several times more through the food chain and organisms at the top of the food chain are worst hit as they contain the maximum quantity of the harmful chemical. Through this route pesticides sprayed on crops accumulate in vegetables, food grains, fish and human tissues (Fig. 8.1).

Bioaccumulation is more marked with some chemicals than others. For example, organochlorine pesticide, DDT is stored in the body fat for a much longer time than methoxychlor. The half-lives of these insecticides in rats are 6-12 months and 12 weeks respectively (Lu and Kacew, 2002).

8.8 BIOMAGNIFICATION

Sometimes the two terms – bioaccumulation and biomagnification are alternatively used. Biomagnification is also called *food chain magnification* (Fig. 8.2) so that the organisms at the top of the food chain contain maximum amount of the harmful chemical. According to Lu and Kacew (2002), biomagnification is the result of bioaccumulation in the organism alone or in conjuction with persistence in the environment. For example, DDT may occur in the fat of body fluids, including milk. While a mother's daily intake of DDT may be about 0.5 μg per kg, a baby feeding on her milk might receive about 11.2 μg per kg This magnification results from the fact that DDT is stored in human body at 10-20 fold chronic daily intake level. Much greater intake of DDT by babies is that they consume essentially only the milk and the caloric intake per kg body weight is higher in babies than adults (Lu *et al* 1965).

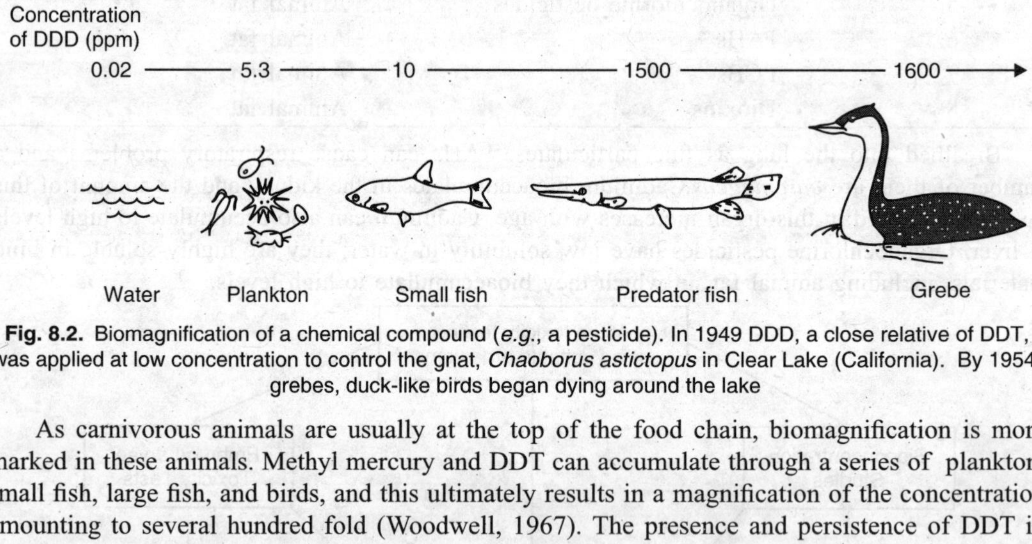

Fig. 8.2. Biomagnification of a chemical compound (*e.g.*, a pesticide). In 1949 DDD, a close relative of DDT, was applied at low concentration to control the gnat, *Chaoborus astictopus* in Clear Lake (California). By 1954, grebes, duck-like birds began dying around the lake

As carnivorous animals are usually at the top of the food chain, biomagnification is more marked in these animals. Methyl mercury and DDT can accumulate through a series of plankton, small fish, large fish, and birds, and this ultimately results in a magnification of the concentration amounting to several hundred fold (Woodwell, 1967). The presence and persistence of DDT in the environment may create other ecological problems also.

8.9 BIOTRANSFORMATION

Many chemicals undergo *biotransformation* (metabolic transformation) while in the tissues and organs such as skin, liver, lungs, stomach, intestine and kidney. Thus, biotransformation may be defined as a process that converts the parent compound into metabolites and then forms conjugates (Fig. 8.3), but it may involve only one of the following reactions. There are two types of biotransformation reactions:

(*a*) Phase I Degradation reactions involving oxidation, reduction, and hydrolysis

(*b*) Phase II Conjugation reactions involving the production of a congugate that is biosynthesized from the toxicant, or its metabolite, plus an endogenous metabolite.

For example, benzene undergoes oxidation (a phase I reaction) to form phenol, which conjugates with sulphate (a phase II reaction). Organophosphorus insecticides such as fenithrothion, chlorofenvinphos, and omethoate can be metabolized through oxidation, dealkylation, desulfuration, or hydrolysis, yielding ten or more metabolites.

However, the rate of *biotransformation* of a toxicant differs from species to species and often the age, sex and exposure to other chemicals may also affect biotransformation. In the process of biotransformation of toxicants a number of highly reactive electrophilic metabolites are formed; some of which can react with cellular constituents and cause cell death, induce tumor formation, or affect immune system. In some cases *Glutathione* reacts with the electrophilic metabolites to prevent their harmful effects on the cells. Wills (1981) described the role of glutathione against covalent binding of acetaminophen to liver proteins. Glutathione conjugation is affected by glutathione S-transferaes and the cofactor glutathione. These ractions play an important role in hepatotoxicity (Commandeur *et al* 1995; De Leve and Kaplowitz, 1991).

Thus, *biotransformation* is a process by which organisms convert absorbed toxic chemicals into other chemicals. The liver is especially active for biotransformation and converts a *xenobiotic* into less toxic chemicals. However, in some cases the result is a more toxic chemical. For example, the liver converts the pollutant benzene into benzene oxide, a reactive chemical that can harm bone marrow. The relative importance of various types of biotransformations of a toxicant depends on several factors including the dose of the toxicant, organism, environmental and chemical factors.

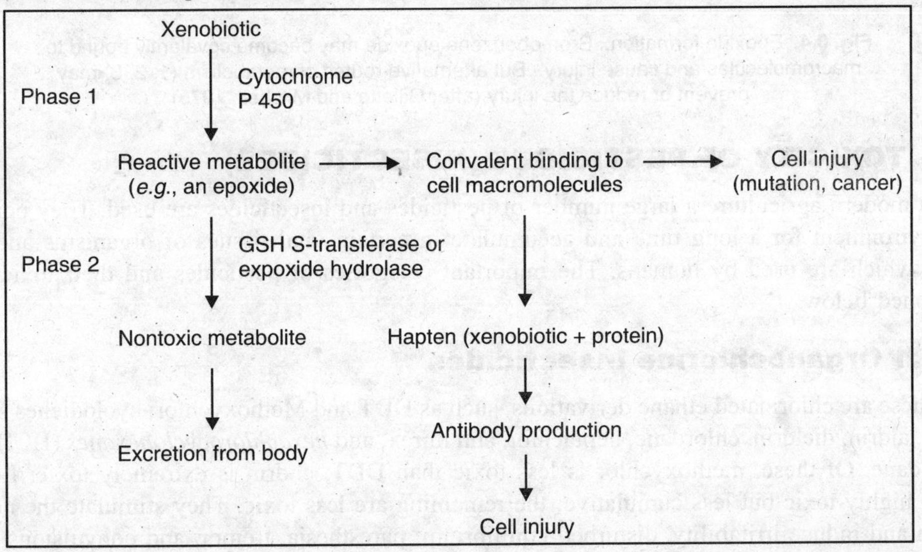

Fig. 8.3. Scheme showing how a xenobiotic can be detoxified and excreted; and how it can result in cell injury, immunologic damage or cancer

8.10 BIOACTIVATION

Bioactivation may be defined as a process whereby certain chemically stable compounds can be converted to chemically reactive metabolites. For example, Parathion, an organophosphorus pesticide, is bioactivated in the liver to Paraoxon, which is a more potent inhibitor of cholinesterase (ChE). The reactive metabolites such as *Epoxides* can form covalent binding to celluar proteins and cause necrosis and cancer. (Fig. 8.3) Other metabolites, such as free radicals, can cause lipid peroxidation resulting in tissue damage. For example carbon tetrachloride forms trichloromethyl radical, which causes peroxidation of polyunsaturated lipid as well as covalently binds to unsturated fat and protein. The herbicide Paraquat is known to produce superoxide radicals (Halliwell *et al* 1992).

Gillette and Mitchell (1975) stated that bromobenzene epoxide from the parent chemical (bromobenzene) may covalently bind to cell macromolecules but such binding may be minimal when the other metabolic routes are predominant (Fig. 8.4). The most important of these routes is the conjugation with Glutathione. This reaction is a protective mechanism against liver necrosis. Bioactivation takes place mainly in the liver and the resulting reactive metabolites induce toxicity through covalent binding with liver proteins, resulting in necrosis or cancer formation.

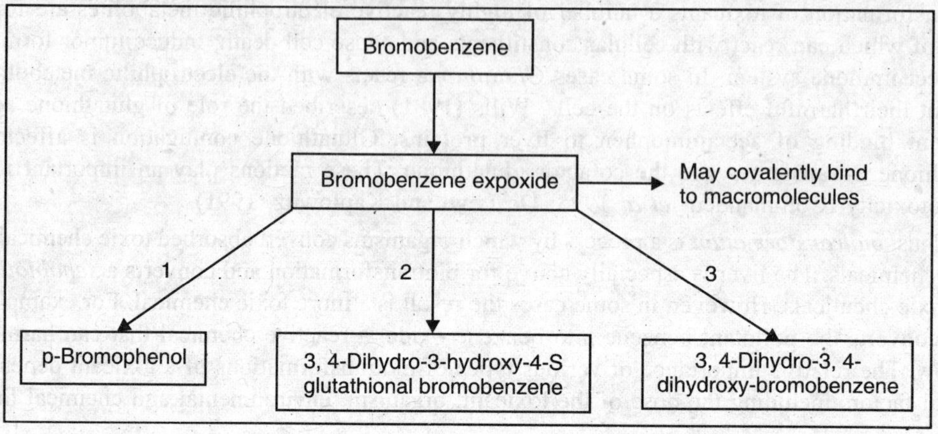

Fig. 8.4. Epoxide formation. Bromobenzene epoxide may become covalently bound to macromolecules and cause injury. But alternative routes of metabolism (1, 2, 3) may prevent or reduce the injury (after Gillette and Mitchell, 1975)

8.11 TOXICITY OF PESTICIDES (INSECTICIDES)

In modern agriculture a large number of pesticides and insecticides are used. They persist in the environment for a long time and accumulate in certain vital tissues of organisms and other items, which are used by humans. The important categories of pesticides and their toxicity are mentioned below.

8.11.1 Organochlorine Insecticides

These are chlorinated ethane derivations, such as DDT and Methoxychlor; cyclodienes such as endrin, aldrin, dieldrin, chlordane, heptachlor, and mirex; and *hexachlorocyclohexanes* (HCH) such as Lindane. Of these, methoxychlor is less toxic than DDT; endrin is extremely toxic; Lindane is also highly toxic but less cumulative; the remaining are less toxic. They stimulate the nervous system and induce irritability, disturbed equilibrium, paresthesia, tremor, and convulsions. Some of these chemicals, such as aldrin, dieldrin and lindane affect neurotransmitter activity. DDT may exert its toxic effects in the nervous system by adversely affecting ion transport across the axon membrane (Doherty, 1979; Narahashi, 1980). Some organochlorine insecticides including DDT are hepatotoxic.

8.11.2 Organophosphorus Insecticides

These are esters of phosphoric acid (Dichlorvos) and theophosphoric acid (Parathion). Other pesticides in this group are Diazinon, Dimethoate, Malathion, Mevinphos and Dipterex (Trichlorfon). Their toxic effects vary over a wide range. They act by inhibiting acetylcholinesterase (AchE). As a result, the accumulated acetylcholine (Ach) induces tremor, incoordination and convulsion. The accumulation of Ach at the neuromuscular synapse will lead to contraction of the muscles, loss of reflexes and paralysis. Several organophosphorus compounds may cause delayed neurotoxicity.

8.11.3 Carbamate Insecticides

Insecticides of this class include Carbaryl (Sevin), Aldicarb (Temik), Carbofuran Methomyl and Propoxur (Baygon). These are esters of N. methylcarbamic acid. They also act by inhibiting AchE. However, inhibition of AchE by a carbamate is readily reversible.

8.11.4 Herbicides

Several types of herbicides are used. Herbicides like 2, 4-D (2, 4-dichlorophenoxy- acetic acid), Paraquat and Diquat have been widely used. Some herbicides retard the growth of weeds by inhibiting photosynthesis, respiration, cell division or protein or lipid synthesis. Their toxicities in animals are relatively low. Paraquat and Diquat exert their toxicity via the formation of free radicals. Paraquat causes lung damage after inhalation and also after ingestion. It also causes hemorrhage and fibrosis.

8.11.5 Rodenticides

A number of rodenticides including Warfarin, Thioureas, Sodium fluoroacetate, and Sodium fluoroacetamide have been used. Warfarin is an antimetabolite to vitamin K, thus acts as an anticoagulant. Thioureas main toxicity is pulmonary edema and pleural effusion. They are highly toxic to rats but moderately toxic to humans. Sodium fluoroacetate exerts its toxic effect through blockage of the citric acid cycle.

8.11.6 Fungicides

Methyl and ethyl mercury are very effective fungicides. However, permanent neurologic damage and deaths have been reported after their use. Fungicides which have been widely used in agriculture are Dicarboximides but they are reported to have carcinogenic effects.

8.12 TOXICITY OF HEAVY METALS

A number of toxic metals are present in the environment. Metals of major toxicological concern are Lead, Mercury and Cadmium.

8.12.1 Lead

Lead (Pb) cannot be completely eliminated from our environment. Small concentrations of lead are present everywhere. In soil, lead levels range from 5-25 mg per kg; in ground water from 1-60 μg per litre (little lower in natural surface water); and in air below 1 μg per m^3. In most people, the lead intake from food is 100-300 μg per day. However, in some countries average lead levels in the blood of children have been found 4-6 μg per 100 ml. Lead poisoning of children is a matter of serious concern. Like calcium, lead is deposited in bones and during pregnancy it is slowly released from the mother's bones to the blood reaching the fetus.

The inhibitory effect of lead to the enzymes involved in the synthesis of heme is well known. A moderate exposure to lead with its level in the blood about 50 μg per dl causes anemia. However, other effects may be observed at lower levels of exposure. The impaired heme synthesis may result in anemia, which may be due partly to the greater fragility of the erythrocyte membrane (WHO, 1977). The nervous system is also a target organ of lead and *encephalopathy* may occur with blood lead level over 80 μg per dl. At lower levels of 40-50 μg per dl children may exhibit hyperactivity, low attention span and decreased IQ. (Ernhardt *et al* 1981) which is due to impaired function of neurotransmitters and Ca ion. Lead also affects the nervous system of occupational workers. Lead is also known to cuase renal failure after long-term exposure. However, some uses of lead, *e.g.*, lead-acid battery used in motor vehicles and lead use in consumer electronics are still going on or are slowly increasing, human population is advised to be cautious against lead poisioning.

8.12.2 Mercury

Mercury, like lead, provokes most concern because it is toxic to the nervous system and the fetus and small babies are more sensitive to its adverse effects. A number of anthropogenic activities are responsible for raising Hg levels in the environment. The Hg level in the ambient air is extremely low; its level in unpolluted water is about 0.1 μg per l. The level of Hg in food, except fish, is also very low (5-20 μg per kg.); but most fishes contain higher levels (200-1000 μg per kg in swordfish and tuna). The consumer population is exposed to mercury toxicity due to food grain treated with mercury fungicide or fish contamenated with methyl mercury (10-40 ppm). About 90% methyl mercury is absorbed across the human gastrointestinal tract. Cases of mercury poisoning due to eating mercury contaminated fish have occurred throughout the world. The toxic effects of Hg are related to the nervous system. The earliest symptom is Paresthesia and at higher levels of exposure, ataxia, deafness, dysarthria and eventually death may occur (Clarkson, 1987). Some studies have concluded that methyl mercury poisoning would be unlikely if the daily intake corresponds to a blood level of 20 by per dl and a hair level of 50 μg per g (WHO, 1976). The divalent mercury salts are also harmful after contact or ingestion. Mercury has also been used to fill up dental cavities. The critics of this practice claim that the amalgam is linked to arthritis, multiple sclerosis, and mental disorders. In view of so many uses of mercury and the problems associated with them, it is advised that the use of mercury should be reduced as much as possible.

8.12.3 Cadmium

Generally environmental cadmium levels are lower than those of lead and people ordinarely have lower exposure. The cadmium level in water is very low, about 1 μg per l, except in polluted areas. Its level in the air is in the range of few nanograms per m^3, but may be several milligrams per m^3 in certain workplaces. The environmental levels are raised by smelting and industrial uses. Foodgrains and cereal products are the main sources of cadmium. Humans are also exposed to Cd through cigarette smoking. Cadmium bioaccumulates in the liver and kidney and the amount of Cd stored in kidney increases with age. Kidney damage is the most prevalent chronic effect of Cd exposure.

The Cd poisoning in Japan has once resulted in *itai itai* disease, which was due to eating of rice produced from soil containing very high levels of Cd (Nomiyama, 1980). Whenever-present, high Cd levels in soil become a matter of worry since plants take up Cd more readily than other metals. Because tobacoo plants are able to concentrate Cd, cigarette smoking increases the risk of Cd exposure. The acute effects of Cd toxicity result mainly from local irritation; after inhalation, the lesions include pulmonary edema and chemical pneumonitis; after ingestion the clinical effects are vomitting and abdominal pain. After long term exposure, Cd causes kidney lesions. Other effects of Cd exposure include hypertension and prostate cancer in occupational workers. In all the developed contries efforts are on to control Cd emissions so that people are not exposed to high levels of Cadmium.

8.12.4 Other Metals

Other metals of concern are Arsenic, Chromium, Beryllium, Nickel and Selenium. Their acute and chronic exposures have been observed to cause serious problems such as skin and lung cancer and black-foot disease (due to *As* poisoning); hypersensitivity, renal problems and lung cancers (due to *Cr* exposure); berylliosus(lesions in lung alveoli due to *Be* exposure); nasal cancer and cancers of the kidney, lung and stomach (due to *Ni* exposure); and hair loss, nail pathology and teeth decay (due to over exposure to *Se*). In animals, and overexposure of *Se* induces more severe effects including liver necrosis, anemia, disorders of reproductive function and enlargement of pancreas.

Excessive intake of cobalt results in *polycythemia*, an overproduction of erythrocytes. Overexposure to *Fe* may cause liver dysfunction. Occupational exposure to *Mn* results in pneumonitis and encephalopathey. Occupational exposure to Zn_2O_3 fume results in 'metal fume fever' (Goyer,1996). Metals such as Co,Cu and Fe are essential metals required in the proper development of erythrocytes. However, certain metals have been used in medicine.

MODEL QUESTIONS

1. Which factors affect the toxicity of a chemical? What do you mean by selective toxicity?
2. Define environmental toxicology. Which characteristics of pesticides determine their mode of action and toxicity ?
3. Give an account of toxicity of heavy metals.
4. Write notes on the following:
   (*a*) Bioaccumulation
   (*b*) Biomagnification
   (*c*) Biotransformation
   (*d*) Bioactivation
   (*e*) Environmental toxicology
   (*f*) Dose-response relation
   (*g*) Toxins
   (*h*) Antibiotics
   (*i*) Vaccines

MULTIPLE CHOICE QUESTIONS

5. Which of the following shows selective toxicity ?
   (A) DDT
   (B) Snake poison
   (C) Parathion
   (D) Scropion poison
6. Most important toxic metals are:
   (A) Lead, Mercury and Cadmium
   (B) Lead and Chromium
   (C) Nickel and Selenium
   (D) Iron and Cobalt
7. The Combined effect of two toxic chemicals may be more than the sum, this type of interaction is called:
   (A) Additive
   (B) Synergistic
   (C) Antagonistic
   (D) All the above
8. Long-term exposure to a toxic chemical produces an effect, which is called:
   (A) Acute
   (B) Highly acute
   (C) Subchronic
   (D) Chronic
9. Carbon monoxide exerts its effect by;
   (A) Destroying haemoglobin molecules
   (B) Destroying WBC
   (C) Blocking the formation of oxyhaemoglobin
   (D) Destroying ATP molecules
10. Which of the following accumulate in body fat ?
    (A) Fluoride
    (B) Lead
    (C) PCBs
    (D) All of them
11. Which of the following bioaccumulates in soft tissues of the body?
    (A) Fluoride
    (B) Cadmium
    (C) PAHs
    (D) Dioxins

12. Which of the of the following metals is inhaled though cigarette smoking:
    (A) Cadmium (B) Lead
    (C) Mercury (D) Iron
13. Which of the following metals was responsible for the disease *itai itai* in Japan?
    (A) Lead (B) Cadmium
    (C) Iron (D) Copper
14. Which of the following aquatic species is not used in aquatic toxicology ?
    (A) Daphnia (B) Chironomus
    (C) Chlorella (D) Mahseer (Tor species)
15. Medico-legal effects of harmful chemicals are a subject-matter of
    (A) Aquatic toxicology (B) Forensic toxicology
    (C) Economic toxicology (D) General toxicology

9 CHAPTER Environmental Biotechnology

9.1 INTRODUCTION

Environment means surroundings. It consists of both abiotic and biotic factors of the medium where an organism occurs. These factors interact among themselves and with the organism and influence the adaptability of the organism. *Biotechnology* has shown explosive growth over the past three decades and microorganisms have been exploited in a number of ways. It may be defined as the industrial exploitation of biological systems or processes and it is largely based upon the expertise of biological systems in recognition and catalysis (Higgins, 1985).

Man has used biotechnology for thousands of years in such activities as wine-making, brewing, bread- making, food preservation and processing by fermentation (*e.g.*, vinegar, cheese) and the manufacture of soap from fats, primitive medications and waste treatment. But recent developments in biotechnology have been made after the discovery of recombinant DNA technology. And scientists all over the world have been attempting the application of biotechnology in agriculture, materials (*e.g.* mineral extraction), medicine (*e.g.* use of microbes and enzymes in manufacture of complex drugs and new antibiotics), food (e.g., new methods of food preservation, food additives and use of enzymes in food processing), energy (*e.g.*, use of biogas and production of ethanol as a liquid fuel), chemicals (*e.g.* organic acids and use of enzymes in detergents) and environment (*e.g.* waste treatment techniques and use of microbes for xenobiotic degradation).

The maintenance of environmental quality has been a continuing problem since the dawn of civilization. *Environmental biotechnology* deals with the use of biotechnology to solve the major environmental problems. For many years environmental biotechnology has been most commonly linked with waste treatment technology and bioremediation. But now there is a paradigm shift in environmental biotechnology and there is a move from bioremediation focus to pollution abatement, and pollution prevention to environmental protection and sustainability. *Environmental sustainability* is operationally defined as utilisation and management of natural resources in a manner that is consistent with their steadystate availability for future generations (Sayler,1997). It is expected that microbial and genetic biodiversity is going to play an everincreasing role in the development of environmental biotechnology for human welfare.As on date the most reliable strategy is the use of eco-friendly microbes for biodegradation, which is generally accepted as an environmentally sound and economically feasible protocol for the treatment of hazardous wasts and effluents (Lalithakumari, 2003). Among other things this chapter covers briefly the following aspects:

(*a*) Sewage and wastewater treatment

(*b*) Microbiological degradation of xenobiotics,

(*c*) Mineral recovery,

(*d*) Pollution control by genetic engineering and biological treatment of air pollutants

9.2 SEWAGE AND WASTEWATER TREATMENT

In most parts of the world, increases in population, the use of huge quantities of fertilizers and pesticides in modern agriculture, the expansion of food processing industry and the growth of other industrial processes contribute to the volume of sewage and wastewater. Therefore, wastewater treatment attempts to remove compounds with a high BOD, pathogenic organisms and harmful chemicals. The wastewater treatment is done as follows:

9.2.1 Primary Treatment

Primary processing is done to remove solids and large objects by passing the wastewater through a series of screens. Rapidly sedimentable solid particles ('grit') then settle out during flow through a grit chamber. The effluent is then allowed to settle in a sedimentation tank (Fig. 9.1). Primary treatment removes about 60% of settleable solids and about 35% of oxygen-demanding waste. In other words, it results in 35% reduction in BOD.

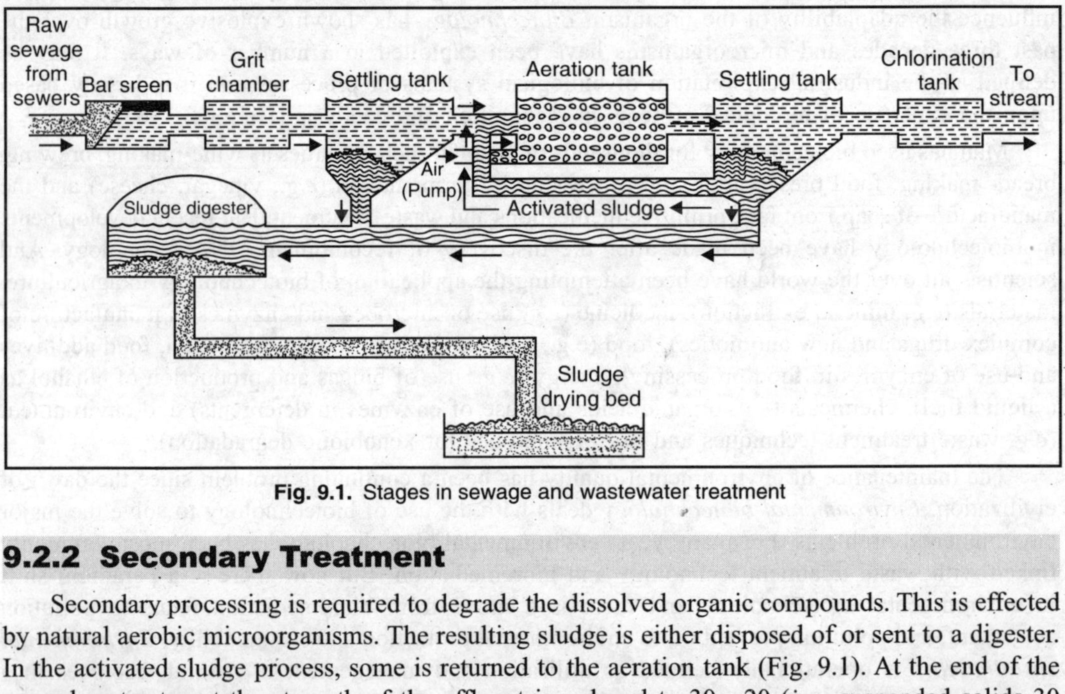

Fig. 9.1. Stages in sewage and wastewater treatment

9.2.2 Secondary Treatment

Secondary processing is required to degrade the dissolved organic compounds. This is effected by natural aerobic microorganisms. The resulting sludge is either disposed of or sent to a digester. In the activated sludge process, some is returned to the aeration tank (Fig. 9.1). At the end of the secondary treatment the strength of the effluent is reduced to 30 : 20 (*i.e.*, suspended solids 30 mg per litre and BOD 20 mg per litre), which can be discharged into water sources. However, major part of the nitrogen and phosphorus compounds still remain in the effluent.

9.2.3 Tertiary Treatment

Tertiary treatment involves chemical precipitation and separation of nitrogen and phosphorus. However, in many cases tertiary treatment is considered optional. The resulting effluent at the end of tertiary treatment is of 10:10 strength (*i.e.*, suspended solids 10 mg per litre and BOD 10 mg per litre). However the effluent cannot be used as drinking water without giving the normal chemical treatment.

The drinking water may be disinfected by chlorine. However, organic material present in water reacts with chlorine to produce disinfection by-products (DPBs) such as chloroform. Some alternate disinfectants that produce smaller amounts of DPBs are ozone, UV light, chlorine dioxide and chloramines.

9.2.4 Digestion Processing

It is used to treat the sludge from the primary and secondary treatments, by an anaerobic process. This treatment reduces the solid volume, the odour and the number of pathogens and also generates the valuable organic fuel, methane.

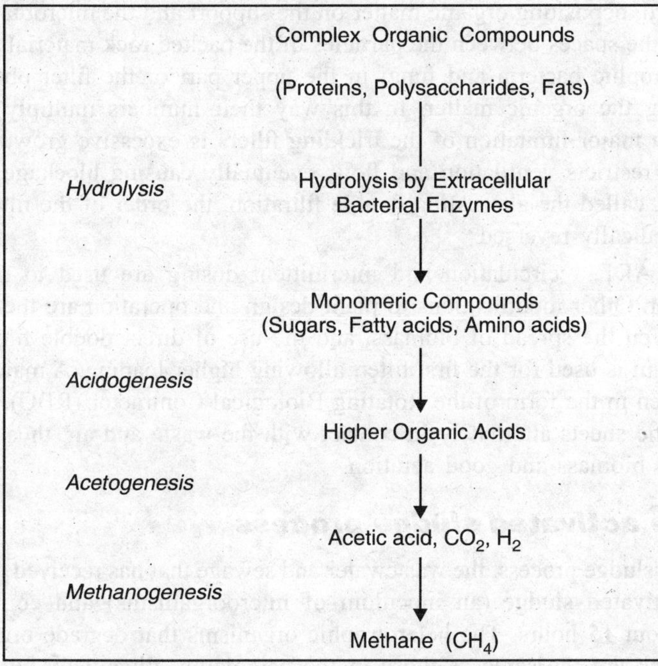

Fig. 9.2. Major metabolic reactions in the anaerobic decomposition of organic wastes into carbon dioxide and methane (after Glazer and Nikaido, 1998)

9.2.5 Biotechnological Improvements

Treatment of industrial wastewater uses processes similar to those described above. Therefore, any biotechnological improvements to these processes are likely to have immediate industrial application. Biotechnological improvements may include —

(*i*) an increase in the capacity of treatment plants;

(*ii*) increased recovery of useful by-products;

(*iii*) replacement of synthetic chemical additives that are currently used; and

(*iv*) removal of metals, recalcitrant compounds and odour.

9.2.6 Aerobic Treatment Systems

The main purpose of secondary treatment is to reduce the BOD of liquid waste. BOD is a measure of the amount of DO consumed by aerobic microorganisms as they metabolize the degradable organic material in the waste (milligrams of DO consumed per litre on incubation for 5 days at 20°C). Aerobic effluent treatment is the largest controlled use of microorganisms in the biotechnological industries. It includes — substrate adsorption to the biological surface; adsorbed solid breakdown by extracellular enzymes; dissolved material absorption into cells; growth and endogenous respiration; release of excretory products; and ingestion of primary population by secondary grazers. These steps should ordinarily result into complete mineralization of the waste. For aerobic processing of waste either trickling (percolating) filter system or activated sludge tanks are used.

9.2.6.1 A trickling filter tank

A trickling filter tank is 3 to 10 feet deep, usually packed with crushed rock on which the microbial population forms a thin film. The liquid waste is applied to the top of the filter and percolates downward, depositing organic matter on the support and the microbial film. An upward flow of air through the spaces between the particles of the packed rock material maintains aerobic conditions. Heterotrophic bacteria and fungi in the upper part of the filter obtain nutrients and energy by oxidizing the organic matter. In this way their numbers multiply and new film is formed. However, a major limitation of the trickling filters is excessive growth of the microbes in the filter, which restricts ventilation and flow, eventually causing blockage and failure. In a recent modification, called the alternating double filtration, the order of the filters first receiving the effluent is periodically reversed.

In addition to ADF, recirculation and intermittent dosing are used to dissipate the load deeper into the filter. Other modifications to plant design and operation are the slowing down of the distributor to even the spread of biomass, and the use of direct double filtration in which a larger size of medium is used for the first filter, allowing higher loading. A major modification to plant design has been in the form of the Rotating Biological Contractor (RBC). This is a rotating honeycomb of plastic sheets alternately in contact with the waste and air, thus providing a large surface area for the biomass and good aeration.

9.2.6.2 In the activated sludge process

In the activated sludge process, the wastewater and sewage that has received primary treatment, is mixed with "activated sludge"(an inoculum of microorganisms) and continuously aerated with oxygen for about 15 hours. The heterotrophic organisms that degrade organic matter in an activated sludge tank are the same as those in the percolating filter tank, but the major group of bacteria involved in the treatment process is *Zoogloea ramigera* , a slime-forming bacteria. As these bacteria grow, they form clumps called flocs to which soluble organic matter, as well as protozoa and other organisms, become attached. When the effluent from an activated sludge tank passes into a sedimentation tank, these flocs with the material they are carrying, sediment out and are transferred to an anaerobic sludge digester.

As in the case of filtration processes, recently a number of biotechnological modifications particularly associated with aeration have been introduced into the activated sludge system, some of them are — (*a*) tapered aeration, which relates aeration capacity to the oxygen demand, which is less at the outlet than the inlet; (*b*) step aeration, which introduces the waste at intervals throughout the length of the tank; (*c*) contact stabilization, in which the returned sludge is aerated for organisms to utilize any stored nutrient; (*d*) the use of pure oxygen in closed tanks, which enables them to operate at higher biomass concentrations; and (*e*) the use of deep-shaft air-lift fermenter, which is more economic than the conventional process through reduced residence times and low running costs of the system.

The efficiency of activated sludge process can be improved through a better understanding of the metabolic control in the microflora of the system. However, it is not easy to control biodegradation, but an appreciation of biochemistry of these pathways may allow manipulation of the control process. For example, low concentration addition of Krebs cycle intermediates, glucose, amino acids and vitamins, such as alanine and nicotinic acid, to the sludge can accelerate the rate of oxidation of specific components. Addition of these intermediates to biomass produces an energy requirement that results in ATP production by increased oxidation of inorganic compounds such as ammonia or sulphur.

9.2.6.3 *Fluidized bed*

This is a combination of the trickling filter and activated sludge systems. The two basic designs are — (*a*) Simon Hartley captor: In this, the biomass is grown in spaces inside polyester foam pads which are retained in the reactor by mesh. The pads are periodically removed from the reactor, the thick biomass (about 15 kg. in each m^3 of fluidized support element) machine squeezed out and the empty pads returned to the reactor. (*b*) Dorr-Oliver oxitron: In this sand particles are used as the support medium. The sand is allowed to overflow the reactor, is cleaned and then recycled.

9.2.7 Anaerobic Treatment Process

The digestion of sewage sludge is the most common anaerobic treatment. All the reactions in the anaerobic secondary treatment process can be divided into two groups – acid- forming and methane- forming (Fig. 9.2). Numerous different bacterial species that participate in acid- forming stage include some that are known to be obligate aerobes, but that may grow by utilizing alternative electron acceptors, such as nitrate, sulphate and carbonate. In the methane-forming stage, anaerobic forms of the genera *Methanobacillus, Methanobacterium, Methanococcus* and *Methanosarcina* convert the acetate hydrogen and carbon dioxide produced by the fermenters to methane (CH_4). All methanogens, microorganisms responsible for methane production, are archaebacteria that derive their energy by reduction of compounds such as CO_2, acetate, or methanol.

Methanogens occur in diverse natural anaerobic habitats. Some reactions are shown below:

$$4H_2 + HCO_3^- + H^+ \rightarrow CH_4 + 3H_2O$$

$$CH_3COO^- + H_2O \rightarrow CH_4 + HCO_3^-$$

There are obstacles to implementing anaerobic digesters for gas fuel production. The commercial application of anaerobic digester systems is increasing for the treatment of farm, industrial and food manufacturing wastes, in addition to the processing of energy crops.

9.3 MICROBIOLOGICAL DEGRADATION OF XENOBIOTICS

Many kinds of xenobiotics (Gk. xenos = foreign) occur in the waste effluents produced by the manufacture and consumption of all the commonly used synthetic products. Xenobiotics that are released into the environment on a large scale are numerous different halogenated aliphatic and aromatic compounds, nitroaromatics, phthalate esters, and polycyclic aromatic hydrocarbons. These compounds enter the environment as components of pesticides, fertilizers, and herbicides. Many toxic xenobiotics are progressively more concentrated in each link of a food chain, a process called biomagnification (Fig. 9.3).

Microbial degradation of xenobiotics has been thoroughly studied. However, the term 'biodegradation' is widely used for — (*i*) primary degradation, in which the characteristic property of the original compound is disappeared; (*ii*) environmentally acceptable biodegradation, in which the minimum alteration of the parent compound necessary to remove properties occurs; and (*iii*) ultimate biodegradation, which involves the complete conversion of the parent compound to the inorganic end-products associated with the microorganism's normal metabolic processes. However, many microorganisms capable of degrading unusual compounds have been isolated and tested. Detoxification of pollutants may involve as little as a single modification of structure to render a potentially hazardous chemical innocuous. The fate of xenobiotic compounds will depend upon many interacting factors, both extrinsic (pH photooxidation, weathering) and intrinsic (water solubility, stability, molecular size and charge, volatility). The transformation of a xenobiotic compound that is available to the microbial community, is determined by its entry into the cell

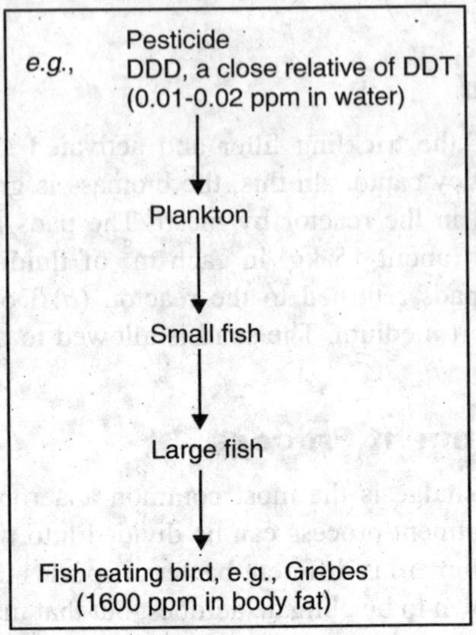

Fig. 9.3. Biomagnification of a pesticide through food chain. Organisms at the peak of food chain contain maximum amount of the pesticide.

and the degree of structural analogy between the synthetic compound and the natural compound for which the biodegradative mechanism exists. Fortuitous metabolism and co-metabolism play important roles in the removal of xenobiotic compounds from the environment. *Gratuitous biodegradation* occurs when an enzyme is able to transform a compound other than its natural substrate. It is done only when the unnatural substrate is able to bind to the active site of the enzyme in such a manner that the enzyme can exert its catalytic activity. Many bacteria and fungi produce enzymes that are able to act on a wide range of organic compounds. *Co-metabolism* is the ability of an organism to transform a nongrowth substrate as long as a growth substrate or other transformable compound is also present. A nongrowth substrate is one that cannot serve as the sole source of carbon and energy for a pure culture of a bacterium and hence cannot support cell division.

9.4 BIOREMEDIATION

Use of microorganisms (probiotic bacteria or friendly bacteria) for improving environmental conditions or restoration of contaminated sites is very well known. Bioremediation has been defined as "a biological response to environmental abuse" (Hamer, 1993). This definition serves to distinguish between the use of microorganisms to remediate contaminated sites and their application in biorecycle/biotreatment processes designated to reduce inorgannic and organic contaminant emissions at source (Colleran,1997). Bioremediation focuses on the former while it is recognized that at present biological treatment processes play a major role in preventing and reducing the organic and inorganic environmental contamination from the municipal, industrial and agricultural sectors. Bioremediation is based on the premise that naturally occurring bacteria in an impacted environment develop the means to degrade or tolerate the presence of organic contaminats.

This indigenous activity can be considered *passive* or *intrinsic bioremediation* (Leavitt, 1997). Intrinsic bioremediation has been recognized most frequently for groundwater (saturated systems) contaminated with hydrocarbons. Sometime a more active remediation system is used. Semi-passive bioremedation systems are defined as those *in-situ* treatments that induce favourable conditions for accelerated biodegradation. The excellent remedial approach is *aggressive bioremediation.* It may be defined as engineered bioremediation to produce optimal treatment in both *in-situ* and *ex-situ* systems. Its advantage is much shorter treatment period, alleviating any construction limitations sooner than with passive system.

Microorganisms are able to degrade a chemical because they are using it as an energy source to promote their own growth. According to Glazer and Nikaido (1998), *bioremedidation* is defined as a spontaneous or managed process in which biological, especially microbiological, catalysis acts on pollutants and thereby remedies or eliminates environmental contamination. Bioremediation is currently being used to decrease the organic chemical waste content of groundwater, soils, effluent from food processing industry and chemical plants, oily sludge from petroleum refineries and oil spills in the ocean. Degradation of pesticides, insecticides and fungicides *in-situ* (in soil) by microbes *Pseudomonas* A3, *P.putida, P.aeruginosa* and *Serratia marinorubra* either singly or as a consortia has been successifully attempted. Some microbial strains such as *S. marinorubra, Bacillus* species YW and YDLK consortia have also been used for decolorization and degradation of industrial effluents and textile mill azo dyes and effluents. Heavy metal detoxification and biosorption by employing the bacterium *bacillus* species YW has been found to be effective in reducing hexavalent chromium to its non-toxic trivalent form and the chromate resistance and reduction was found to be plasmid mediated process (Lalithakumari, 2003). The removal of less toxic trivalent chromim through biosorption using the EPS *Azotobacter* species as the biomatrix has also been successfully attempted. Bioremediation techniques fall into four categoies – *in situ* treatment, composting, landfarming, and above-ground reactors.

9.4.1 In-situ Bioremediation Techniques

In-situ, in Latin, means "in the orginal place". Thus, *in-situ* bioremediation means bioremediation based on the degradative activities of endogenous microbial populations. In other words, it relies on the indigenous microbial flora of subsurface soils and groundwater. It depends on the premise that the microorganisms already present in a contaminated site have adapted to the organic chemical wastes found there and are able to degrade some or all of the components of these wastes. The degradation by these so - called adapted microorganisms goes on until some nutrient or electron acceptor attains a limiting concentration. In most cases, oxygen level is the limiting factor, but phosphate and nitrate may also become limiting factor. The stimulation of natural biotransformation by nutrient addition (N, P) to the environment is called *enhanced in-situ bioremediation*. The addition of nutrients to contaminated soils and also to hydrocarbon-contaminated marine environments has been investigated.

One of the best examples of enhanced *in-situ* bioremediation was the clean-up of the Exxon Valdez oil spill (1989) in Alaska. The spill, about 11 million gallons of crude oil, severely affected 350 miles of shoreline in Bligh Reef in Prince William Sound. In this case, fertilizers were used to accelerate the removal of oil from the beaches, supplying extra nutrients that were in limiting concentrations. A single application of inorganic fertilizer was shown to speed up the disappearance of oil by a factor of two to three over its rate of disappearance on untreated site. Samples of oil taken at the end of that time from surfaces of treated beaches showed changes in composition consistent with extensive biodegradation. Thus enhanced *in-situ* bioremediation offers several potential advantages in the elimination of hazardous wastes.

Besides *Pseudomonas* species, HC1 and *Raulstonia* species have also been successfully used for oil degradation. Most of the *Psedumonas* species strains harboured a catabolic plasmid, which encodes the genes for hidorcarbon degradation. The biotransformation of this plasmid to various bacteria in natural soil and marine water has been carried out, indicating the horizointal transfer of catabolic genes from one bacterium to another, paving way to create "Superbugs" for bioremediation in differing and metamorphosing ecosystems.

Bioventing is one of the most widely used methods of remediating soils contaminated by petroleum hydrocarbons. Bioinventing supplies air to an unsaturated soil zone by using a combination of pumps and blowers that apply a vacuum to the target area, while continuously injecting low volumes of air. *In-situ* bioremediation under anaerobic conditions may also be enhanced by providing electron acceptors such as sulphate or nitrate. However, *in-situ* methanogenic bio-remediation has not, so far, been investigated thoroughly, although it is recognised that degradation of organic pollutants in anaerobic microniches in soil,sediment and groundwater environments contributes significantly to overall *in-situ* bioremediation rates.

9.4.2 Ex-situ Bioremediation

Ex-situ bioremediation techniques are usually aerobic and involve treatment of contaminated soils or sediments using solid or slurry-phase systems. Solid-phase systems include compost heaps, landfarming (soil-treatment units) and engineered biopiles.

9.4.2.1 Composting

Compost is a mixture of soil, partially decayed plants, and sometimes manure and commercial fertilizer. It is very rich in microorganisms. It has long been used by farmers and gardners to make soils more fertile and to improve crop yields. Compost heaps consist of contaminated soil or sediment supplemented with composting material in order to enhance water and air-holding capacity and improve physical handling properties. Composting techniques are used for remediation of highly contaminated sites and have proved successful for military sites contaminated with explosives such as 2, 4, 6-trinitrotoluene (TNT), hexahydro-1, 3, 5-trinitro-1, 3, 5-triazine (RDX) and N-methyl-1, 2, 4, 6-tetranitroaniline (Tetryl). Biodegradation or biotransformation of over 90% of these explosives was achieved within 80 days in a compost pile maintained at 55°C. After 150 days, a starting concentration of 18,000 mg of explosives per kg of soil was reduced to 74 mg perkg.

9.4.2.2 Landfarming

Landfarming is used to dispose of oily sludges from petroleum refinery operations. The oily sludge from refinery wastes is mixed with soil and subjected to enhanced *in-situ* bioremediation. The sludges may be pretreated or not. Biological pretreatment of refinery effluents partially mineralizes the organic waste components; the residual solid waste (sludge) then has a high content of aromatic hydrocarbon compounds. In contrast, untreated settled solids, such as the solids from tank bottoms, contain high amounts of aliphatic hydrocarbons and silt. The terrain of a landfarm must be flat to minimize runoff; the soil should be light and loamy for proper aeration; and a clay layer should underlie the porous surface soil to reduce the possibility of groundwater contamination through seepage. The optimal temperature range for biodegradation is 20-30°C. Inorganic fetilizer is applied to the site to provide fixed nitrogen, and phosphate and Ca CO_3 are added to raise the pH of the soil-waste mixture to about 7.8. For untreated sludge, maximal oil biodegradation rates in soil are achieved at a hydrocarbon load of 5-10% by weight — that is , 100-200 metric tones of hydrocarbon per ha. In such landfarms, about 50-70% of the applied organic waste is degraded before the next batch of sludge is applied. However, a disadvantage of landfarming is that the process is quite slow. Moreover, the heavy metal constituents of the sludge gradually accumulate in the landfarm soil. As a result, a plot of land used intensively as a landfarm cannot be used later for grazing livestock or growing crops.

9.4.2.3 Above-ground Bioreactors

The above-ground bioreactors are based on the same technology as fermenters. There are four common types of bioreactors — the stirred- tank reactor, the bubble column, the air lift, and the packed bed (or fixed bed). They are used for the treatment of either excavated soil or groundwater containing high levels of contaminants. Contaminated soil is mixed with water and introduced into the reactor as a slurry. Granulated charcoal, plastic spheres, glass beads, or diatomaceous earth provide a large surface area for microbial growth in bioreactors. The large surface area of the microbial biofilm that forms on such supports leads to a rapid rate of biodegradation. The microbial inoculum may be taken from activated sludge from a sewage treatment plant, an indigenous population at the contaminated site, or from a pure culture of appropriate microbes.

Such bioreactors can be used in series to accomplish different kinds of degaradation. For example, the first reactor can be operated in an anaerobic mode and its effluent transferred to a second reactor operated in an aerobic mode. Mineralization requires aerobic, conditions whereas biotrasformations such as dehalogenations of certain compounds require anaerobic conditions.

9.4.3 Environmental Factors Influencing *in-situ* Bioremediation

Both the rate and the extent of microbial remediation of organic contaminats *in-situ* are affected by a number of environmental factors, some of which may be mainpulated, whereas others are difficult to modify within the contaminated site. Important environmental factors which affect *in-situ* bioremediation are:

PH: Most laboratory based bioremediation studied have been carried out in about neutral pH since the majority of bacteria show optimal growth at this pH range. In many studies, adjustment of pH enhances the rate of biodegrdation.

Temperature: Temperature directly affects the metabolism and growth of bacteria. The vast majority of *in-situ* bioremediation applications have been carried out under mesophilic conditions (between 20 and 40°C). However, the thermophilic or thermotrophic species are capable of degrading a diverse range of organic compounds in wastes or wastewaters at higher temperature (60-70°C). A variety of techniques have been utilized to increase the temperature in *in-situ* soil remediation applications. All these studies indicated that even modest increase in temperature may significantly increase bioremediation rates.

Water content: Water content in soils or sediments in an important factor affecting bioremediation rates. Microbes generally require water activity values (a_w) of 0.9-1.0 in order to metabolize and grow. The majority of bacteria grow optimally at a_w values in the upper limit of this range.

Soil type: In general, *in-situ* bioremediation rates are enhanced when the soil is granular or porous. *In-situ* degradation rates are slowed down under unfavourable geological characteristics which include low permeability of soil, faractured rock, and water-logged or arid conditions.

Nutrient availability: In almost all cases nutrient supplementation (addition of N,P) significantly increases bioremediation rates. It has shown very good results in hydrocarbon bioremediation of soils and groundwaters when N and P levels have been shown to be limiting.

External Electron availability: Many *in-situ* bioremediation techniques use some provision of oxygen supply to enhance aerobic respiratory breakdown of organic contaminats. Addition of hydrogen peroxide is used to introduce oxygen. Hydrogen peroxide is about 7 times more soluble in water than oxygen and its decomposition in soil yields 0.5 mol of O_2 per mol of H_2O_2 introduced to contaminated site ($2H_2O_2 \rightarrow 2H_2O + O_2$).

Bioavailability of organic pollutants: This is an important factor governing the rate of *in-situ* bioremediation. Improvements in bioremediation rates have been achieved by the addition of biosurfactants or synthetic detergents to the contaminated zone. Addition of biodegradable solvents, which assist in desorption and dissolution rates also exhibits increase in the biodegradation of the adsorbed pollutants (*e.g.*, adsorption of PAHs by soil particles)

***Cometabolism*:** It is a process whereby microbes involved in the metabolism of a growth promoting substrate also transform other organic contaminants (cosubstrates) that are not growth supporting if supplied as sole carbon and energy source. Cometabolic transformation of organic pollutants is an important process in both aerobic and anaerobic environments. Bacterial transformation of DDT, PCBs provide examples of both aerobic and anaerobic cometabolic biodegradation. Provision of a readily metabolizable substrate may also promote pollutant transformation by enhancing the growth of associated microbes involved in the overall microbial activity.

***Gene Expression*:** The ability of indigenous microorganisms to degrade organic pollutants is dependent on expression of the genes encoding the required uptake and degradative enzyme systems. If the bioavailable concentration of a pollutant is too low, the expression of inducible operons may not occur.

9.5 USE OF PLANTS IN BIOREMEDIATION

The idea that plants can be used for environmental remediation is not new. Extensive research on using plants for treating radionuclide contamination was conducted in Russia in the early 1960s. (Timofeev-Resovsky *et al* 1962). Since then, there have been a number of reports that aquatic plants such as water hyacinth, duckweed and water velvet can accumulate heavy metals from contaminated water (Mo *et al.*, 1989; Jackson *et al.*, 1990). Recently the value of terrestrial plants for environmental remediation has been recognized (Cunningham and Berti, 1993). Crop plants such as Indian mustard have been used to extract heavy metals from soil and translocate them to the leaves and stalks of plants (Raskin *et al* 1994, Duchenkov *et al* 1995). A list of plants used in phytoremediation is given in Table 9.1. Aquatic plants can be used in bioremediation (Phytoremediation) especially to remediate sites contaminated with heavy metals. For example, water hyacinth (*Eichhornia crassipes*) has been used for purifying not only domestic wastewater but also industrial wastewater. It can readily absorb, accumulate and concentrate heavy metals such as Pb, Fe, Cu, Cd, Hg and Ni. Other studies have evaluated the potential of water hyacinth to absorb various organic chemicals — phenols and toxaphene and to remove even radioactive metals from effluents. Cell suspension cultures of *Datura innoxia* have been found to remove a wide variety of metal ions from solutions.

Table 9.1. List of some plants used in Bioremediation (phytoremediation)

S.No.	*Aquatic/Semiaquatic plants*	*Crop/Terrestrial plants*
1.	Water hyacinth (*Eichhornia crassipes*)	Indian mustard (*Brassica juncea*)
2.	Duckweed (*Lemna minor*)	New Caledonian tree (*Sebertia acuminata*)
3.	Pennywort (*Hydrocotyle umbellata*)	*Thlaspi caerulescens*
4.	Water velvet (*Azolla pinnata*)	*Alyssum*
5.		Sunflower (*Helianthus annuus*)

Like animals, most plants can be poisoned by heavy metals. But certain plants can accumulate as much as 40% of their weight as heavy metals without harm to themselves. According to Ensley *et al* (1997), further advances in phytoremediation require a multidisciplinary approach, covering fields as diverse as agricultural engineering, plant biology, agronomy, soil science, microbiology and genetic engineering. A better understanding of the biochemical processes involved in plant heavy metal uptake, transport and accumulation will certainly improve phytoremediation using modern genetic approaches. One strategy for improving the phytoremediation potential of high biomass plant species is the introduction of genes responsible for metal accumulation and resistance from the wild metal accumulators. In the absence of known "phytoremediation genes" this may be accomplished via somatic and sexual hybridization followed by extensive screening and backcrossing of progeny. Mutagenesis of selected high biomass plant species may also produce improved phytoremediating cultivars. Phytoremediation of heavy metals is designed to concentrate metals in plant tissues, thus minimizing the amount of hazardous waste, which needs to be treated and deposited at hazardous waste sites. But an economical method of reclaiming metals from plant residue should be developed. As present, methods for further concentration of metals in plant tissues include — sun, heat air drying, environmentally safe ashing or incineration, composting, pressing and compacting, and acid leaching. These plants can then be harvested and burned to recover the metals. Plants can be useful in treating other chemicals also. One aquatic plant reduced the concentration of TNT from 128 ppm to 10 ppm. Submerged roots of sunflowers can concentrate radioactive wastes from the water. Because the redioactive chemicals accumulate only in the roots, only they require special disposal.

9.6 BIOAUGMENTATION

In case when a competent degradative population either does not exist, or in which the time needed for acclimation/adaptation may be too long, microbial inoculants may be added to promote or enhance bioremediation rates. This technique is referred to as *bioaugmentation* and may involve the addition of natural isolates or *genetically engineered microorganisms* (GEMs). However, GEM augmentation studies are confined to contained laboratory-based bioreactors. The success, or otherwise, of bioaugmentation depends both on the degradative capacity of the introduced inoculant and on its environmental competence with respect to the site or system under investigation.

Both *in-situ* and *ex-situ* bioremediation protocols are designed keeping in view the catabolic pathways used by aeorbic and anaerobic bacteria for degradation of complex aromatic and aliphatic compouds. Consequently the utilization of sequenced anaerobic-aerobic treatment strategies presents an exciting new approach for the bioremediation of aromatic xenobiotics. The development and application of efficient, cost effective and environmentally sound bioremediation technologies requires a multidisciplinary approach (integration of many different disciplines), as well as on advances in our knowledge of the physico-chemical, biological and ecological factors that govern microbial degradation of organic contaminants both *in-situ* and *ex-situ* (Colleran,1997).

9.7 RECOVERY OF RESOURCES

One of the main objectives of environmental biotechnology is the conservation of resources via the recycling of waste materials. The recoveries of more valuable products such as metals, oils, vitamins are important aspects of this technology. Use of microorganisms in recovery of minerals of commercial interest is also an interesting area. Reclaiming organically polluted water, application of microbes to degrade recalcitrant compounds, use of animal waste as fertilizer, recycling of microbial protein as an animal feed (Fig. 9.4), and removal of heavy metals found in sewage sludges, are examples of this type of technology.

Use of Bactrial ore leaching helps in recovering metals from low-grade ores. Of late, the contribution of bacterial leaching to copper recovery is increasing. Solubilization of metals by bacteria from ores proceeds by either 'indirect leaching' or 'direct leaching'. The former depends on the ability of various species of acidophilic microorganism *Thiobacillus ferrooxidants* to generate metabolic energy by oxidizing either ferrous iron or sulphide leading to the production of ferric sulphate–

$$FeS_2 + 3.5\ O_2 \rightarrow FeSO_4 + H_2SO_4 + H_2O$$

$$2FeSO_4 + 0.5\ O_2 \rightarrow Fe_2(SO_4)_3 + H_2SO_4 + H_2O$$

(ferrous sulphate) (ferric sulphate)

Ferric sulphate is a strong oxidizing agent that is able to dissolve several copper suphide minerals. Thus, leaching by $Fe_2(SO_4)_3$ is called indirect leaching because it is independent of the presence of oxygen or microbial action. It depends upon *T. ferrooxidan's* ability to supply the necessary ferric sulphate by oxidizing Fe^{2+} to Fe^{3+}. However, in the direct leaching process the microorganism becomes attached to the mineral particles. Then the enzymes associated with the cell membrane catalyze oxidation on the crystal lattice of the metal sulphide. The overall reactions for oxidation of the mineral are —

$$CuS + 0.5\ O_2 \rightarrow Cu^{2+} + S° + 2H + H_2O$$

$$S° + 1.5\ O_2 + H_2O \rightarrow H_2\ SO_4$$

Thus, microorganisms have played a key role in the leaching of metals from mineral-bearing rocks. Besides *T. ferrooxidans* other bacteria involved in the leaching process are *T. thiooxidans, Leptospirillum ferrooxidans* and *Sulfolobus* species. Some other species of *Thiobacillus* such as *T.thiooxidans, T. acidophilus* and *T.organoparus* also play a role in leaching. Recovery of copper is done by dump leaching. *Thiobacilli* living in dumps are largely confined to the top metre in densities upto 10^8 bacteria per gram of ore. Once leaching is well under way in a dump or heap, little change is observed in its microbial population. A dump can be viewed as a continuous- flow reactor in which solubilization of metals is performed by the bacteria attached to the ore particles.

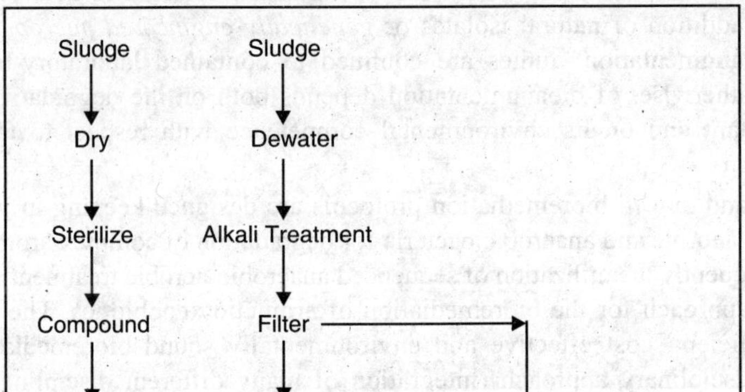

Fig. 9.4. Extraction of protein from activated sludge (from Best *et al,* 1985)

Microorganisms are also used in the removal of heavy metals from aqueous effluents. Microorganisms immobilize metal ions by both active and pasive process. For example, bacteria that use sulphate as a terminal electron acceptor actively produce and excrete sulphide ion that forms an insoluble complex with metal ions present in solution, causing these ions to precipitate.

The H_2S produced by the sulphate-reducing bacteria forms highly insoluble metal sulphides such as CdS, CuS, FeS, HgS and ZnS, which remain in the bottom sediments. On the other hand, *biosorption* is a passive process seen with both living and dead cells. It may be defined as strong binding of metal ions to bacterial cells and to polymeric substances secreted by the cells. The *biosorbents* prepared from bacterial biomass are similar to synthetic ion exchange resins in their capacity for loading metal ions. Microorganisms can remove chloride and sulphate from water bodies. Some wetlands artificially enriched by appropriate organic nutrients have been shown to act as "biofilters" for the heavy metals present in acid mine drainage. Microbes are also able to remove Hg from waste water by volatilization as dimethyl mercury $(CH_3)_2Hg$.

$$Hg^{2+} \rightarrow CH_3\,Hg \rightarrow (CH_3)_2\,Hg$$

By methylating the metal, many microorganisms protect themselves from the toxic effects of Hg, but the resulting alkylmercury compounds are very toxic to higher organisms.

With the help of microorganisms cyanide may also be removed from wastewater by converting cyanide to nitrate. The first stage of the process consists of conversion of cyanide and thiocyanate to ammonia and bicarbonate by *Pseudomonas* spp. In the second stage, nitrifying bacteria, *Nitrosomonas* and *Nitrobacter*, convert ammonia to nitrate.

9.8 POLLUTION CONTROL BY GENETIC ENGINEERING

Many toxic compounds produced by human activities cannot be degraded by microorganisms available in natural water bodies and soil sediments. But genetic engineering is being applied to generate microorganisms that may be capable of degrading them. Some examples of such toxic compounds are: polychlorinated biphenyls (PCBs), chlorinated benzenes, and chlorinated phenols. Recently two strategies have been applied to the development of novel degradative pathways.

The first strategy, termed as *pathway restructuring,* is used when the resistant compound is similar in structure to a chemical compound that is degradable. It consists of determining which engyme-catalyzed steps of the known degradation pathway are unable to degrade these related compounds and therefore an attempt is made to modify the enzymes.

The seond strategy, termed *pathway assembly*, is used when it is possible to plan a pathway that would utilize enzymes from several microorganisms and by means of genetic engineering those enzymes are assembled within a single host or in a consortium and placed under the appropriate metabolic regulation.

Pollution control by genetic engineering works best when pollutants are a known mixture of relativily concentrated organic compounds that are related to each other in structure, when conventional alternative organic nutrients are absent or perhaps added in controlled amounts only, and when there is no competition from indigenous microorganisms. Thus, genetic engineering makes it possible to alter the properties of existing degradative enzymes, to modify regulatory mechanisms, and to assemble within a single organism degradative enzymes from various different or phylogenetically distant organisms. However, the task of engineering an organism for the degradation of a particular target compound is not simple. The easiest way to create an appropriate genetically engineered strain is to begin with an organism that already possesses much of the necessary degradative enzymatic machinery. An example of horizontal expansion of a catabolic pathway by genetic engineering is the following:

Psuedomonas putida carrying the TOL plasmid is able to grow on a variety of alkylbenzoates such as 3- and 4- methylbenzoate (3MB and 4MB), 3,4 dimethylbenzoate (3,4 DMB) and 3-ethylbenzoate (3-EB). However, the organism cannot grow on the very closely related compound 4- ethylbenzoate (4EB). To redesign the pathway to enable it to degrade 4 EB, it was necessary to broaden the effector specificity of the *xylS* regulator protein in a way that it would be activated by 4 EB, and to generate catechol 2,3- dioxygenase mutant enzymes capable of degrading 4- ethyl catechol. A genetic engineering strategy was applied for selection of mutants whose *xylS* protein would form a complex with 4 EB that could bind at *Pmeta* promoter and allow the transcription of genes controlled by this promoter (Ramos and Timmis, 1987).

9.9 BIOLOGICAL TREATMENT OF AIR POLLUTANTS

Biological treatment of air pollutants is now attempted in many countries. The gases must be treated at the end of pipe to prevent release to the atmosphere. The treatment is done by oxidation. The *volatile organic compounds* (VOCs) are oxidized to CO_2, H_2S and SO_2 to sulphate and Nitrogen oxides (NO_x) to nitrate-

$$VOC + O_2 \rightarrow CO_2 + H_2O$$

The bioreactions require no fuel, or chemicals create no secondary pollutants and the reaction rates increase with pollutant emission rates. The microorganisms in soil and water carry out the reaction shown above for their metabolic energy. The following biological methods of air pollution control are used (Fig. 9.5).

9.9.1 Biofilter

Biofilteration may be defined as the oxidation of air pollutants as contaminated air flows slowly through porous media such as soils, peat, compost, activated carbon, plastic packing

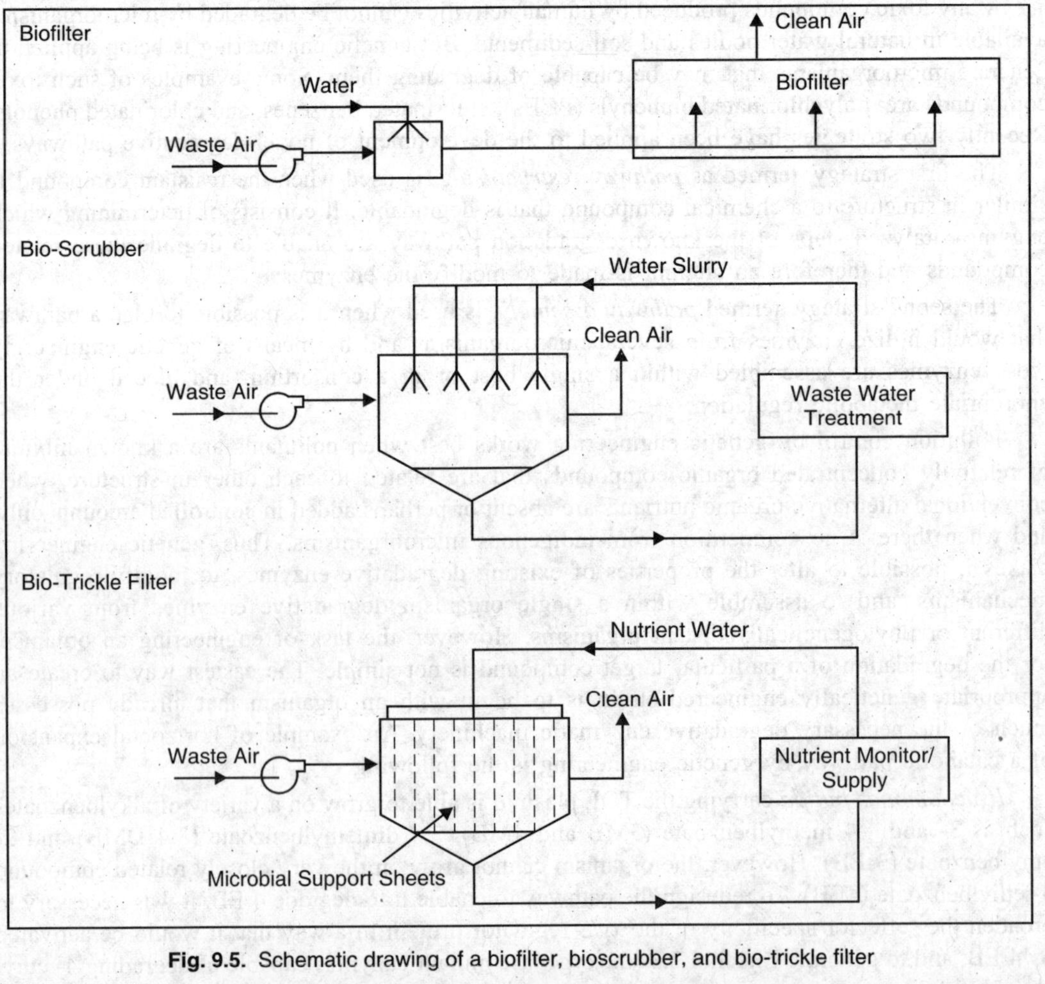

Fig. 9.5. Schematic drawing of a biofilter, bioscrubber, and bio-trickle filter

material and ceramic packing material that supports a degradative microbial population attached to the walls of the pores. Air pollutants adsorb on the pore walls and flow more slowly than the air. The high O_2 concentration and the intimate mixing of oxygen with the VOCs ensures complete oxidation.

However, the various media have both advantages and disadvantages. Soil is inexpensive, neutralizes acids well, its moisture content may be easily controlled. Biofiltration does not pollute the soil because soils cannot accumulate gases or their metabolic protects. In soil, organic gases are converted to CO_2 and water; inorganic gases are converted to their calcium oxy salts, *e.g.*, SO_2 and H_2S to Ca SO_4, and NO_x to Ca $(NO_3)_2$. Among other media, compost is also not expensive. It is lighter than soil and requires less area. It neutralises acids poorly, is hydrophobic when dry and has to be replaced within 2-5 years. The inorganic media such as plastic packing materials, ceramics, and activated carbon offer promise of smaller size because uniform air flow is easier to attain. Overcoming the size disadvantage of biofilters means increasing the VOC loading rate. Biofiltration requires only a fan and some means of maintaining a high moisture content in the biofilter bed. An advantage of biofiltration is that it is quite inexpensive compared to physico-chemical air pollution control methods.

9.9.2 Bioscrubber

Bioscrubber is a modified water scrubber. The spray into the gas stream is a sludge of suspended microbes which increases the aqueous solubility of the VOCs and removes them more effectively than water alone. The sludge then flows to a water treatment plant to biodegrade the dissolved VOCs and generate a new sludge.

This method appears best suited for large air flows because of their low backpressure and small size. Their disadvantage is operating a wastewater treatment plant in addition to one that destroys the VOCs, rather than just stripping VOCs from the water, generating a nonodorous sludge.

Thus, bioscrubbing is helpful to deal with toxic and odorous gases. Microbiological methods for scrubbing of waste effluents are available, although, in general, their performance at the point of waste discharge remains to be determined.

9.9.3 Biotrickle Filter

Biotrickle filter consists of a sheet of a plastic or other microbial support medium hung in the contaminated air stream. The sheet is bathed continuously by a recirculating stream of water containing the nutrients and vitamins required by the microbes. Biooxidation rates per unit volume are high so biotrickle filters can be as small as physico-chemical units. Their disadvantages are that at high loading rates, biological reactions approach zero order kinetics and are incapable of responding to peak loads, accidental discharges and explosion incidents. Bio-trickle filters would seem to be susceptible to microbially produced bactericides and to invasion of less-desirable microbial populations.

Biological air pollution control is related to wastewater treatment and is only slowly being accepted on an industrial scale. Wastewater usually contains a suitable mixture of nitrogen, phosphate and other chemicals as microbial growth factors. In contrast, industrial waste air might contain only a single carbon-, hydrogen molecule. Therefore, the biofilter medium must supply the other growth factors including water so that the microorganisms may metabolize the "the empty calories" in the contaninated air.

9.10 SURFACTANT

A *surfactant* works on the premise of "like dissolving like" (polar solutes tend to dissolve in polar solvents, nonpolar solutes tend to dissolve in nonpolar solvents). It is recently proposed that CO_2 should be used as a soluent in place of other industrial solvents (Cann and Connelly, 2000). The solvent that are currently used are — VOCs, CFCs HCFCs (hydrochlorofluoro carbons) and PERC (perchloro- ethylene). The VOCs create air pollution problems because they are some of the "ingredients" needed for photochemical smog —

$$VOCs + NO + sunlight \rightarrow O_3 + HNO_3 + \text{organic compouds}$$

The halogenated organic compounds are also problematic, for example, CFCs and HCFCs are ozone- depleting substances (see the reactions given below) and many organic compounds such as PERC contaminate groundwater and are potential carcinogens.

$$CF_2\ Cl_2 + UV\text{-}C \rightarrow CF_2\ Cl + Cl.$$

$$O_3 + Cl \rightarrow O_2 + ClO.$$

$$ClO + O \rightarrow Cl + O_2$$

$$O_3 + O \rightarrow 2O_2$$

But surfactants designed by J.M.De Simoni have made the use of supercritical and liquid CO_2 as an industrial solvent possible. These surfactants increase the solubility of various materials in CO_2 that would otherwise be insoluble in CO_2. Thus, by replacing many of the current industrial solvents with CO_2, the risk of environmental pollution is reduced.

9.10.1 Biosurfactant

Biosurfactants are a chemically unique class of compounds produced by many bacteria and fungi and are used in bioremediation. In terms of chemical structure, biosurfactants can be divided into several broad groups - glycolipids, lipopeptides, lipopolysaccharides, phospholipids and neutral lipids/ fatty acids. Generally their molecular weights range from about 500-1500, although some may have molecular weight upto 14300. Few biosurfactants have been well studied with respect to remediation. The best - studied biosurfactant is *rhamnolipid* produced by *Pseudomonas* spp. Some information is also available regarding *Bacillus*-surfaction, *Rhodococcus* — trehalose lipid, and *Candida-sophrorose* lipid systems. Procedures for testing the effect of any surfactant (either synthetic or biological) on solubilization and biodegradation of model hydrocarbons have been described (Miller and Zhag, 1997). Biosurfactants are not yet available commercially and therefore must be produced in the laboratory. For example, *rhamnolipids* can be produced and recovered from many *Pseudomonas aeruginosa* strains. Protocols for production, purification and use of a number of biosurfactants have been developed.

Hydrocarbon biodegradation can be determined by measurement of hydrocarbon mineralization or by measurement of protein increase as an indication of cell growth. The solubilization of hydrocarbons by surfactants is markedly affected by environmental conditions. One common way to measure biodegradation is to measure substrate disappearance. Surfactants do not confer degradative ability; instead, they aid in increasing mass transfer of hydrocarbons into the aqueous phase and they may possibly aid in delivery of hydrocarbons to cells. However, the behaviour of surfactants in soil systems needs to be investigated further. In this case, the effect of biosurfactant sorption by soil must be considered because sorption can significantly reduce the effective amount of biosurfactant available in solution. Thus, much higher biosurfactant concentrations may be required in such cases.

9.11 SUSTAINABLE DEVELOPMENT

It is possible through the use of environmental biotechnology and traditional ecological knowledge (TEK). The term sustainable development was brought into vogue by the World Commission on Environment and Development in 1987 in its report entitled '*our common future*'. It means the development without destruction of natural resource. In other words, sustainable development means improving the quality of human life while living within the carrying capacity of supporting ecosystems.

In all the developing countries developmental plans have been carried out at the expense of natural resources and environment. This has led to many problems like deforestation, over exploitation of resources and the resulting ecological disasters (as seen in Uttarakhand in June, 2013) like the flash floods and landslides in uplands and drought and forest fires at other places. But now it has been realized that sustainable development is the only answer to our problems.

Sustainable development is possible only through the *rational use* of resources and by checking environmental degradation. In a recent National Symposium on Agriculture and Environment–issues in sustainable development (1995) it was concluded that a holistic approach for development was essential where integrated development principles are followed. Trends to global sustainability indicators (after *Brown*, 1998) are given in Table 9.2. These trends do indicate tremendous environmental changes, but we can pin our hopes on practical application of environmental biology. Our model of sustainable development may be in these form of a triangle having three equal arms representing *education, resource management* and *economic development*, based on sound ecological principles.

The traditional ecological knowledge (TEK) is the best way to achieve sustainable development. It teaches us *sustainable use* of renewable resources. It means using them at rates within their capacity for renewal. A '*sustainable economy*' is the product of sustainable development. It maintains its natural resource use. It can continue to develop by adapting and through improvements in knowledge, technical efficiency on wisdom. Leopold believes that the greatest obstacle towards achieving environmental ethics is the economic mind set. He has offered a principle which brings into focus the broader ethical concerns of environment. 'A thing is right when it tends to preserve the integrity, stability and beauty of the biotic community. It is wrong when it tends otherwise. However, we try to manage and simplify nature with too little knowledge about its working.

Table 9.2. Trends to global sustainability indicators (from Brown, 1998)

Parameter	*1950*	*1990*
Ozone depletion, Antarctic	110 Dobson units	80 Dobson units
Global warming	14.89° C	15.39° C
Irrigation	136 million hectares	248 million hectares
Forests	40% Earth cover	27% Earth cover
Productive land (grain)	0.23 ha per person	0.12 ha per person
Biodiversity (*e.g.* Australian frog species)	294	224
Population (World total)	2.55 billion	5.73 billion
Annual increase	37 million	87 million
Energy (Oil)	518 Mt p.a.	3031 Mt p.a.
Natural gas	187 Mt p.a.	2114 Mt p.a.
Wind power	< 10 MW p.a.	4880 MW p.a.

Measures agreed by UNDP, World Watch Institute and World Resources Institute (1995).

9.12 FUTURE OF ENVIRONMENTAL BIOTECHNOLOGY

Environmental biotechnology has made rapid strides during the past three or four decades. Many problems pertaining to accumulation of hazardous wastes and toxic metals in water, wastewater and soil have been solved. We now use biological methods not only for wastewater treatment but also for air pollution control. Bioremediation, whether it be by natural attenuation, bioaugmentation or use of genetically engineered microorganisms (GEMs) has been used for environmental protection. Use of both aquatic and terrestrial plants in phytoremediation and recovery of metals from plant residues has been a promising area. Biological wastewater treatment facilities have played a major role in preventing water pollution and environmental degradation. *Green chemistry* and *industrial ecology* have suggested techniques and pathways for manufacture of alternative chemicals and a new paradigm about the relationship between the industrial system and the natural environment.

In spite of all these achievements and problems environmental biotechnology holds many promises for more and more understanding of the use of microorganisms and other biological techniques for environmental protection and restoration. A better understanding of the conditions and methods required for environmental cleaning and waste disposal can lead to more efficient remediation and waste treatment processes. The development of more biodegradable products will also become an improtant part of environmental protection. In the light of all such hopes there are many reasons to be optimistic that with more advances in environmental biotechnology we can achieve a *sustainable environment*.

Role of micro-organism in *aquaculture* has also been recognized. They may have positive or negative effects on the outcome of aquaculture operations. Positive microbial activities include elimination of toxic materials such as ammonia, nitrite and hydrogen sulphide, degradation of uneaten food, and also providing nutrition to aquatic animals such as fish and shrimp. These and other similar function make micro-organisms as key players in the sustainable development of aquaculture.

MODEL QUESTIONS

1. Give an account of four basic operations of sewage treatment plants.
2. Give an account of microbiological degradation of xenobiotics.
3. Define the term 'bioremediation'. Give an account of bioremediation techniques.
4. Write short notes on the following:
   (*a*) Phytoremediation (*b*) Pollution control by genetic engineering
   (*c*) Sustainable development

MULTIPLE CHOICE QUESTIONS

5. One of the broader aims of environmental biotechnology is the conservation of resources via:
   (A) Recycling of waste materials (B) Producing less toxic waste materials
   (C) Producing less harmful pesticides (D) All of the above
6. Microbial treatment of oil pollution is done by:
   (A) Degradation of oil (B) Biodegradation of oil
   (C) Biodegradation of oil after addition of nutrients like glucose
   (D) Biodegradation of oil after addition of nutrients such as nitrogen and phosphorus
7. Majority of pesticides can be degraded by:
   (A) Aquatic plants (B) Aquatic animals
   (C) Bacteria and fungi (D) By all of them
8. Activated sludge treatment is done in:
   (A) Primary treatment (B) Secondary treatment
   (C) Tertiary treatment (D) In none of them
9. Who gave the statement "*the land ethic simply enlarges the boundaries of the community to include soils, waters, plants and animals*"?
   (A) Aldo Leopold (B) Odum
   (C) Barry Commoner (D) Lester Brown
10. Bioremediation includes:
   (A) Use of aquatic plants (B) Aquatic animals
   (C) Use of bacteria (D) All of them

ANSWERS TO MCQs

Chapter 1

5. (D)	6. (D)	7. (B)	8. (D)	9. (D)	10. (D)

Chapter 2

11. (D)	12. (D)	13. (D)	14. (B)	15. (C)	16. (D)
17. (C)	18. (A)	19. (C)	20. (D)	21. (C)	22. (D)
23. (C)	24. (A)	25. (B)	26. (C)	27. (D)	28. (D)
29. (A)	30. (B).				

Chapter 3

11. (C)	12. (D)	13. (D)	14. (A)	15. (B)	16. (C)
17. (D)	18. (D)	19. (D)	20. (B)	21. (C)	22. (A)
23. (A)	24. (A)	25. (B)	26. (D)	27. (A)	

Chapter 4

17. (D)	18. (C)	19. (B)	20. (A)	21. (B)	22. (D)
23. (C)	24. (B)	25. (C)	26. (D)	27. (B)	28. (D)
29. (D)	30. (B)	31. (B)	32. (B)	33. (A)	34. (C)
35. (D)					

Chapter 5

10. (B)	11. (A)	12. (D)	13. (B)	14. (C)	15. (D)
16. (C)	17. (A)	18. (D)	19. (B)	20. (A)	21. (A)
22. (D)	23. (B)	24. (A)	25. (C)		

Chapter 6

12. (A)	13. (B)	14. (C)	15. (A)	16. (C)	17. (D)
18. (C)	19. (B)	20. (D)	21. (D)	22. (D)	23. (D)
24. (C)	25. (A)	26. (C)	27. (D)	28. (D)	29. (D)
30. (C)					

Chapter 7

9. (B)	10. (B)	11. (A)	12. (C)	13. (D)	14. (A)
15. (C)	16. (A)	17. (B)	18. (C)	19. (D)	20. (A)

Chapter 8

5. (B)	6. (A)	7. (B)	8. (D)	9. (C)	10. (C)
11. (B)	12. (A)	13. (B)	14. (D)	15. (B)	

Chapter 9

5. (A)	6. (D)	7. (C)	8. (B)	9. (A)	10. (C)

REFERENCES

Abrams, P.A. 1991. The relationship between food availability and foraging effort: Effects of life history and time scale. Ecology 72:1242-1252.

Agnihotri, A.1998.Biotechnological approaches for conservation and sustainable utilization of biodiversity. In (ed.P.S.Srivastava) *Plant Tissue Culture and Molecular Biology*, pp. 598-606. Narosa Publishing House, New Delhi.

Andrewartha,H.G. and L.C.Birch 1954. *The Distribution and Abundance of Animals*. University of Chicago Press, Chicago.

Arif, M.1993. *Animal life in Antarctica. The seventh continent*. J.Appl.zool.Res. 4:13-20.

Armbrust, E.J. ad and G.G.Gyrisco. 1982. Forage crops insect pest management. In (eds. R.L. Metcalf and W. Luckman) *Introduction to Insect Pest Management*, pp. 443-46, Wiley, New York.

Barcley, G.W. 1958. *Techniques of Population Analysis*. Wiley, New York.

Batzli, G.O., R.G. White, S.F. MacLean, F.A. Pitelka and B.D. Collier 1980. The herbivore based trophic system. In (eds. J.Brown *et al*) An Arctic Ecosystem, The Coastal (tundra at Barrow, Alaska) pp. 335-410, Dowden, Hutchinson & Ross, Stroudsburg, PA.

Begon, M., J.L. Harper ad and C.R. Townsend.1996. *Ecology: Individuals, populations and Communities*. pp. 1068, Blackwell Science Ltd.

Berner, E.K and R.A. Berner 1987. *The Global Water Cycle: Geochemistry and Environment*. Prentice-Hall, New Jersey, USA.

Best, D.J., J.Jones and D.Stafford 1985. *The Environment and Biotechnology*. In (eds. I.J.Higgins, D.J.Best and J.Jones) *Biotechnology: Principles and Applications*. Blackwell Scientific Publications.

Bhatt, S.D. and J. K. Pathak 1991. *Himalayan Environment, Water Quality of the drainage basins*. pp. 318, Shree Almora Book Depot, Almora.

Bhatt, S. D. and R.K. Pande 1991. *Ecology of the Mountain Waters*. pp. 389, Ashish Publishing House, New Delhi.

Cann, M.C. and M.E. Connelly 2000. *Real-World Cases in Green Chemistry*. American Chemical Society (ICS)

Caraco, N.F. 1993. Disturbance of the phosphorus cycle: A case of indirect effects of human activity. Trends Ecol. Evol. 8:51-54.

Chapman, J.L. and M.J. Reiss 1995. *Ecology, Principles and Applications*. pp. 294, Cambridge University Press.

Chittay, D.1960. Population processes in the vole and their relevance in general theory. Canadian Journal of Zoology 38:99-113.

Clapham Jr., W.B., 1981, *Human Ecosystems*, MacMillan, New York.

Clarke, G.L. 1954. *Elements of Ecology*. John Wiley, New York.

Clarkson. T.W. 1987. Metal toxicity in the central nervous system. Environ. Health Perspect. 75:59-64.

Clements, F.E.1916. Plant succession: Analysis of development of vegetation. Carnegie Inst. Wash. Publ. 242:1-512.

Clements, F.E. 1936. Nature and structure of the climax. J.Ecol. 24:252-284.

Colleran, E.1997. Uses of bacteria in bioremediation. In (ed.D. Sheehan) Methods in Biotechnology, Vol. 2 Bioremediation Protocols. Humana Press Inc. Totowa. NJ.

Commander, J.N.M. , G.J. Stijntjes and N.P.E. Vermeulen 1995. Enzymes and transport systems involved in the formation and disposition of glutathione S-conjugates. Pharmacol.Revs. 47: 271-330.

Connell, J.H. and Slatyer R.O. 1977. Mechanisms of succession in natural communities and their role in community stability and organization. Amer. Naturalist 111:1119-1144.

Cowles, H.C.1899. The ecological relations of the vegetation on the sand dunes of Lake Michigan. Bot.Gaz. 27:95-117, 167-202, 281-308, 361-391.

Cowles, H.C. 1901. The physiographic ecology of Chicago and vicinity. Bot. Gaz. 31:73-108, 145-181.

Cunningham, S.D. and W.R. Berti 1993. Remediation of contaminated soils with green plants: An overview. *In Vitro* Cell. Dev. Biol. 29P: 207-212.

Dagani, R. 1980. Aquatic toxicology matures ,gains importance. Chern. Eng. News June 30: 18-23.

Dahl, K. 1919. Studies of trout and trout waters in Norway.Salmon and Trout Magazine 18: 16-33.

Daubenmire, R. 1966. Vegetation; Identification of tipical communites. Science 151:291-298

Daubenmire, R. 1968. *Plant Communities: A Textbook of Plant Synecology*. Harper & Row, New York.

DeLeve, L.D. ad and N. Kaplowitz 1991. Glutathione metabolism and its role in hepatotoxicity. Pharmacol. Ther. 52:287-305.

Dasmann, R.F. 1968. *Environmental Conservation*, John Wiley.

Dix, H.M. 1981. *Environmental Pollution*, John Wiley, Pub., New York.

Dobriyal, A.K. and H.R. Singh 1988. Ecological basis for ichthyofaunal variation in two hillstreams of Garhwal Himalaya In (ed. M.M. Joseph) The First Indian Fisheries Forum, Proc. Asian Fisheries Society Indian Branch Mangalore, 313-317.

Dobson, M. and C. Frid 1998. *Ecology of Aquatic systems*. pp. 222, Longman.

Doherty, J.D. 1979. Insecticides affecting ion transport. Pharmacol. Therap. 7:123-151.

Donovan, T.M. ad and C.W. Welden 2002. *Spreadsheet Exercises in Ecology and Evolution*. Sinauer Associates, Inc. Publishers Sunderland Massachusetts, USA, pp. 556.

Egler, F.E. 1954. Vegetation science concepts .I. Initial floristic composition, a factor in old-field vegetation development. Vegetatio.14:412-417.

Elton, C. 1927. *Animal Ecology*. MacMillan New York (2/e 1935, 3/e 1947)

Elton, C. 1933. *The Ecology of Animals*. Methuen, London.

Elton, C. 1966. *The Pattern of Animal Communities*. Methuen, London.

Enold, G.L. 1984. 'the significance of healthy animal in toxicity studies. Concepts Toxicol. 1:34-41.

Ernhardt, C.B, B. Landa and N.B. Schnell 1981. Subclinical levels of lead and developmental deficits: A multivariate follow-up reassessment. Pediatrics 67:911-919.

Ensley, B.D., I. Raskin and D.E. Salt 1997. Phytoremediation applications for removing heavy metal contamination from soil and water. In (eds. G.S. Sayler, J. Sanseverino and K.L. Davis) *Biotechnology in the Sustainable Environment*. Plenum Press, New York and London.

Farlow, J .O. 1975. A consideration of the trophic dynamics of a late Cretaceous large dinosaur community (Oldham Formation). Ecology 57:841-857.

Fauth, J.E., J. Bernando, M. Camara, W.J. Ressetartis Jr, J. Van Buskirk, and S.A. McCollum 1996. Simplifying the jargon of Community Ecology: A conceptual approach. Am. Nat. 147: 282-286.

Feisthaur, N.P. Sibley, S. Burke, and N. Kaushik 2002. A review of the toxicity of detergents and its formulation components on aquatic organisms. International Journal of Ecology and Environmental Sciences 28:223-297.

Flint, M. and R. van den Bosch 1981. *Introduction to Integrated Pest Management*. Plenum Press, New York.

Freedman, H.I.1980. *Deterministic Mathematical Models in Population Ecology*. Marcel Dekker Inc. New York and Basel, pp. 254.

Gadgil, M.2000. Conservation: Taking care of all life, pp. 69-75, The Hindu survey of the Environment.

Gadgil, M. and V. M - Meher Homji 1986a. Localities of great significance to conservation of India: A biological diversity. Proc. Indian Acad. Sci. (Suppl) B265:180

Gadgil, M and V. M. Meher - Homji 1986b. Role of protected areas in conservation. In (eds. V.L. Chopra and T.N. Khoshoo Conservation of Productive Agriculture. pp. 144-59, Indian Council Agric. Res. (ICAR) New Delhi.

Gadgil, M.*et al* 1998. Conservation : Where are the people? pp. 107-17. The Hindu Survey of the Environment.

Gaston, K.G. and J.I. Spicer 1998. *Biodiversity, an Introduction*, p. 113, Blackwell.

Gause, G.F. 1934. *The Struggle for Existence*. Williams and Wilkins, Balimore.

Glazer, A.N. and H. Nikaido 1998. *Microbial Biotechnology, Fundamentals of Applied Microbiology*. W.H. Freeman and Company, New York.

Gooday, G.W. 1988. The potential of the microbial cell and its interaction with other cells. In (eds. J.M. Lynch and J.E. Hobbie) Micro-organisms in action: Concepts and Applications in Microbial Ecology. pp. 7-32, Blackwell Scientific Publications, Oxford.

Gopal, B. 1982. Ecology and and management of freshwater wetlands in India. pp. 127-62. Proc. Int. Sci. Workshop Ecosystem Dynamics in FreshWater Wetlands and Shallow Water Bodies. SCOPE-UNEO, Moscow.

Gopal B. 2003. *Understanding Wetlands, our natural heritage*. NIE, New Delhi.

Gopal, B and U.Goel 1993. Competition and allelopathy in aquatic plant communities. Bot. Rev. 59:155-210.

Gaffney, P.M. 1975. Roots of the niche concept.Amer.Nat. 109:490.

Graves, J and D. Reavey 1996. *Global Environmental Change*. pp. 226, Longman.

Grinnell, J. 1904. The origin and distribution of the chestnut-backed chickdee. Auk .21:364-382.

Hadley, N.F. 1970. Micrometeorology and energy exchange in two desert arthropods. Ecology 51:434-444.

HaIl, D.C., R.B. Nogard and P.K. True 1975. The performance of independent pest management consultants. California Agriculture 29:12-14.

Halliwell, B., J.M.C. Gutteridge and C.E. Cross 1992. Free radicals, antioxidants and human diseases, where are we now? J. Lab. Clin. Med. 119:598-620.

Hamilton. W.A. 1988. Microbial energetics and metabolism. In (eds. J.M. Lynch and J.E. Hobbie) *Micro-organisms in Action: Concepts and Applications in Microbial Ecology*. pp. 75-100, Blackwell Scientific Publications. Oxford.

Hardin, G. 1960. The competitive exclusion principle. Science 131:1292-1297.

Harper, D.M. and J.D. Ferguson 1982. Zooplankton and their relationships with water quality and fisheries. Hydrobiologia 88:135-145.

Hassell, M.P. 1976. *The Dynamics of Competition and Predation*. Edward Arnold, London.

Hallawell, J.M. 1986. Biological indicators of freshwater pollution and Environmental Management. Elsevier.

Hssell, M.P. 1978. *The Dynamics of Arthropod Predator-Prey Systems*. Princeton University Press, Princeton, N.J.

Hassell, M.P. 1981. Arthropod predator-prey systems. In (ed.R.M.May) Theoretical Ecology, pp. l05-31, Blackwell, Oxford, England.

Heal, O.W. and S.F. MacLean 1975. Comparative productivity in ecosystems, secondary productivity. In (eds. W.H., vanDobben and R.,H., Lowe-McConnell) Unifying Concepts in Ecology. pp. 89-108, Junk, The Hague.

Higgins, I.J. 1985. What is biotechnology? In (eds. I.J. Higgins, D.J. Best and J. Jones) *Biotechnology, Principles and Applications*. Blackwell Scientific Publications.

Holland, M.M.-1988. SCOPE/MAB technical consultations on landscape boundaries. Biol. Intl. 17:47-106.

Holling, C.S. 1959. The components of predation as revealed by a study of small mammal predation of the European pine sawfly. Can. Entornol. 91:293-320.

Huffaker, C.B. and P.S. Messenger (eds) 1976. *Theory and Practice of Biological Control*. Academic Press, New York.

Hunter, M.D. and P.W. Price 1992. Playing chutes and ladders: Heterogeneity and the relative roles of bottom-up and top-down forces in natural communities. Ecology 73:724-732.

Henschel, J.R.1990. Spiders wheel to escape. South African Journal of Science 886:151-152.

Howard, L.O. and W.F. Fiske 1911. The importation into the United States of the parasites of the gypsy-moth and the brown-tail moth. U.S Department of Agriculture, Bureau of Entomology Bulletin, 91.

Holling, C.S. 1995. Sustainability: The Cross-scale Dimension. In (eds. M.Munasinghe and W. Shearer) Defining and Measuring Sustainability. The Biophysical Foundations, U.N.Univ., Tokyo and The World Bank, Washington, D.C. pp. 65-75.

Hutchinson, G.E. 1957. Concluding Remarks. Cold Spring Harbor Symp. on Quant. Biol. 25: 415-427.

Jantsch, E. 1970. Inter- and transdisciplinary university: A systems approach to education and innovation. Policy Sciences 1:203-428.

Johal, M.S. 2002. Ecology of hillstreams of Himachal Pradesh and Garhwal region with special reference to fish communities. Final Report U.S. Fish and Wildlife Service.

Johnston, R.H. 1910. Determinate evolution in color pattern of the lady- beetles. Carengie Inst. Wash. Publ. 122 : p. 104.

Kak, A. and T. R. Rao 1998. Does the evasive behaviour of *Hexarthra* influence its competition with cladocerans? Hydrobiologia 387/388:409-419.

Kendeigh S.C., 1961. *Animal Ecology*, Prentice-Hall

Kendeigh, S.C. 1980. *Ecology with Special reference to animal and man*, Prentice-Hall, New Delhi.

Khan, M.A.Q. 1981. *Pesticides in Aquatic Environments*, Rleum, Khoshoo, T.N.

Khoshoo, T.N. 1986. *Environmental priorities in India and sustainable development*, Presidential Address, of 3rd Indian Science Congress, New Delhi.

Kormondy, E.J. 1976. *Concepts of Ecology*. Prentice Hall, Englewood Cliffs, N.J. p. 238.

Krebs, C.J. 1994. *Ecology, the experimental analysis of distribution and abundance*, 4th edn, Addison-Wesley Educational Publishers, Inc. p. 801.

Kumar, H.D. 2000. *Plant-Animal Interactions*. p. 400, Affiliated East- West Press Pvt. Ltd., New Delhi.

Kumar. H.D. 2001. *Sustainable Human Ecology*. p. 307, Affiliated East-West Press Pvt. Ltd., New Delhi.

Kumar, N.1992. Ecological studies on the macrozoobenthic communities of the Alaknanda river system. D. Phil. Thesis, H.N., B. Garhwal University, Srinagar Garhwal.

Kumar, R. and T.R. Rao 1999*a*. Effect of algal food on animal prey consumption rates in the omnivorous copepod, *Mesocyclops thermocyclopoides*. Internat. Rev. Hydrobiol. 84:419-426.

Kumar, R. and T. R. Rao 1999*b*. Demographic responses of adult *Mesocyclops thermocyclopoides* (Copepoda,Cyclopoida) to different plant and animal diets. Freshwater Biology 42:487-501.

Lack, D. 1944. Ecological aspects of species formation in passerine birds. Ibis 86:260-286.

Lalithakumari, D. 2003. Microbes; "a tribute" to clean environment. In (eds. Manuswamy *et al*) EMCB-ENVIS NODE newsletter, Micro-organism and Environment Management. Vol 1 (3-5), Department of Zoology, University of Madras, Chennai, India.

Lambert, B. and M. Peferoen 1992. Insecticidal promise of *Bacillus thurengiesis*: facts and mysteries about a successful biopesticide. Bioscience 42:112-121.

Leavitt, M. 1997. Bioremediation. 'the Green Thumb in Brownfields Management. In (eds. G.S. Sayler, J. Sanseverino and K.L. Davis) *Biotechnology in the sustainable environment*. Plenum Press, New York and London.

Lieth, H. 1975. Primary productivity in ecosystems: Comparative analysis of global patterns. In (eds. W.H.vanDobben and R.H ~ Lowe-McConnell) unifying Concepts in Ecology. pp. 67-88, Junk, The Hague.

Lindeman, R.L. 1942. The trophic-dynamic aspect of ecology. Ecology 23:399-417.

Lovelock, J.E. 1988. *The Ages of Gaia*. Cambridge: Cambridge University Press.

Lovelock, J.E. and L. Margulis 1974. Atmospheric homeostasis, by and for the biosphere: The Gaia Hypothesis. Tellus 26:1-10.

Lu, F.C., D.C. Jessup and A. Lavallee 1965. Toxicity of pesticides to young versus adult rats. Food Cosmet. Toxicol. 5:591-596.

Lu, F.C. and S.Kacew 2002 *Lu's Basic Taxicology*, 4th edn, p. 392, Taylor and Francis, London and New York.

Magurran, A.E. 1988. *Ecological diversity and its measurement*. p. 119, Croom Helm, London

Mann, C. 1991. Lynx Margulis: Science's unruly Earth Mother. Science 252:378-381.

Maxwell F.G. and P.R. Jenings 1980. *Breeding Plants Resistant to Insects*. Wiley, New York.

Messerli, B. 1985. Stability and instability of mountain ecosystems, an interdisciplinary approach. In (eds. T.V. Singh and J. Kaur) *Integrated Mountain Development*. Himalayan Books, New Delhi.

Miller, R.M. and Y. Zhang 1997. Measurement of biosurfactant-enhanced solubilization and biodegradation of hydrocarbons. In (ed.D. Sheehan) *Methods in Biotechnology,* Vol. 2, Bioremediation Protocols. Humana Press, Inc., Totowa, N.J.

Misra, V. 1982. Toxicology and environmental problems of detergent industry. Journal of Scientific and Industrial Research 41:48-53.

Morris, R.F. 1957. The interpretation of mortality data in studies on population dynamics. Canadian Entomologist 89:49-69.

Mo, S.C. D.S. Choi and J.W. Robinson 1989. Uptake of mercury from aqueous solutions by duckweed: The effect of pH, Copper and Humic acid. J. Env. Sci. Health. A 24:135-146.

Mukherjee, B. and G.V.N. Pankajakshi 1995. The impact of detergents on plankton diversity in freshwaters. Journal of Environmental Biology 16: 211-218.

Murdoch, W.W. 1971. The developmental response of predators to changes in prey density. Ecology 52:132-137.

Murdoch, W.W., J. Chessonand P.L. Chesson 1985. Biological control in theory and practice. American Naturalist 125:344-366.

Nagdali, S.S. and P.K. Gupta 2002. Impact of mass mortality of a mosquito fish, *Gambusia affinis* on the ecology of a freshwater eutrophic lake (Lake Nainital, India). Hydrobiologia 468 :45-52.

Nomiyama, K. 1980. Recent progress and perspectives in cadmium health effects studies. Sci. Total Environ. 14:199-232.

Narahashi, T. 1980. Nerve membrane as a target of environmental toxicants. In (eds. P. Spencer and H.H. Schaumberg) *Neurotoxicology*. Baltimore, M.D: Williams & Wilkins.

Nautiyal, R., P. Nautiyal and H.R. Singh 1996*a*. Impact of sewage on the diatom communities of the river Alaknanda (Srinagar Garhwal) J. International Journal of Ecology and Environmental Sciences 22:289-296.

Nautiyal, R., P. Nautiyal and H.R. Singh 1996*b*. Community structure of coldwater epilithic diatoms in relation to substrate and flow conditions in Himalayan river Alaknanda. J. Freshwater Biol. 8:1-5.

Nautiyal, R., P. Nautiyal and H.R. Singh 2000. Species richness and diversity of epilithic diatom communities on different natural substrates in a coldwater river Alaknanda. Tropical Ecology 41:255-258.

Nicholson, A.J. 1933. The balance of animal populations. Journal of Animal Ecology 2:131-178.

Nicholson, A.J. and V.A. Bailey 1935. The balance of animal populations. Proc. Zool. Soc. London 3:551-598.

Nicol, J.A., C. 1989. *The Eyes of Fishes*. Clarenden, Oxford.

Odum, H.T. 1957. Trophic structure and productivity of Silver Springs Florida. Ecological Monographs 27:55-112.

Odum, E.P. 1971. *Fundamentals of Ecology*. W.B. Saunders Company, USA First Indian edition 1996, Natraj Publishers Dehradun. p. 574.

Odum, E.P. 1983. *Basic Ecology*. Saunders College Publishing, Philadelphia.

Odum, E.P. 1993. *Ecology and our Endangered Life-support Systems*. 2nd edn, Sinauer Associates,

Inc. Publishers Sunderland, Massachusetts.

Oglive, D.M. and J.M. Anderson 1965. Effects of DDT on temperature selection by young Atlantic salmon, *Salmo salar*. J. Fish Res. Bd. Canada 22:503-512.

Omkar, 1994. *Concepts of Toxicology*. Shobanlal Nagin Chand & Co. Jalandhar-4, pp. 228.

Owen, O.S. 1985. *Natural Resource Conservation, An ecological approach*, MacMillan, New York.

Paine, R.T. 1966. Food web complexity and species diversity. Am. Nat. 100:65-75.

Park, T. 1948. Experimental studies of interspecies competition. I. Competition between populations of the flour beetles *Tribolium confosum* Duval and *Tribolium castaneum* Herbst. Ecological Monographs 18:265-307.

Park, T. 1954. Experimental studies of interspecies competition. II. Temperature,humidity and competition in two species of *Tribolium.* Physiological Zoology 27:177-238.

Park, T. 1962. Beetles,competition and populations. Science 138:1369-1375.

Patterson, R.S., D.E. Weidhass, H.R. Ford and C.S. Lofgren 1970. Suppression and elimination of an island population of *Culex pipens quinquefasciatus* with sterile males. Science 168: 1368-1370.

Peakall, D.B. 1974. DDE: its presence in peregrine eggs in 1948. Science 183:673-674.

Pimentel, D. (ed.) 1991. *Handbook of Pest Management in Agriculture*. CRC Press, Boca Raton, Fl.

Postgate, J. 1988. The ghost in the laboratory. New Scientist 4:49-52.

Pough, F.H., C.M. Janis and J.B. Heiser 1999. *Vertebrate Life*. 5th edn. Prentice Hall, p. 798.

Putman, R.J. and S.D. Wratten 1984. *Principles of Ecology*. p. 388, University of California Press Berkeley and Los Angeles.

Qasim, S.Z. 1979. Primary production in some tropical marine environments. In (ed. M.J. Dunbar) *Marine Production Mechanisms*. pp. 31-69, University Press, Cambridge, (U.K.)

Ramakrishnan, P.S. 2000. Ecology teaching in India and in developing countries. Biology International 39:33-34.

Ramakrishnan, P.S. 2001. *Ecology and Sustainable Development*. National Book Trust, India, New Delhi, p. 198.

Ramos, J.L. and K.N. Timmis 1987. Experimental evolution of catabolic pathways of bacteria. Microbial Sci. 4:228-237.

Raskin, I., N. Kumar, S. Dushenkov and D. Salt 1994. Current Opinion in Biotechnology. 5:285-290.

Resetarits, W.J.Jr. 1997. Interspecific competition and qualitative competitive asymmetry between two benthic stream fish. Oikos 78:429-439.

Ricklefs, R.E. and G.L. Miller 2000. *Ecology*, 4th edn. W.H. Freeman and Company New York, p. 822.

Risser, P.G. 1990. Landscape processes in the vegetation of the North American grassland. In (eds. S.L. Collins and L.L. Wallace) Fire in North American Tallgrass Prairies. University of Oklahoma Press, Norman.

Rosenzweig, M.L. 1968. Net primary productivity of terrestrial communities: Prediction from climatological data. Amer. Nat. 102:67-74.

Saha, S.K. 1993. *Limnology of Thermal Springs* p. 176. Narendra Publishing House, Delhi.

Saha, S.K. and J.S. Datta Munshi 1982. Primary production, standing crop and efficiency of net production in thermal stream. Trop. Ecol. 23:300-309.

Saha, S.K. and J.S. Datta Munshi 1983. Algal flora of Sitakund and Bhimbandh thermal springs of Bihar. Biol. Bull. India 5:257-261.

Salisbury, F.B. and C.W. Ross 1985. *Plant Physiology*, 3rd edn. Wadsworth Bel. mont, California.

Samways, M.J. 1995. *Insect Conservation Biology*. p. 358, Chapman & Hall.

Savage, J.M. 1966. *Evolution* p. 126. Holt, Rinehart and Winston, Inc.

Sayler, G.S. 1997. Challenges and opportunities in the area of environmental biotechnology. In (eds. G.S. Sayler J. Sanseverino and K.L. Davis) *Biotechnology in the Suatainable Environment*. Plenum Press New York and London.

Schleginer, W.H. 1991. *Biogeochemistry: An Analysis of Global Change*. Academic Press, San Diego.

Schmidt-Nielsen, K. 1964. *Desert Animals, Physiological Problems of Heat and water*. Oxford University Press, Oxford.

Schoener, T.W. 1986. Overview : Kinds of ecological communities–ecology becomes pluralistic. In (eds. J. Diamond and T. Case) *Community Ecology* 467-479, Harper & Row New York.

Schoener, T.W. 1983. Field experiments on interspecific competition. Am. Nat. 122:240-285.

Schwerdtferger, F. 1941. Uber die Ursachen des Massenweschels der Insekten. Zeischrift fuer angewandte Entomologie 28:254-303.

Seigler, D.S. 1996. Chemistry and mechanisms of allelopathic interactions. Agron. J. 88:876-885.

Sharma, R.C. 1985. Effect of physico-chemical factors on benthic fauna of Bhagirathi river. Indian J. Ecol. 13:133-137.

Sharma, R.C. and H.R. Singh 1980. Impact of Bhagirathi blockade on its riverine ecosystem. Journal of Himalayan Studies and Regional Development 4:35-37.

Shukla, O.P., Omkar and A.K. Kulshrestha (eds) 1998. *Pesticides, Man and Biosphere* p. 424. A.P.H. Publishing Corporation, New Delhi.

Sih, A. 1982. Foraging strategies and the avoidance of predation by an aquatic insect, *Notonecta hoffmani*. Ecology 63:786-796.

Sih, A. and R.D. Moore 1990. Interacting effects of predator and prey behaviour in determining diets. In (ed. R.N.Hughes) *Behavioural mechanisms of food selection*, pp. 771-96, Springer-Verlag, Berlin.

Sinclair, A.R.E. and P. Arcese 1995. Population consequences of predation-sensitive foraging: The Serengeti wildebeest. Ecology 76:882-891.

Singh, H.R. 1989. *Humankind and Biosphere*. In (eds. B.Dev, R.Singh and U. D. Sharma) *Perspectives in Zoology*, Vol. l, 193-209, C.B.S. Publishers and Distributors, Delhi.

Singh, H. R. 1991. The Life Support System. In : 2nd Public Hearing (16.2.91, Hardwar) on Environmental Problems and Developmental Strategies. pp. 11-15. The People's Commission on Environment and Development, India, New Delhi.

Singh, H. R. 1999. Environmental Ethics. pp. 37-41. Proceedings of Third Conference of International Academy of Physical Sciences, Allahabad.

Singh, H. R. 1995. Water Pollution and Fisheries. In (ed. R.K.Trivedi) *Encyclopaedia of Environmental Pollution and Control*, Vol 2, pp. 403-21, Enviro Media, Karad, India.

Singh, H.R. 2002. Ecological diversity of hillstreams. In (ed. A. Kumar) *Ecology and Coservation of Lakes, Reservoirs and Rivers*. A.B.D. Publishers, Jaipur.

Singh, H.R. 2004. Fish diversity of India and some related issues. Invited Lecture (20.3.04) at Symposium on Advances in Fish Physiology: Ecological Considerations, B.H.U, Varanasi.

Singh, H.R. 2004. Pollution Monitoring and Related Issues. Invited Lecture (24.4.04) at National Workshop on Environment and Pollution Management for Sustainable Development, LNCT, Bhopal.

Singh, H.R. and S.P. Badola 1980. Ecological study of the fish and fishery of Garhwal waters with a note on some applied problems. In (ed. Director Zoological Survey of India, Calcutta) *Wildlife Ecology. Proc. Wildlife Workshop*, 173-177.

Singh, H.R. and S. N. Bahuguna 1983. Gross morphology of the alimentary canal and seasonal variation in feeding of *Noemacheilus montanus* (McClelland). Anat. Anz. Jena 154:119-124.

Singh, H. R. and N. Kumar 1995. River Ecology and Water Pollution. In (ed. R.K. Trivedi) *Encyclopedia of Environmental Pollution and Control*, Vol. 2, pp. 255-72. Enviro Media, Karad (India).

Singh, H.R. and N. Kumar 2003. Biodiversity of mountain streams and some new areas for research. In (ed.P. Nautiyal) *Emerging Trends in Freshwater Biology*. Transmedia, Srinagar Garhwal.

Singh, H.R. and P. Nautiyal 1990. Altitudinal changes and the impact of municipal sewage on the community structure of macrobenthic insects in the torrential reaches of the river Ganga in Garhwal Himalaya (India). Acta Hydrobiol. 32:407-421.

Singh, H.R., P. Nautiyal, A.K. Dobriyal, R.C. Pokhriyal, M. Negi, V. Baduni, R. Nautiyal, N.K. Agarwal, P. Nautiyal and A. Gautam 1994. Water quality of the river Ganga (Garhwal Himalaya). Acta Hydrobiol 36:3-15.

Singh, J. S. and M.C. Joshi 1979. Primary Production. In (ed. R.T. Coupland) Grassland Ecosystems of the World: Analysis of Grasslands and their Uses. pp. 197-18. Cambridge University Press, London.

Singh, J.S., Y. Hanxi, P.E. Sajise 1985*a*. Structural and Functional Aspects of Indian and Southeast Asian Savanna Ecosystems. (eds. J.C. Tothill and J.J. Mott) Ecology and Management of the World's Savannas. pp. 34-51, Austr. Acad. Sci., Canberra.

Singh, J. S, S.P. Singh, A.K. Saxena and Y.S. Rawat 1985*b*. The Forest Vegetation of Silent Valley in India. In (A.C. Chadwick and S.L. Sutton) The Tropical Rain Forests: Ecology and Resource Management. pp. 25-52. Leeds Philosophical and Literary Soc.

Singh, J.S. and K. P. Singh 1995. *Ecology*. In (ed. B.M. Johri) *Botany in India*: History and Progress. pp. 407-43. Oxford & I.B.H. Publishing Co. Pvt. Ltd. New Delhi.

Singh J. S. 2004. Biodiversity. Prof. D.D. Pant Memorial Lecture, Department of Botany, University of Allahabad.

Shimwell, D.W. 1971. Description and Classification of Vegetation. University of Washington Press, Seattle.

Sladecek, V., 1983. Rotifers as indicators of water quality. Hydrobiologia 100:169-201.

Smith, H.S. 1935. The role of biotic factors in the determination of population densities. Journal of Economic Entomology 28:873-898.

Stephens D.W. and J.R. Krebs1986. Foraging Theory. Princeton University Press Princeton, N.J.

Stern, V.M., R.F. Smith, R. van den Bosch and K.S. Hagen 1959. The integration of chemical and biological control of the spotted alfalfa aphid. Hilgardia 29:81-154.

Swaminathan, M.S..2000. Biodiversity: Equity in benefit sharing. pp: 45-51 The Hindu Survey of the Environment.

Swaminathan, M.S. 2002. Agriculture. pp. 33-39, The Hindu Survey of the Environment.

Tansley, A.G. 1939. *The British Islands and their Vegetation*: Cambridge University Press, Cambridge.

The Hindu 2001. State of Forests. pp. 199-2004. The Hindu Survey of the Environment.

Tilman, D.1988. Plant Strategies and the Dynamics and Structure of Plant Communities. Princeton University Press, Princeton.

Townsend, C.R., J.L. Harper and M. Begon 2000. *Essentials of Ecology*. pp. 552, Blackwell.

Trivedy, R.K., 1995. *Encyclopedia of Environmental Pollution and Control* (2 vols), Environment Media, Karad (India).

Uvarov, B.P. 1931. Insects and Climate. Transactions of the Royal Entomological Society of London 79:1-247.

Volterra, V. 1926*a*. Variazioni e fluttuazioni del nurnero d'individui in specie animali conviventi. Mem. Acad. Lincei. 2:31-113.

Volterra, V. 1926*b*. Variations and fluctuation of the numbers of individuals in animal species living together. Reprinted 1931 in R.N. Chapman, Animal Ecology, McGraw-Hill, New York.

Warming, J.E.B. 1896. Lehrbuch der oekologischen Pflanzengeographie. Berlin.

Warming , J.E.B. 1909. *Oecology of Plants*. Oxford University Press, New York.

Wayne, D. and R. Nimmo 1984. Aquatic Toxicology, an evolving science. Concepts Toxicol. 1:200-212.

Weaver, J.E. and F.E.Clements 1938. *Plant Ecology*. McGraw-Hill, New York.

Westselaar, R. and G.D. Farquhar 1980. Nitrogen losses from tops of plants. Advances in Agronomy 33:263-302.

Whittaker, R.H. 1953. A consideration of climax theory : The climax as a population and pattern. Ecol. Monogr. 22:1-44.

Whittaker, R.H. 1975. *Communities and Ecosystems*, 2nd edn. Macmillan, New York.

WHO. 1976. Mercury. Environmental Health Criteria. 1. Geneva: World Health Organization.

WHO. 1977. Lead. Environmental Health Criteria. 3. Geneva: World Health Organization.

WHO. 1972. Evaluation of Certain Food Additives and the Contaminants Mercury, Lead and Cadmium. Tech. Rep. Ser. 505, Geneva: World Health Organization.

WHO. 1974. Pesticide Resistence in Food. Report of the 1973 Joint FAO/WHO Meeting. Tech. Rep. Sere 545. Geneva: World Health Organization.

Willmer, P., G. Stone and I. Johnston 2000. *Environmental Physiology of Animals* p. 644. Blackwell Science Ltd. Oxford,

Wills, E.D. 1981. The role of glutathione in drug metabolism and the protection of the liver against toxic metabolites. In (ed. J.W. Gorrod) Testig for Toxicity, London: Taylor and Francis.

Wilson, E.O. 1985. The biological diversity crisis, a challenge to science. BioScience 35:700-706.

Wit, C.T. de 1960. On competition. Versl. Landbouwk. Onderz. 66:1-82.

Wynne-Edwards, V.C 1962 *Animal Dispersion in Relation to Social Behaviour*. Oliver and Boyd, Edinburgh.

Zutshi, D.P. 1985. The Himalayan Lake Ecosystems. ln (ed. J.S. Singh) *Environmental Regeneration in the Himalaya: Concepts and Strategies*. pp. 325-38, Gyanodaya Prakashan, Nainital (India).

Zutshi, D.P. 1989. Twenty-five years of ecological research on the lakes of North-West Himalayas. ln (eds J.S. Singh and B. Gopal) *Perspectives in Ecology*, pp. 49-66. Jagmander Book Agency, New Delhi.

INDEX

M

N

O

P

Q

R

S

T

U

V

W

X

Y

Z